T0226181

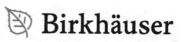

George Grätzer

The Congruences of a Finite Lattice

A "Proof-by-Picture" Approach

Second Edition

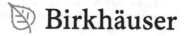 Birkhäuser

George Grätzer
Toronto, ON, Canada

ISBN 978-3-319-81748-4   ISBN 978-3-319-38798-7   (eBook)
DOI 10.1007/978-3-319-38798-7

Mathematics Subject Classification (2010): 06B10, 06D05, 06C05, 06C10

Printed on acid-free paper

This book is published under the trade name Birkhäuser
The registered company is Springer International Publishing AG Switzerland
(www.birkhauser-science.com)

To László Fuchs,
my thesis advisor, my teacher,
who taught me to set the bar high;

and to my coauthors,
who helped me raise the bar.

Short Contents

Contents

VII   Congruence Structure   289

24   Prime Intervals and Congruences   291

25   Some Applications of the Swing Lemma   315

Glossary of Notation

Symbol	Explanation	Page
0_I and 1_I	the zero and unit of the interval I	14
$\mathrm{Atom}(U)$	set of atoms of the ideal U	94
$\mathrm{Aut}\,L$	automorphism group of L	12
B_n	boolean lattice with n atoms	5
$\mathrm{C_l}(L), \mathrm{C_{ll}}(L), \mathrm{C_{ul}}(L)$	left boundary chains of a planar lattice	47
C_n	n-element chain	4
$\mathrm{con}(a,b)$	smallest congruence under which $a \equiv b$	16
$\mathrm{con}(c)$	principal congruence for a color c	41
$\mathrm{con}(H)$	smallest congruence collapsing H	16
$\mathrm{con}(\mathfrak{p})$	principal congruence generated by a prime interval $\mathfrak{p}$	39
$\mathrm{Con}\,L$	congruence lattice of L	15, 58
$\mathrm{Con_J}\,L$	ordered set of join-irreducible congruences of L	39
$\mathrm{Con_M}\,L$	ordered set of meet-irreducible congruences of L	81
$\mathrm{C_r}(L), \mathrm{C_{lr}}(L), \mathrm{C_{ur}}(L)$	right boundary chains of a planar lattice L	47
$\mathrm{Cube}\,K$	cubic extension of K	81
$\mathbf{D}$	class (variety) of distributive lattices	25
$\mathrm{Diag}(K)$	diagonal embedding of K into $\mathrm{Cube}\,K$	81
$\mathrm{Down}\,P$	ordered set of down sets of the ordered set P	5, 9, 242
$\mathrm{ext}\colon \mathrm{Con}\,K \to \mathrm{Con}\,L$	for $K \leq L$, extension map: $\boldsymbol{\alpha} \mapsto \mathrm{con}_L(\boldsymbol{\alpha})$	42
$\mathrm{fil}(a)$	filter generated by the element a	15

Symbol	Explanation	Page
$\mathrm{fil}(H)$	filter generated by the set H	15
$\mathrm{Free}_{\mathbf{D}}(3)$	free distributive lattice on three generators	26
$\mathrm{Free}_{\mathbf{K}}(H)$	free lattice generated by H in a variety $\mathbf{K}$	26
$\mathrm{Free}_{\mathbf{M}}(3)$	free modular lattice on three generators	28
$\mathrm{Frucht}\,C$	Frucht lattice of a graph C	190
$\mathrm{hom}_{\{\vee,0\}}(X,Y)$	the set of $\{\vee,0\}$-homomorphisms of X into Y	261
$\mathrm{id}(a)$	ideal generated by the element a	14
$\mathrm{id}(H)$	ideal generated by the set H	14
$\mathrm{Id}\,L$	ideal lattice of L	14, 59
(Id)	condition to define ideals	14, 59
Isoform	class of isoform lattices	153
$\mathrm{J}(D)$	ordered set of join-irreducible elements of D	19
$\mathrm{J}(\gamma)$	$\mathrm{J}(\gamma)\colon \mathrm{J}(E) \to \mathrm{J}(D)$, the "inverse" of $\gamma\colon D \to E$	32
$\mathrm{J}(a)$	set of join-irreducible elements below a	19
$\ker(\gamma)$	congruence kernel of γ	17
$\mathbf{L}$	class (variety) of all lattices	25
$L_{\mathrm{bottom}}, L_{\mathrm{top}}$	bottom and top of a rectangular lattice L	53
$\mathrm{lc}(L)$	left corner of a rectangular lattice L	53
$L_{\mathrm{left}}, L_{\mathrm{right}}$	left and right of a rectangular lattice L	53
$\mathbf{M}$	class (variety) of modular lattices	25
Max	maximal elements of an ordered set	59
$\mathbf{mcr}(n)$	minimal congruence representation function	97
$\mathbf{mcr}(n, \mathbf{V})$	$\mathbf{mcr}$ for a class $\mathbf{V}$	97
$\mathrm{M}(D)$	ordered set of meet-irreducible elements of D	32
M_3	five-element modular nondistributive lattice	xxiii, 11, 30
$\mathrm{M}_3[L]$	ordered set of boolean triples of L	68
$\mathrm{M}_3[L,a]$	interval of $\mathrm{M}_3[L]$	73
$\mathrm{M}_3[L,a,b]$	interval of $\mathrm{M}_3[L]$	75
$\mathrm{M}_3[a,b]$	ordered set of boolean triples of the interval $[a,b]$	68

Symbol	Explanation	Page
$M_3[\alpha]$	restriction of α^3 to $M_3[L]$	70
$M_3[\alpha, a]$	restriction of α^3 to $M_3[L, a]$	74
$M_3[\alpha, a, b]$	restriction of α^3 to $M_3[L, a, b]$	xxiii, 77
N_5	five-element nonmodular lattice	xxiii, 11, 20, 30
$N_{5,5}$	seven-element nonmodular lattice	102
$N_6 = N(p, q)$	six-element nonmodular lattice	xxiii, 20, 90
$N_6[L]$	2/3-boolean triple construction	209
$N(A, B)$	a lattice construction	144
$O(f)$	Landau big O notation	xxxiv
Part A	partition lattice of A	7, 9
Pow X	power set lattice of X	5
Pow$^+ X$	ordered set of nonempty subsets of X	228
Princ L	ordered set of principal congruences L	38
rc(L)	right corner of a rectangular lattice	52
Prime(L)	set of prime intervals of L	39
re: Con $L \to$ Con K	restriction map: $\alpha \mapsto \alpha]K$	41
SecComp	class of sectionally complemented lattices	97
SemiMod	class of semimodular lattices	122
Simp K	simple extension of K	81
$(SP_\vee)$	join-substitution property	15, 58
$(SP_\wedge)$	meet-substitution property	15, 58
sub(H)	sublattice generated by H	13
S_7	seven-element semimodular lattice	xxiii, 34
S_8	eight-element semimodular lattice	34
Swing, $\curvearrowleft$	$p \curvearrowleft q$, p swings to q	295
T	class (variety) of trivial lattices	25
Uniform	class of uniform lattices	153
Traj L	set of all trajectories of L	48

Symbol	Explanation	Page
Relations and		
Congruences		
A^2	set of ordered pairs of A	3
$\varrho, \tau, \pi, \ldots$	binary relations	
$\alpha, \beta, \ldots$	congruences	
$\mathbf{0}$	zero of Part A and Con L	7
$\mathbf{1}$	unit of Part A and Con L	7
$a \equiv b \pmod \pi$	a and b in the same block of π	7
$a \varrho b$	a and b in relation ϱ	3
$a \equiv b \pmod{\alpha}$	a and b in relation α	3
a/π	block containing a	7, 15
H/π	blocks represented by H	7
$\alpha \circ \beta$	product of α and β	21
$\alpha \overset{r}{\circ} \beta$	reflexive product of α and β	30
$\alpha\rceil_K$	restriction of α to the sublattice K	15
L/α	quotient lattice	16
β/α	quotient congruence	17
π_i	projection map: $L_1 \times \cdots \times L_n \to L_i$	21
$\alpha \times \beta$	direct product of congruences	21
Ordered sets		
$\leq, <$	ordering	3
$\geq, >$	ordering, inverse notation	3
$K \leq L$	K a sublattice of L	13
$\leq_Q$	ordering of P restricted to a subset Q	4
$a \parallel b$	a incomparable with b	3
$a \prec b$	a is covered by b	5
$b \succ a$	b covers a	5
0	zero, least element of an ordered set	4
1	unit, largest element of an ordered set	4
$a \vee b$	join operation	9
$\bigvee H$	least upper bound of H	4
$a \wedge b$	meet operation	9
$\bigwedge H$	greatest lower bound of H	4
P^d	dual of the ordered set (lattice) P	4, 10
$[a, b]$	interval	14
$\downarrow H$	down set generated by H	5
$\downarrow a$	down set generated by $\{a\}$	5
$P \cong Q$	ordered set (lattice) P isomorphic to Q	4, 12

Symbol	Explanation	Page

Constructions

$P \times Q$	direct product of P and Q	5, 20
$P + Q$	sum of P and Q	6
$P \dotplus Q$	glued sum of P and Q	17
$A[B]$	tensor extension of A by B	256
$A \otimes B$	tensor product of A and B	253
$U \circledast V$	modular lattice construction	131

Prime intervals

$\mathfrak{p}, \mathfrak{q}, \ldots$	prime intervals	
$\mathrm{con}(\mathfrak{p})$	principal congruence generated by $\mathfrak{p}$	39
$\mathrm{Prime}(L)$	set of prime intervals of L	39
$\mathrm{Princ}\, L$	the ordered set of principal congruences	38

Perspectivities

$[a, b] \sim [c, d]$	$[a, b]$ perspective to $[c, d]$	32
$[a, b] \overset{\mathrm{up}}{\sim} [c, d]$	$[a, b]$ up-perspective to $[c, d]$	33
$[a, b] \overset{\mathrm{dn}}{\sim} [c, d]$	$[a, b]$ down-perspective to $[c, d]$	33
$[a, b] \approx [c, d]$	$[a, b]$ projective to $[c, d]$	33
$[a, b] \twoheadrightarrow [c, d]$	$[a, b]$ congruence-perspective onto $[c, d]$	36
$[a, b] \overset{\mathrm{up}}{\twoheadrightarrow} [c, d]$	$[a, b]$ up congruence-perspective onto $[c, d]$	35
$[a, b] \overset{\mathrm{dn}}{\twoheadrightarrow} [c, d]$	$[a, b]$ down congruence-perspective onto $[c, d]$	35
$[a, b] \Rightarrow [c, d]$	$[a, b]$ congruence-projective onto $[c, d]$	35
$\mathfrak{p} \overset{\mathrm{p}}{\longrightarrow} \mathfrak{q}$	$\mathfrak{p}$ prime-perspective to $\mathfrak{q}$	292
$\mathfrak{p} \overset{\mathrm{p\text{-}up}}{\longrightarrow} \mathfrak{q}$	$\mathfrak{p}$ prime-perspective up to $\mathfrak{q}$	292
$\mathfrak{p} \overset{\mathrm{p\text{-}dn}}{\longrightarrow} \mathfrak{q}$	$\mathfrak{p}$ prime-perspective down to $\mathfrak{q}$	292
$\mathfrak{p} \overset{\mathrm{p}}{\Longrightarrow} \mathfrak{q}$	$\mathfrak{p}$ prime-projective to $\mathfrak{q}$	292
$\mathfrak{p} \curvearrowright \mathfrak{q}$	$\mathfrak{p}$ swings to $\mathfrak{q}$	295
$\mathfrak{p} \overset{\mathrm{in}}{\curvearrowright} \mathfrak{q}$	internal swing	295
$\mathfrak{p} \overset{\mathrm{ex}}{\curvearrowright} \mathfrak{q}$	external swing	295

Miscellaneous

$\overline{x}$	closure of x	12
$\varnothing$	empty set	4

Picture Gallery

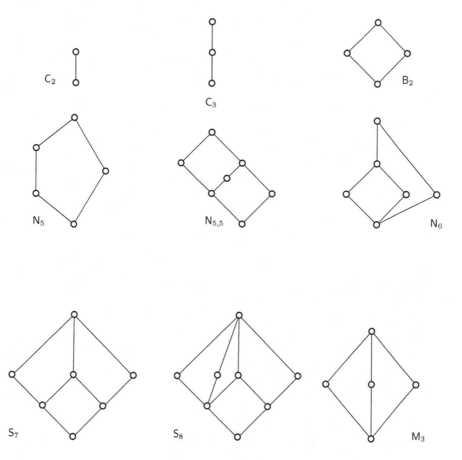

Preface to the Second Edition

A few years after the publication of the first edition of this book, I submitted a paper on congruence lattices of finite lattices to a journal. The referee expressed surprise that I have more to say on this topic, after all, I published a whole book on the subject.

This Second Edition, will attest to the fact the we, indeed, have a lot more to say on this subject. In fact, we have hardly started.

The new topics include

- The minimal representation theorem for rectangular lattices, a combinatorial result, see Chapter 11; this is joint work with E. Knapp.

- A new field investigating the ordered set $\operatorname{Princ} K$ of principal congruences of a lattice K, see Part VI. I started this field; it was continued by G. Czédli and myself.

- Congruences and prime intervals, a field concerned with the congruence structure of finite lattices, especially of SPS lattices, see Part VII. For SPS lattices this field was started by G. Czédli; the results are due to him and myself.

Part I has been extended to also provide an introduction to the new topics.

LTF refers to the book *Lattice Theory: Foundation*, [62], and LTS1 to the book *Lattice Theory: Special Topics and Applications, Volume 1* [152].

We use now the notation as in LTF, in particular, the functional notation for maps and bold Greek letters for congruences. This introduced $\aleph_0$ incorrect formulas and diagrams. I think I have been able to catch 99% of the errors. However, if you find a map $\varphi\psi$ difficult to figure out, just change it to $\psi\varphi$.

I would like to thank all those who sent me corrections, especially, G. Czédli, K. Kaarli, H. Lakser, R. Padmanabhan, and F. Wehrung.

Note All the unpublished papers in the references can be found in the
ResearchGate Website, `researchgate.net`; since `Gratzer` will not find my
publications, use the link

`https://www.researchgate.net/profile/George_Graetzer`

Toronto, ON, Canada   George Grätzer

Introduction

The topic of this book, congruences of finite lattices, naturally splits into three fields of research:

- congruence lattices of finite lattices;

- the ordered set of principal congruences of finite lattices;

- the congruence structure of finite lattices.

Congruence lattices of finite lattices

This is the major topic of this book, it covers over 70 years of research and almost 200 papers.

The congruences of a finite lattice L form a lattice, called the *congruence lattice of L*, denoted by $\operatorname{Con} L$. According to a 1942 result of N. Funayama and T. Nakayama [53], the lattice $\operatorname{Con} L$ is a finite *distributive* lattice. The converse is a result of R. P. Dilworth from 1944 (see [11]):

Dilworth Theorem. *Every finite distributive lattice D can be represented as the congruence lattice of a finite lattice L.*

This result was not published until 1962, see G. Grätzer and E. T. Schmidt [117]. In the almost 60 years since the discovery of this result, a large number of papers have been published, strengthening and generalizing the Dilworth Theorem. These papers form two distinct fields:

(i) Representation theorems of finite distributive lattices as congruence lattices of lattices with special properties.

(ii) The Congruence Lattice Problem (CLP): Can congruence lattices of lattices be characterized as distributive algebraic lattices?

A nontrivial finite distributive lattice D is determined by the ordered set $J(D)$ of join-irreducible elements. So a representation of D as the congruence lattice of a finite lattice L is really a representation of a finite ordered set P ($= J(D)$), as the ordered set, $\text{Con}_J L$, of join-irreducible congruences of a finite lattice L. A join-irreducible congruence of a nontrivial finite lattice L is exactly the same as a congruence of the form $\text{con}(a, b)$, where $a \prec b$ in L; that is, the smallest congruence collapsing a prime interval. Therefore, it is enough to concentrate on such congruences, and make sure that they are ordered as required by P.

The infinite case is much different. There are really only two general positive results:

1. The ideal lattice of a distributive lattice with zero is the congruence lattice of a lattice—see E. T. Schmidt [177] (also P. Pudlák [168]).

2. Any distributive algebraic lattice with at most $\aleph_1$ compact elements is the congruence lattice of a lattice—A. P. Huhn [159] and [160] (see also H. Dobbertin [45]).

The big breakthrough for negative results came in 1999 in F. Wehrung [190], based on the results of F. Wehrung [189].

This book deals with the finite case. For a detailed review of the infinite case, see F. Wehrung [192]–[194], Chapter 7–9 of the book LTS1.

The two types of representation theorems

The basic representation theorems for the finite case are all of the same general type. We represent a finite distributive lattice D as the congruence lattice of a "nice" finite lattice L. For instance, in the 1962 paper (G. Grätzer and E. T. Schmidt [117]), we already proved that the finite lattice L for the Dilworth Theorem can be constructed as a *sectionally complemented lattice*.

To understand the second, more sophisticated, type of representation theorem, we need the concept of a congruence-preserving extension.

Let L be a lattice, and let K be a sublattice of L. In general, there is not much connection between the congruence lattice of L and the congruence lattice of K. If they happen to be naturally isomorphic, we call L a *congruence-preserving extension* of K. (More formally, see Section 3.3.)

For sectionally complemented lattices, the congruence-preserving extension theorem was published in a 1999 paper, G. Grätzer and E. T. Schmidt [129]: *Every finite lattice K has a finite, sectionally complemented, congruence-preserving extension L.* It is difficult, reading this for the first time, to appreciate how much stronger this theorem is than the straight representation theorem. While the 1962 theorem provides, for a finite distributive lattice D, a finite sectionally complemented lattice L whose congruence lattice is isomorphic to D, the 1999 theorem starts with an *arbitrary* finite lattice K, and builds a sectionally complemented lattice L extending it with the "same" congruence structure.

The ordered set of principal congruences of a finite lattice

A large part of this book investigates the congruence lattice, Con L, of a finite lattice L. But Con L is not the only interesting congruence construct we can associate with a finite lattice L. A new one is Princ L, the ordered set of principal congruences of L. We discuss this topic in Part VI.

It turns out that only a tiny consequence of finiteness (being bounded) is of importance for this topic. So we will phrase the results for bounded lattices.

We start discussing this new field in Chapter 22, characterizing Princ K of a bounded lattice K as a bounded ordered set, see G. Grätzer [65]. Utilizing this result, we prove in Section 22.3 that for a bounded lattice[1] L, the two related structures Princ L and Aut L are independent, see G. Czédli [28].

If K and L are bounded lattices and φ is a bounded homomorphism[2] of K into L, then there is a natural bounded isotone map of Princ K into Princ L. The main result of Chapter 23 is the converse: every bounded isotone map of Princ K into Princ L can be so represented, see G. Czédli [26], [27] and G. Grätzer [75], [77].

Congruence structure of finite lattices

The spreading of a congruence from a prime interval to another prime interval involves intervals of arbitrary size. Can we describe such a spreading with prime intervals only?

We can indeed, by introducing the concept of prime-projectivity, see Section 24.2, and obtaining Prime-Projectivity Lemma, see G. Grätzer [69].

Then, in Section 24.3, we develop much sharper forms of this result for SPS (slim, planar, semimodular) lattices. The main result is the Swing Lemma, G. Grätzer [70], from which we derive many of the known results of G. Czédli and myself concerning congruences of SPS lattices.

Proof-by-Picture

In 1960, trying with E. T. Schmidt to prove the Dilworth Theorem (unpublished at the time) we came up with the construction—more or less—as presented in Section 8.2. In 1960, we did not anticipate the 1968 result of G. Grätzer and H. Lakser [89], establishing that the construction of a chopped lattice solves the problem. So we translated the chopped lattice construction to a closure space, as in Section 8.4, proved that the closed sets form a

---

[1] A lattice is *bounded*, if it has 0 and 1, see Section 1.1.1.

[2] A homomorphism is *bounded*, if it preserves 0 and 1, see Section 1.3.1.

sectionally complemented lattice L, and based on that, we verified that the congruence lattice of L represents the given finite distributive lattice.

When we submitted the paper [117] for publication, it had a three-page section explaining the chopped lattice construction and its translation to a closure space. The referee was strict: "You cannot have a three-page explanation for a two-page proof." I believe that in the 50 plus years since the publication of that article, few readers have developed an understanding of the idea behind the published proof.

The referee's dictum is quite in keeping with mathematical tradition and practice. When mathematicians discuss new results, they explain the constructions and the ideas with examples; when these same results are published, the motivation and the examples are largely gone. We publish definitions, constructions, and formal proofs (and conjectures, Paul Erdős would have added).

Tradition has it, when Gauss proved one of his famous results, he was not ready to publicize it because the proof gave away too much as to how the theorem was discovered. "I have had my results for a long time: but I do not yet know how I am to arrive at them", Gauss is quoted in A. Arber [3].

I try to break with this tradition in this book. In many chapters, after stating the main result, I include a section: *Proof-by-Picture*. This is a misnomer. A *Proof-by-Picture* is not a proof. The Pythagorean Theorem has many well known *Proofs-by-Picture*—sometimes called "Visual Proofs"; these are really proofs. My *Proof-by-Picture* is an attempt *to convey the idea of the proof*. I trust that if the idea is properly understood, the reader should be able to provide the formal proof, or should at least have less trouble reading it. Think of a *Proof-by-Picture* as a lecture to an informed audience, concluding with "the formal details now you can provide."

Outline

In the last paragraph, I call an audience "informed" if they are familiar with the basic concepts and techniques of lattice theory. Part I provides this. I am very selective as to what to include. There are no proofs in this part—with a few exceptions; they are easy enough for the reader to work them out on his own. For proofs, lots of exercises, and a more detailed exposition, I refer the reader to my book LTF.

Most of the research in this book deals with representation theorems; lattices with certain properties are constructed with prescribed congruence structures. The constructions are *ad hoc*. Nevertheless, there are three basic techniques to prove representation theorems:

- chopped lattices, used in almost every chapter in Parts III–V;

- boolean triples, used in Chapters 12, 14, and 18, and generalized in Chapter 21; also used in some papers that did not make it in this book, for instance, G. Grätzer and E. T. Schmidt [131];

- cubic extensions, used in most chapters of Part IV.

These are presented in Part II with proofs.

Actually, there are two more basic techniques. *Multi-coloring* (and its recent variant, *quasi-coloring*, see many recent papers of G. Czédli) is used in several relevant papers; however, it appears in the book only in Chapter 19, so we introduce it there. *Pruning* is utilized in Chapters 13 and 16—it would seem to qualify for Part II; however, there are only concrete uses of pruning, there is no general theory to discuss in Part II.

Part III contains the representation theorems of congruence lattices of finite lattices, requiring only chopped lattices from Part II. I cover the following topics:

- The Dilworth Theorem and the representation theorem for *sectionally complemented* lattices in Chapter 8 (G. Grätzer and E. T. Schmidt [117], P. Crawley and R. P. Dilworth [13]; see also [11]).

- *Minimal representations* in Chapter 9; that is, for a given $|J(D)|$, we minimize the size of L representing the finite distributive lattice D (G. Grätzer, H. Lakser, and E. T. Schmidt [102], G. Grätzer, Rival, and N. Zaguia [114]).

- The *semimodular* representation theorem in Chapter 10 (G. Grätzer, H. Lakser, and E. T. Schmidt [105]).

- Another *minimal representations* in Chapter 11 for rectangular lattices (G. Grätzer and E. Knapp [87]).

- The representation theorem for *modular* lattices in Chapter 12 (E. T. Schmidt [174] and G. Grätzer and E. T. Schmidt [134]); we are forced to represent with a countable lattice L, since the congruence lattice of a finite modular lattice is always boolean.

- The representation theorem for *uniform* lattices (that is, lattices in which any two congruence classes of a congruence are of the same size) in Chapter 13 (G. Grätzer, E. T. Schmidt, and K. Thomsen [141]).

Part IV is mostly about congruence-preserving extension. I present the congruence-preserving extension theorem for

- *sectionally complemented* lattices in Chapter 14 (G. Grätzer and E. T. Schmidt [129]);

- *semimodular* lattices in Chapter 15 (G. Grätzer and E. T. Schmidt [132]);

- *isoform* lattices (that is, lattices in which any two congruence classes of a congruence are isomorphic) in Chapter 16 (G. Grätzer, R. W. Quackenbush, and E. T. Schmidt [113]).

These three constructions are based on cubic extensions, introduced in Part II.

In Chapter 17, I present the congruence-preserving extension version of the Baranskiĭ-Urquhart Theorem (V. A. Baranskiĭ [4], [5] and A. Urquhart [188]) on the independence of the congruence lattice and the automorphism group of a finite lattice (see G. Grätzer and E. T. Schmidt [126]).

Finally, in Chapter 18, I discuss two congruence "destroying" extensions, which we call "magic wands" (G. Grätzer and E. T. Schmidt [135], G. Grätzer, M. Greenberg, and E. T. Schmidt [82]).

What happens if we consider the congruence lattices of two related lattices, such as a lattice and a sublattice? I take up three variants of this question in Part V.

Let L be a finite lattice, and let K be a sublattice of L. As we discuss it in Section 3.3, there is a map ext from $\operatorname{Con} K$ into $\operatorname{Con} L$: For a congruence relation α of K, let the image ext α be the congruence relation $\operatorname{con}_L(\alpha)$ of L generated by α. The map ext is a $\{0\}$-separating join-homomorphism.

Chapter 19 proves the converse, a 1974 result of A. P. Huhn [158] and a stronger form due to G. Grätzer, H. Lakser, and E. T. Schmidt [103].

We deal with ideals in Chapter 20. Let K be an ideal of a lattice L. Then the restriction map re: $\operatorname{Con} L \to \operatorname{Con} K$ (which assigns to a congruence α of L, the restriction $\alpha\rceil_K$ of α to K) is a bounded homomorphism. We prove the corresponding representation theorem for finite lattices—G. Grätzer and H. Lakser [90].

We also prove two variants. The first is by G. Grätzer and H. Lakser [97] stating that this result also holds for *sectionally complemented* lattices. The second is by G. Grätzer and H. Lakser [95] stating that this result also holds for *planar* lattices.

The final chapter is a first contribution to the following class of problems. Let $\circledast$ be a construction for finite lattices (that is, if D and E are finite lattices, then so is $D \circledast E$). Find a construction $\circledcirc$ of finite distributive lattices (that is, if K and L are finite distributive lattices, then so is $K \circledcirc L$) satisfying $\operatorname{Con}(K \circledast L) \cong \operatorname{Con} K \circledcirc \operatorname{Con} L$.

If the lattice construction is the direct product, the answer is obvious since $\operatorname{Con}(K \times L) \cong \operatorname{Con} K \times \operatorname{Con} L$.

In Chapter 21, we take up the construction defined as the distributive lattice of all isotone maps from $\operatorname{J}(E)$ to D.

In G. Grätzer and M. Greenberg [78], we introduced another construction: the *tensor extension*, $A[B]$, for nontrivial finite lattices A and B. In Chapter 21, we prove that $\operatorname{Con}(A[B]) \cong (\operatorname{Con} A)[\operatorname{Con} B]$.

In 2013, I raised the question whether one can associate with a finite lattice L a structure of some of its congruences? We could take the ordered set of the principal congruences generated by prime intervals, but this is just $\operatorname{Con}_J L$, which is "equivalent" with $\operatorname{Con} L$ (see Section 2.5.2 for an explanation). We proposed to consider the ordered set $\operatorname{Princ} K$ of principal congruences of a lattice K. In Part VI, we describe the first few results of this field.

It turns out that the class of bounded (not necessarily finite) lattices is the natural setting for this topic.

The highlights include

- The Representation Theorem, characterizing $\operatorname{Princ} K$ of a bounded lattice K as a bounded ordered set, see Chapter 22 (G. Grätzer [65]).

- For a bounded lattice L, the two related structures $\operatorname{Princ} L$ and $\operatorname{Aut} L$ are independent, see Section 22.3 (G. Czédli [28], see also G. Grätzer [76]).

- If K and L are bounded lattices and γ is a bounded homomorphism of K into L, then there is a natural bounded isotone map of $\operatorname{Princ} K$ into $\operatorname{Princ} L$. The main result of Chapter 23 is the converse, every bounded isotone map of $\operatorname{Princ} K$ into $\operatorname{Princ} L$ can be so represented (G. Czédli [26], [27]) and G. Grätzer [75]).

Finally, Part VII deals with another new topic: the congruence structure of finite lattices, focusing on prime intervals and congruences.

The two main results are

- The Prime-Projectivity Lemma, which verifies that, indeed, we can describe the spreading of a congruence from a prime interval to another prime interval involving only prime intervals, see G. Grätzer [69].

- The Swing Lemma, a very strong form of the Prime-Projectivity Lemma, for SPS (slim, planar, and semimodular) lattice, see G. Grätzer [70].

Each chapter in Parts III–VII concludes with an extensive discussion section, giving the background for the topic, further results, and open problems. This book lists more than 80 open problems, hoping to convince the reader that, indeed, we have hardly started. There are almost 200 references and a detailed index.

This book is, as much as possible, visually oriented. I cannot stress too much the use of diagrams as a major research tool in lattice theory. I did not include in the book the list of figures because there is not much use to it; it would list about 150 figures.

Notation and terminology

Lattice-theoretic terminology and notation evolved from the three editions of G. Birkhoff's *Lattice Theory*, [10], by way of my books, [54]–[57], [59], [60], LTF, and R. N. McKenzie, G. F. McNulty, and W. F. Taylor [165], changing quite a bit in the process.

Birkhoff's notation for the congruence lattice and ideal lattice of a lattice changed from $\Theta(L)$ and $I(L)$ to $\operatorname{Con} L$ and $\operatorname{Id} L$, respectively. The advent of LaTeX promoted the use of operators for lattice constructions. I try to be consistent: I use an operator when a new structure is constructed; so I use $\operatorname{Con} L, \operatorname{Id} L, \operatorname{Aut} L$, and so on, without parentheses, unless required for readability, for instance, $\operatorname{J}(D)$ and $\operatorname{Con}(\operatorname{Id} L)$. I use functional notation when sets are constructed, as in $\operatorname{Atom}(L)$ and $\operatorname{J}(a)$. "Generated by" uses the same letters as the corresponding lattice construction, but starting with a lower case letter: $\operatorname{Con} L$ is the congruence lattice of L and $\operatorname{con}(H)$ is the congruence generated by H, while $\operatorname{Id} L$ is the ideal lattice of L and $\operatorname{id}(H)$ is the ideal generated by H.

New concepts introduced in more recent research papers exhibit the usual richness in notation and terminology. I use this opportunity, with the wisdom of hindsight, to make their use more consistent. The reader will often find different notation and terminology when reading the original papers. The detailed Table of Notation and Index may help.

In combinatorial results, I will use Landau's big O notation: for the functions f and g, we write $f = O(g)$ to mean that $|f| \leq C|g|$ for a suitable constant C. Natural numbers start at 1.

Toronto, ON, Canada   George Grätzer
Summer 2016
Homepage: http://www.maths.umanitoba.ca/homepages/gratzer.html/

Part I

A Brief Introduction to Lattices

Basic Concepts

In this chapter we introduce the most basic order theoretic concepts: ordered sets, lattices, diagrams, and the most basic algebraic concepts: sublattices, congruences, products.

1.1. Ordering

1.1.1 Ordered sets

A *binary relation* ϱ on a nonempty set A is a subset of A^2, that is, a set of ordered pairs (a, b), with $a, b \in A$. For $(a, b) \in \varrho$, we will write $a \varrho b$ or $a \equiv b \pmod{\varrho}$.

A binary relation $\leq$ on a set P is called an *ordering* if it is *reflexive* ($a \leq a$ for all $a \in P$), *antisymmetric* ($a \leq b$ and $b \leq a$ imply that $a = b$ for all $a, b \in P$), and *transitive* ($a \leq b$ and $b \leq c$ imply that $a \leq c$ for all $a, b, c \in P$). An *ordered set* $(P, \leq)$ consists of a nonempty set P and an ordering $\leq$.

$a < b$ means that $a \leq b$ and $a \neq b$. We also use the "inverse" relations: $a \geq b$ defined as $b \leq a$ and $a > b$ for $b < a$. If more than one ordering is being considered, we write $\leq_P$ for the ordering of $(P, \leq)$; on the other hand if the ordering is understood, we will say that P (rather than $(P, \leq)$) is an ordered set. An ordered set P is *trivial* if P has only one element.

The elements a and b of the ordered set P are *comparable* if $a \leq b$ or $b \leq a$. Otherwise, a and b are *incomparable*, in notation, $a \parallel b$.

Let $H \subseteq P$ and $a \in P$. Then a is an *upper bound* of H iff $h \leq a$ for all $h \in H$. An upper bound a of H is the *least upper bound* of H iff $a \leq b$ for

© Springer International Publishing Switzerland 2016
G. Grätzer, *The Congruences of a Finite Lattice*,
DOI 10.1007/978-3-319-38798-7_1

any upper bound b of H; in this case, we will write $a = \bigvee H$. If $a = \bigvee H$ exists, then it is unique. By definition, $\bigvee \varnothing$ exist ($\varnothing$ is the empty set) iff P has a smallest element, *zero*, denoted by 0. The concepts of *lower bound* and *greatest lower bound* are similarly defined; the latter is denoted by $\bigwedge H$. Note that $\bigwedge \varnothing$ exists iff P has a largest element, *unit*, denoted by 1. A *bounded ordered set* is one that has both 0 and 1. We often denote the 0 and 1 of P by 0_P and 1_P. The notation $\bigvee H$ and $\bigwedge H$ will also be used for families of elements.

The adverb "similarly" (in "similarly defined") in the previous paragraph can be given concrete meaning. Let $(P, \leq)$ be an ordered set. Then $(P, \geq)$ is also an ordered set, called the *dual* of $(P, \leq)$. The dual of the ordered set P will be denoted by P^d. Now if Φ is a "statement" about ordered sets, and if in Φ we replace all occurrences of $\leq$ by $\geq$, then we get the *dual* of Φ.

Duality Principle for Ordered Sets. *If a statement Φ is true for all ordered sets, then its dual is also true for all ordered sets.*

For $a, b \in P$, if a is an *upper bound* of $\{b\}$, then a is an *upper bound* of b. If for all $a, b \in P$, the set $\{a, b\}$ has an upper bound, then the ordered set P is *directed*.

A *chain* (*linearly ordered set*, *totally ordered set*) is an ordered set with no incomparable elements. An *antichain* is one in which $a \parallel b$ for all $a \neq b$.

Let $(P, \leq)$ be an ordered set and let Q be a nonempty subset of P. Then there is a natural ordering $\leq_Q$ on Q induced by $\leq$: for $a, b \in Q$, let $a \leq_Q b$ iff $a \leq b$; we call $(Q, \leq_Q)$ (or simply, $(Q, \leq)$, or even simpler, Q) an *ordered subset* (or *suborder*) of $(P, \leq)$.

A *chain C in an ordered set P* is a nonempty subset, which, as a suborder, is a chain. An *antichain C in an ordered set P* is a nonempty subset which, as a suborder, is an antichain.

The *length* of a finite chain C, length C, is $|C| - 1$. An ordered set P is said to be *of length n* (in formula, length $P = n$), where n is a natural number iff there is a chain in P of length n and all chains in P are of length $\leq n$.

The ordered sets P and Q are *isomorphic* (in formula, $P \cong Q$) and the map $\psi \colon P \to Q$ is an *isomorphism* iff ψ is one-to-one and onto and

$$a \leq b \text{ in } P \quad \text{iff} \quad \psi a \leq \psi b \text{ in } Q.$$

Let C_n denote the set $\{0, \ldots, n-1\}$ ordered by

$$0 < 1 < 2 < \cdots < n - 1.$$

Then C_n is an n-element chain. Observe that length $C_n = n - 1$. If $C = \{x_0, \ldots, x_{n-1}\}$ is an n-element chain and $x_0 < x_1 < \cdots < x_{n-1}$, then $\psi \colon i \mapsto x_i$ is an isomorphism between C_n and C. Therefore, the n-element chain is unique up to isomorphism.

Let B_n denote the set of all subsets of the set $\{0, \ldots, n-1\}$ ordered by containment. Observe that the ordered set B_n has 2^n elements and length $\mathsf{B}_n = n$. In general, for a set X, we denote by $\mathrm{Pow}\, X$ the *power set* of X, that is, the set of all subsets of X ordered by set inclusion.

For an ordered set P, call $A \subseteq P$ a *down set* iff $x \in A$ and $y \le x$ in P, imply that $y \in A$. For $H \subseteq P$, there is a smallest down set containing H, namely, $\{\, x \mid x \le h,\ \text{for some } h \in H \,\}$; we use the notation $\downarrow H$ for this set. If $H = \{a\}$, we write $\downarrow a$ for $\downarrow \{a\}$. Let $\mathrm{Down}\, P$ denote the set of all down sets ordered by set inclusion. If P is an antichain, then $\mathrm{Down}\, P \cong \mathsf{B}_n$, where $n = |P|$.

The map $\psi \colon P \to Q$ is an *isotone map* (resp., *antitone map*) of the ordered set P into the ordered set Q iff $a \le b$ in P implies that $\psi a \le \psi b$ (resp., $\psi a \ge \psi b$) in Q. Then ψP is a suborder of Q. Even if ψ is one-to-one, the ordered sets P and ψP need not be isomorphic. If both P and Q are bounded, then the map $\psi \colon P \to Q$ is *bounded* or a *bounded map*, if it preserves the bounds, that is, $\psi 0_P = 0_Q$ and $\psi 1_P = 1_Q$. Most often, we talk about bounded isotone maps (and bounded homomorphisms, see Section 1.3.1).

1.1.2 Diagrams

In the ordered set P, the element a *is covered by* b or b *covers* a (in formula, $a \prec b$ or $b \succ a$) iff $a < b$ and $a < x < b$ for no $x \in P$. The binary relation $\prec$ is called the *covering relation*. The covering determines the ordering:

Let P be a finite ordered set. Then $a \le b$ iff $a = b$ or if there exists a finite sequence of elements $x_1, x_2, \ldots, x_n$ such that

$$a = x_1 \prec x_2 \prec \cdots \prec x_n = b.$$

A *diagram* of an ordered set P represents the elements with small circles $\bigcirc$; the circles representing two elements x, y are connected by a line segment iff one covers the other; if x is covered by y, then the circle representing x is placed lower than the circle representing y.

The diagram of a finite ordered set determines the order up to isomorphism.

In a diagram the intersection of two line segments does not indicate an element. A diagram is *planar* if no two line segments intersect. An ordered set P is *planar* if it has a diagram that is planar. Figure 1.1 shows three diagrams of the same ordered set P. Since the third diagram is planar, P is a planar ordered set.

1.1.3 Constructions of ordered sets

Given the ordered sets P and Q, we can form the *direct product* $P \times Q$, consisting of all ordered pairs (x_1, x_2), with $x_1 \in P$ and $x_2 \in Q$, ordered componentwise, that is, $(x_1, x_2) \leq (y_1, y_2)$ iff $x_1 \leq y_1$ and $x_2 \leq y_2$. If $P = Q$, then we write P^2 for $P \times Q$. Similarly, we use the notation P^n for $P^{n-1} \times P$ for $n > 2$. Figure 1.2 shows a diagram of $C_2 \times P$, where P is the ordered set with diagrams in Figure 1.1.

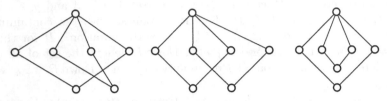

Figure 1.1: Three diagrams of an ordered set.

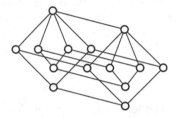

Figure 1.2: A diagram of $C_2 \times P$.

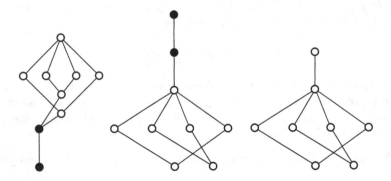

Figure 1.3: Diagrams of $C_2 + P$, $P + C_2$, and $P \dotplus C_2$.

Another often used construction is the (ordinal) *sum* $P + Q$ of P and Q, defined on the (disjoint) union $P \cup Q$ and ordered as follows:

$$x \leq y \quad \text{iff} \quad \begin{cases} x \leq_P y & \text{for } x, y \in P; \\ x \leq_Q y & \text{for } x, y \in Q; \\ x \in P, \ y \in Q. \end{cases}$$

Figure 1.3 shows diagrams of $C_2 + P$ and $P + C_2$, where P is the ordered set with diagrams in Figure 1.1. In both diagrams, the elements of C_2 are black-filled. Figure 1.3 also shows the diagram of $P \dotplus C_2$.

A variant construction is the *glued sum*, $P \dotplus Q$, applied to an ordered set P with largest element 1_P and an ordered set Q with smallest element 0_Q; then $P \dotplus Q$ is $P + Q$ in which 1_P and 0_Q are identified (that is, $1_P = 0_Q$ in $P \dotplus Q$).

1.1.4   Partitions

We now give a nontrivial example of an ordered set. A *partition* of a nonempty set A is a set π of nonempty pairwise disjoint subsets of A whose union is A. The members of π are called the *blocks* of π. The block containing $a \in A$ will be denoted by a/π. A singleton as a block is called *trivial*. If the elements a and b of A belong to the same block, we write $a \equiv b$ (mod π) or $a \pi b$ or $a/\pi = b/\pi$. In general, for $H \subseteq A$,

$$H/\pi = \{\, a/\pi \mid a \in H \,\},$$

a collection of blocks.

An *equivalence relation* ε on the set A is a reflexive, symmetric ($a \varepsilon b$ implies that $b \varepsilon a$, for all $a, b \in A$), and transitive binary relation. Given a partition π, we can define an equivalence relation ε by $(x, y) \in \varepsilon$ iff $x/\pi = y/\pi$. Conversely, if ε is an equivalence relation, then $\pi = \{\, a/\varepsilon \mid a \in A \,\}$ is a partition of A. There is a one-to-one correspondence between partitions and equivalence relations; we will use the two terms interchangeably.

Part A will denote the set of all partitions of A ordered by

$$\pi_1 \leq \pi_2 \quad \text{iff} \quad x \equiv y \pmod{\pi_1} \text{ implies that } x \equiv y \pmod{\pi_2}.$$

We draw a picture of a partition by drawing the boundary lines of the (non-trivial) blocks. Then $\pi_1 \leq \pi_2$ iff the boundary lines of π_2 are also boundary lines of π_1 (but π_1 may have some more boundary lines). Equivalently, the blocks of π_2 are unions of blocks of π_1; see Figure 1.4.

Part A has a zero and a unit, denoted by $\mathbf{0}$ and $\mathbf{1}$, respectively, defined by

$$x \equiv y \pmod{\mathbf{0}} \qquad \text{iff } x = y;$$
$$x \equiv y \pmod{\mathbf{1}} \qquad \text{for all } x, y \in A.$$

Figure 1.5 shows the diagrams of Part A for $|A| \leq 4$. The partitions are labeled by listing the nontrivial blocks.

A *preordered set* is a nonempty set Q with a binary relation $\leq$ that is reflexive and transitive. Let us define the binary relation $a \approx b$ on Q as $a \leq b$ and $b \leq a$. Then $\approx$ is an equivalence relation. Define the set P as $Q/\approx$, and on P define the binary relation $\leq$:

$$a/\approx \; \leq \; b/\approx \quad \text{iff} \quad a \leq b \text{ in } Q.$$

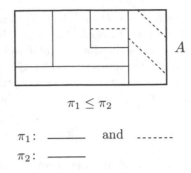

$$\pi_1 \leq \pi_2$$

$\pi_1:$ ——— and - - - - - -

$\pi_2:$ ———

Figure 1.4: Drawing a partition.

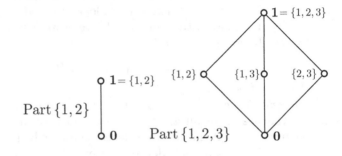

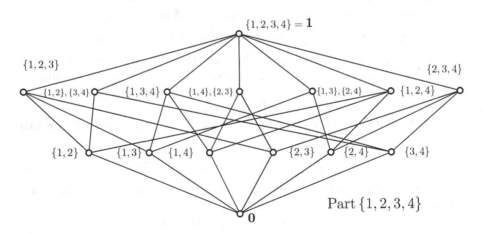

Figure 1.5: Part A for $|A| \leq 4$.

It is easy to see that the definition of $\leq$ on P is well-defined and that P is an ordered set. We will call P the *ordered set associated with the preordered set Q*.

Starting with a binary relation $\prec$ on the set Q, we can define the *reflexive-transitive closure* $\leq$ of $\prec$ by the formula: for $a, b \in Q$, let $a \leq b$ iff $a = b$ or if $a = x_0 \prec x_1 \prec \cdots \prec x_n = b$ for elements $x_1, \ldots, x_{n-1} \in Q$. Then $\leq$ is a preordering on Q. A *cycle* on Q is a sequence $x_1, \ldots, x_n \in Q$ satisfying $x_1 \prec x_2 \prec \cdots \prec x_n \prec x_1$ $(n > 1)$. The preordering $\leq$ is an ordering iff there are no cycles.

1.2. Lattices and semilattices

1.2.1 Lattices

We need two basic concepts from Universal Algebra. An (n-ary) *operation* on a nonempty set A is a map from A^n to A. For $n = 2$, we call the operation *binary*. An *algebra* is a nonempty set A with operations defined on A.

An ordered set $(L, \leq)$ is a *lattice* if $\bigvee\{a, b\}$ and $\bigwedge\{a, b\}$ exist for all $a, b \in L$. A lattice L is *trivial* if it has only one element; otherwise, it is *nontrivial*.

We will use the notations

$$a \vee b = \bigvee\{a, b\},$$
$$a \wedge b = \bigwedge\{a, b\},$$

and call $\vee$ the *join* and $\wedge$ the *meet*. They are both *binary operations* that are *idempotent* ($a \vee a = a$ and $a \wedge a = a$), *commutative* ($a \vee b = b \vee a$ and $a \wedge b = b \wedge a$), *associative* (($a \vee b) \vee c = a \vee (b \vee c)$ and $(a \wedge b) \wedge c = a \wedge (b \wedge c)$), and *absorptive* ($a \vee (a \wedge b) = a$ and $a \wedge (a \vee b) = a$). These properties of the operations are also called the *idempotent identities, commutative identities, associative identities*, and *absorption identities*, respectively. (Identities, in general, are introduced in Section 2.3.) As always in algebra, associativity makes it possible to write $a_1 \vee a_2 \vee \cdots \vee a_n$ without using parentheses (and the same for $\wedge$).

For instance, for $A, B \in \operatorname{Pow} X$, we have $A \vee B = A \cup B$ and $A \wedge B = A \cap B$. So $\operatorname{Pow} X$ is a lattice.

For $\alpha, \beta \in \operatorname{Part} A$, if we regard α and β as equivalence relations, then the meet formula is trivial: $\alpha \wedge \beta = \alpha \cap \beta$, but the formula for joins is a bit more complicated:

$x \equiv y \pmod{\alpha \vee \beta}$ *iff there is a sequence* $x = z_0, z_1, \ldots, z_n = y$ *of elements of A such that* $z_i \equiv z_{i+1} \pmod{\alpha}$ *or* $z_i \equiv z_{i+1} \pmod{\beta}$ *for* $0 \leq i < n$.

So $\operatorname{Part} A$ is a lattice; it is called the *partition lattice* on A.

For an ordered set P, the order $\operatorname{Down} P$ is a lattice: $A \vee B = A \cup B$ and $A \wedge B = A \cap B$ for $A, B \in \operatorname{Down} P$.

To treat lattices as algebras, define an algebra $(L, \vee, \wedge)$ a *lattice* iff L is a nonempty set, $\vee$ and $\wedge$ are binary operations on L, both $\vee$ and $\wedge$ are idempotent, commutative, and associative, and they satisfy the two absorption identities. A lattice as an algebra and a lattice as an ordered set are "equivalent" concepts: Let the order $\mathfrak{L} = (L, \leq)$ be a lattice. Then the algebra $\mathfrak{L}^a = (L, \vee, \wedge)$ is a lattice. Conversely, let the algebra $\mathfrak{L} = (L, \vee, \wedge)$ be a lattice. Define $a \leq b$ iff $a \vee b = b$. Then $\mathfrak{L}^p = (L, \leq)$ is an ordered set, and the ordered set $\mathfrak{L}^p$ is a lattice. For an ordered set $\mathfrak{L}$ that is a lattice, we have $\mathfrak{L}^{ap} = \mathfrak{L}$; for an algebra $\mathfrak{L}$ that is a lattice, we have $\mathfrak{L}^{pa} = \mathfrak{L}$.

Note that for lattices as algebras, the Duality Principle takes on the following very simple form.

Duality Principle for Lattices. *Let Φ be a statement about lattices expressed in terms of $\vee$ and $\wedge$. The dual of Φ is the statement we get from Φ by interchanging $\vee$ and $\wedge$. If Φ is true for all lattices, then the dual of Φ is also true for all lattices.*

If the operations are understood, we will say that L (rather than $(L, \vee, \wedge)$) is a lattice. The dual of the lattice L will be denoted by L^d; the ordered set L^d is also a lattice.

A finite lattice L is *planar* if it is planar as an ordered set (see Section 1.1.2). We have quite a bit of flexibility to construct a planar diagram for an ordered set, but for a lattice, we are much more constrained because L has a zero, which must be the lowest element and a unit, which must be the highest element—contrast this with Figure 1.1. All lattices with five or fewer elements are planar; all but the five chains are shown in the first two rows of Figure 1.6.

The third row of Figure 1.6 provides an example of "good" and "bad" lattice diagrams; the two diagrams represent the same lattice, C_3^2. Planar diagrams are the best. Diagrams in which meets and joins are hard to figure out are not of much value.

In the last row of Figure 1.6 there are two more diagrams. The one on the left is not planar; nevertheless, it is very easy to work with: joins and meets are easy to see (the notation $\mathsf{M}_3[\mathsf{C}_3]$ will be explained in Section 6.1). The one on the right is not a lattice: the two black-filled elements have no join.

In this book, we deal almost exclusively with finite lattices. Some concepts, however, are more natural to introduce in a more general context. An ordered set $(L, \leq)$ is a *complete lattice* if $\bigvee X$ and $\bigwedge X$ exist for all $X \subseteq L$. All finite lattices are complete, of course.

1.2.2   Semilattices and closure systems

A *semilattice* $(S, \circ)$ is an algebra: a nonempty set S with an idempotent, commutative, and associative binary operation $\circ$. A join-semilattice $(S, \vee, \leq)$ is a structure, where $(S, \vee)$ is a semilattice, $(S, \leq)$ is an ordered set, and $a \leq b$ iff $a \vee b = b$. In the ordered set $(S, \leq)$, we have $\bigvee\{a, b\} = a \vee b$. As conventional, we write $(S, \vee)$ for $(S, \vee, \leq)$ or just S if the operation is understood.

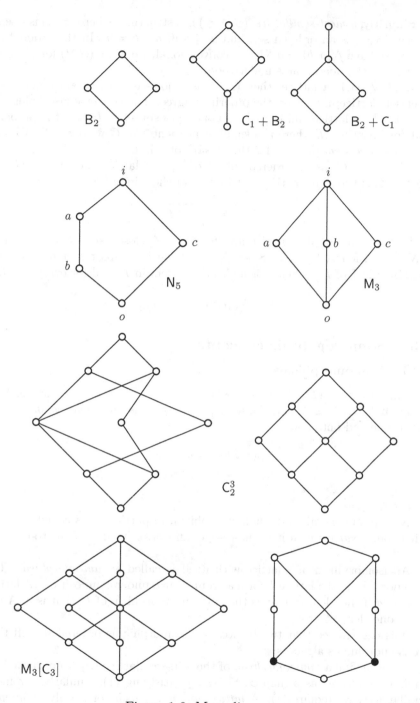

Figure 1.6: More diagrams.

Similarly, a *meet-semilattice* $(S, \wedge, \leq)$ is a structure, where $(S, \wedge)$ is a semi-lattice, $(S, \leq)$ is an ordered set, and $a \leq b$ iff $a \wedge b = a$. In the ordered set $(S, \leq)$, we have $\bigwedge \{a, b\} = a \wedge b$. As conventional, we write $(S, \vee)$ for $(S, \vee, \leq)$ or just S if the operation is understood.

If $(L, \vee, \wedge)$ is a lattice, then $(L, \vee)$ is a join-semilattice and $(L, \wedge)$ is a meet-semilattice; moreover, the orderings agree. The converse also holds.

Let L be a lattice and let C be a nonempty subset of L with the property that for every $x \in L$, there is a smallest element $\overline{x}$ of C with $x \leq \overline{x}$. We call C a *closure system* in L, and $\overline{x}$ the *closure* of x in C.

Obviously, C, as an ordered subset of L, is a lattice: For $x, y \in C$, the meet in C is the same as the meet in L, and the join is

$$x \vee_C y = \overline{x \vee_L y}.$$

Let L be a complete lattice and let C be $\bigwedge$-closed subset of L, that is, if $X \subseteq C$, then $\bigwedge X \in C$. (Since $\bigwedge \varnothing = 1$, such a subset is nonempty and contains the 1 of L.) Then C is a closure system in L, and for every $x \in L$,

$$\overline{x} = \bigwedge (y \in C \mid x \leq y).$$

1.3. Some algebraic concepts

1.3.1 Homomorphisms

The lattices $\mathfrak{L}_1 = (L_1, \vee, \wedge)$ and $\mathfrak{L}_2 = (L_2, \vee, \wedge)$ are *isomorphic* as algebras (in symbols, $\mathfrak{L}_1 \cong \mathfrak{L}_2$), and the map $\varphi \colon L_1 \to L_2$ is an *isomorphism* iff φ is one-to-one and onto and

(1) $$\varphi(a \vee b) = \varphi a \vee \varphi b,$$
(2) $$\varphi(a \wedge b) = \varphi a \wedge \varphi b$$

for $a, b \in L_1$.

A map, in general, and a homomorphism, in particular, is called *injective* if it is onto, *surjective* if it is one-to-one, and *bijective* if it is one-to-one and onto.

An isomorphism of a lattice with itself is called an *automorphism*. The automorphisms of a lattice L form a group Aut L under composition. A lattice L is *rigid* if the identity map is the only automorphism of L, that is, if Aut L is the one-element group.

It is easy to see that two lattices are isomorphic as ordered sets iff they are isomorphic as algebras.

Let us define a *homomorphism* of the join-semilattice $(S_1, \vee)$ into the join-semilattice $(S_2, \vee)$ as a map $\varphi \colon S_1 \to S_2$ satisfying (1); similarly, for meet-semilattices, we require (2). A *lattice homomorphism* (or simply, *homomorphism*) φ of the lattice L_1 into the lattice L_2 is a map of L_1 into L_2 satisfying both (1) and (2). A homomorphism of a lattice into itself is called an *endomorphism*. A one-to-one homomorphism will also be called an *embedding*.

Note that meet-homomorphisms, join-homomorphisms, and (lattice) homomorphisms are all isotone.

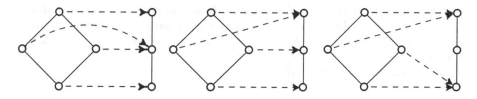

Figure 1.7: Morphisms.

Figure 1.7 shows three maps of the four-element lattice B_2 into the three-element chain C_3. The first map is isotone but it is neither a meet- nor a join-homomorphism. The second map is a join-homomorphism but is not a meet-homomorphism, thus not a homomorphism. The third map is a (lattice) homomorphism.

Various versions of homomorphisms and embeddings will be used. For instance, for lattices and join-semilattices, there are also $\{\vee, 0\}$-homomorphism, and so on, with obvious meanings. An onto homomorphism φ is also called *surjective*, while a one-to-one homomorphism is called *injective*; it is the same as an *embedding*. For bounded lattices, we often use *bounded homomorphisms* and *bounded embeddings*, that is, $\{0, 1\}$-homomorphisms and $\{0, 1\}$-embeddings. (In the literature, bounded homomorphisms sometimes have a different definition; this is unlikely to cause any confusion.)

It should always be clear from the context what kind of homomorphism we are considering. If we say, "let φ be a homomorphism of K into L", where K and L are lattices, then φ is a lattice homomorphism, unless otherwise stated.

1.3.2 Sublattices

A *sublattice* $(K, \vee, \wedge)$ of the lattice $(L, \vee, \wedge)$ is defined on a nonempty subset K of L with the property that $a, b \in K$ implies that $a \vee b, a \wedge b \in K$ (the operations $\vee, \wedge$ are formed in $(L, \vee, \wedge)$), and the $\vee$ and the $\wedge$ of $(K, \vee, \wedge)$ are restrictions to K of the $\vee$ and the $\wedge$ of $(L, \vee, \wedge)$, respectively. Instead of "$(K, \vee, \wedge)$ is a sublattice of $(L, \vee, \wedge)$", we will simply say that "K is a sublattice of L"—in symbols, $K \leq L$. Of course, a sublattice of a lattice is again a lattice. If K is a sublattice of L, then we call L an *extension* of K—in symbols, $L \geq K$.

For every $H \subseteq L$, $H \neq \varnothing$, there is a smallest sublattice $\mathrm{sub}(H) \subseteq L$ containing H called the *sublattice of L generated by H*. We say that H is a *generating set* of $\mathrm{sub}(H)$.

For a bounded lattice L, the sublattice K is *bounded* (also called a $\{0,1\}$-sublattice) if the 0 and 1 of L are in K. Similarly, we can define a $\{0\}$-sublattice, bounded extension, and so on.

The subset K of the lattice L is called *convex* iff $a, b \in K$, $c \in L$, and $a \le c \le b$ imply that $c \in K$. We can add the adjective "convex" to sublattices, extensions, and embeddings. A sublattice K of the lattice L is *convex* if it a convex subset of L. Let L be an extension of K; then L is a *convex extension* if K is a convex sublattice. An embedding is *convex* if the image is a convex sublattice.

For $a, b \in L$, $a \le b$, the *interval*

$$I = [a, b] = \{ x \mid a \le x \le b \}$$

is an important example of a convex sublattice. We will use the notation 1_I for the largest element of I, that is, b and 0_I for the smallest element of I, that is, a.

An interval $[a, b]$ is *trivial* if $a = b$. The smallest nontrivial intervals are called *prime*; that is, $[a, b]$ is *prime* iff $a \prec b$. Another important example of a convex sublattice is an ideal. A nonempty subset I of L is an *ideal* iff it is a down set with the property:

(Id) $a, b \in I$ implies that $a \vee b \in I$.

An ideal I of L is *proper* if $I \ne L$. Since the intersection of any number of ideals is an ideal, unless empty, we can define $\mathrm{id}(H)$, the *ideal generated by a subset H* of the lattice L, provided that $H \ne \varnothing$. If $H = \{a\}$, we write $\mathrm{id}(a)$ for $\mathrm{id}(\{a\})$, and call it a *principal ideal*. Obviously, $\mathrm{id}(a) = \{ x \mid x \le a \} = {\downarrow} a$. So instead of $\mathrm{id}(a)$, we could use ${\downarrow} a$; many do, who work in categorical aspects of lattice theory—and use id for the identity map.

The set $\mathrm{Id}\, L$ of all ideals of L is an ordered set under set inclusion, and as an ordered set it is a lattice. In fact, for $I, J \in \mathrm{Id}\, L$, the lattice operations in $\mathrm{Id}\, L$ are $I \vee J = \mathrm{id}(I \cup J)$ and $I \wedge J = I \cap J$. So we obtain the formula for the ideal join:

$x \in I \vee J$ *iff* $x \le i \vee j$ *for some* $i \in I$, $j \in J$.

We call $\mathrm{Id}\, L$ the *ideal lattice* of L. Now observe the formulas: $\mathrm{id}(a) \vee \mathrm{id}(b) = \mathrm{id}(a \vee b)$, $\mathrm{id}(a) \wedge \mathrm{id}(b) = \mathrm{id}(a \wedge b)$. Since $a \ne b$ implies that $\mathrm{id}(a) \ne \mathrm{id}(b)$, these yield:

The map $a \mapsto \mathrm{id}(a)$ *embeds L into* $\mathrm{Id}\, L$.

Since the definition of an ideal uses only $\vee$ and $\le$, it applies to any join-semilattice S. The ordered set $\mathrm{Id}\, S$ is a join-semilattice and the same join formula holds as the one for lattices. Since the intersection of two ideals could be empty, $\mathrm{Id}\, S$ is not a lattice, in general. However, for a $\{\vee, 0\}$-semilattice (a join-semilattice with zero), $\mathrm{Id}\, S$ is a lattice.

For lattices (join-semilattices) S and T, let $\varepsilon\colon S \to T$ be an embedding. We call ε an *ideal-embedding* if εS is an ideal of T. Then, of course for any ideal I of S, we have that εI is an ideal of T. Ideal-embeddings play a major role in Chapter 20.

By dualizing, we get the concepts of *filter*, fil(H), the *filter generated by a subset* H of the lattice L, provided that $H \neq \varnothing$, *principal filter* fil(a), and so on.

1.3.3 Congruences

An equivalence relation $\boldsymbol{\alpha}$ on a lattice L is called a *congruence relation*, or *congruence*, of L iff $a \equiv b \pmod{\boldsymbol{\alpha}}$ and $c \equiv d \pmod{\boldsymbol{\alpha}}$ imply that

$$(\text{SP}_\vee) \qquad\qquad a \wedge c \equiv b \wedge d \pmod{\boldsymbol{\alpha}},$$
$$(\text{SP}_\wedge) \qquad\qquad a \vee c \equiv b \vee d \pmod{\boldsymbol{\alpha}}$$

(*Substitution Properties*). Trivial examples are the relations $\mathbf{0}$ and $\mathbf{1}$ (introduced in Section 1.1.4). As in Section 1.1.4, for $a \in L$, we write $a/\boldsymbol{\alpha}$ for the *congruence class* (*congruence block*) containing a; observe that $a/\boldsymbol{\alpha}$ is a convex sublattice.

If L is a lattice, $K \leq L$, and $\boldsymbol{\alpha}$ a congruence on L, then $\boldsymbol{\alpha}\rceil K$, the *restriction of $\boldsymbol{\alpha}$ to K*, is a congruence of K. Formally, for $x, y \in K$,

$$x \equiv y \pmod{\boldsymbol{\alpha}\rceil K} \quad \text{iff} \quad x \equiv y \pmod{\boldsymbol{\alpha}} \text{ in } L.$$

We call $\boldsymbol{\alpha}$ *discrete* on K if $\boldsymbol{\alpha}\rceil K = \mathbf{0}$.

Sometimes it is tedious to compute that a binary relation is a congruence relation. Such computations are often facilitated by the following result (G. Grätzer and E. T. Schmidt [116] and F. Maeda [164]), referred to as the Technical Lemma in the literature.

Lemma 1.1. *A reflexive binary relation $\boldsymbol{\alpha}$ on a lattice L is a congruence relation iff the following three properties are satisfied for $x, y, z, t \in L$:*

(i) $x \equiv y \pmod{\boldsymbol{\alpha}}$ *iff* $x \wedge y \equiv x \vee y \pmod{\boldsymbol{\alpha}}$.

(ii) $x \leq y \leq z$, $x \equiv y \pmod{\boldsymbol{\alpha}}$, *and* $y \equiv z \pmod{\boldsymbol{\alpha}}$ *imply that* $x \equiv z \pmod{\boldsymbol{\alpha}}$.

(iii) $x \leq y$ *and* $x \equiv y \pmod{\boldsymbol{\alpha}}$ *imply that* $x \wedge t \equiv y \wedge t \pmod{\boldsymbol{\alpha}}$ *and* $x \vee t \equiv y \vee t \pmod{\boldsymbol{\alpha}}$.

Let $\operatorname{Con} L$ denote the set of all congruence relations on L ordered by set inclusion (remember that we can view $\boldsymbol{\alpha} \in \operatorname{Con} L$ as a subset of L^2).

We use the Technical Lemma to prove the following result.

Theorem 1.2. Con L *is a lattice. For* $\alpha, \beta \in \mathrm{Con}\, L$,

$$\alpha \wedge \beta = \alpha \cap \beta.$$

The join, $\alpha \vee \beta$*, can be described as follows:*
$x \equiv y \pmod{\alpha \vee \beta}$ *iff there is a sequence*

$$x \wedge y = z_0 \leq z_1 \leq \cdots \leq z_n = x \vee y$$

of elements of L *such that* $z_i \equiv z_{i+1} \pmod{\alpha}$ *or* $z_i \equiv z_{i+1} \pmod{\beta}$ *for every* i *with* $0 \leq i < n$.

Remark. For the binary relations γ and δ on a set A, we define the binary relation $\gamma \circ \delta$, the *product of* γ *and* δ, as follows: for $a, b \in A$, the relation $a\, \gamma \circ \delta\, b$ holds iff $a\, \gamma\, x$ and $x\, \delta\, b$ for some $x \in A$. The relation $\alpha \vee \beta$ is formed by repeated products. Theorem 1.2 strengthens this statement.

The integer n in Theorem 1.2 can be restricted for some congruence joins. We call the congruences α and β *permutable* if $\alpha \vee \beta = \alpha \circ \beta$. A lattice L is *congruence permutable* if any pair of congruences of L are permutable. The chain C_n is congruence permutable iff $n \leq 2$.

Con L is called the *congruence lattice* of L. Observe that Con L is a sublattice of Part L; that is, the join and meet of congruence relations as congruence relations and as equivalence relations (partitions) coincide.

If L is nontrivial, then Con L contains the two-element sublattice $\{\mathbf{0}, \mathbf{1}\}$. If Con $L = \{\mathbf{0}, \mathbf{1}\}$, we call the lattice L *simple*. All the nontrivial lattices of Figure 1.5 are simple. Of the many lattices of Figure 1.6, only M_3 is simple.

Given $a, b \in L$, there is a smallest congruence $\mathrm{con}(a, b)$—called a *principal congruence*—under which $a \equiv b$. The formula

$$(3) \qquad \alpha = \bigvee (\,\mathrm{con}(a, b) \mid a \equiv b \pmod{\alpha}\,)$$

is trivial but important. For $H \subseteq L$, the smallest congruence under which H is in one class is formed as $\mathrm{con}(H) = \bigvee (\,\mathrm{con}(a, b) \mid a,\ b \in H\,)$.

Homomorphisms and congruence relations express two sides of the same phenomenon. Let L be a lattice and let α be a congruence relation on L. Let $L/\alpha = \{\, a/\alpha \mid a \in L \,\}$. Define $\wedge$ and $\vee$ on L/α by $a/\alpha \wedge b/\alpha = (a \wedge b)/\alpha$ and $a/\alpha \vee b/\alpha = (a \vee b)/\alpha$. The lattice axioms are easily verified. The lattice L/α is the *quotient lattice* of L modulo α.

Lemma 1.3. *The map*

$$\varphi_\alpha : x \mapsto x/\alpha, \qquad \text{for } x \in L,$$

is a homomorphism of L *onto* L/α.

The lattice K is a *homomorphic image* of the lattice L iff there is a homomorphism of L *onto* K. Theorem 1.4 (illustrated in Figure 1.8) states that any quotient lattice is a homomorphic image. To state it, we need one more concept: Let $\varphi\colon L \to L_1$ be a homomorphism of the lattice L into the lattice L_1, and define the binary relation α on L by $x\,\alpha\,y$ iff $\varphi x = \varphi y$; the relation α is a congruence relation of L, called the *kernel* of φ, in notation, $\ker(\varphi) = \alpha$.

Theorem 1.4 (Homomorphism Theorem). *Let L be a lattice. Any homomorphic image of L is isomorphic to a suitable quotient lattice of L. In fact, if $\varphi\colon L \to L_1$ is a homomorphism of L onto L_1 and α is the kernel of φ, then $L/\alpha \cong L_1$; an isomorphism (see Figure 1.8) is given by $\psi\colon x/\alpha \mapsto \varphi x$ for $x \in L$.*

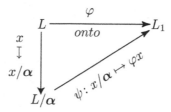

Figure 1.8: The Homomorphism Theorem.

We also know the congruence lattice of a homomorphic image:

Theorem 1.5 (Second Isomorphism Theorem). *Let L be a lattice and let α be a congruence relation of L. For any congruence β of L such that $\beta \geq \alpha$, define the relation β/α on L/α by*

$$x/\alpha \equiv y/\alpha \quad (\text{mod } \beta/\alpha) \qquad \textit{iff} \qquad x \equiv y \quad (\text{mod } \beta).$$

Then β/α is a congruence of L/α. Conversely, every congruence γ of L/α can be (uniquely) represented in the form $\gamma = \beta/\alpha$ for some congruence $\beta \geq \alpha$ of L. In particular, the congruence lattice of L/α is isomorphic with the interval $[\alpha, 1]$ of the congruence lattice of L.

Let L be a bounded lattice. A congruence α of L *separates* 0 if $0/\alpha = \{0\}$, that is, $x \equiv 0 \pmod{\alpha}$ implies that $x = 0$. Similarly, a congruence α of L *separates* 1 if $1/\alpha = \{1\}$, that is, $x \equiv 1 \pmod{\alpha}$ implies that $x = 1$. We call the lattice L *non-separating* if 0 and 1 are not separated by any congruence $\alpha \neq 0$.

Similarly, a homomorphism φ of the lattices L_1 and L_2 with zero is *0-separating* if $\varphi 0 = 0$, but $\varphi x \neq 0$ for $x \neq 0$.

Chapter

2

Special Concepts

In this chapter we introduce special elements, constructions, and classes of lattices that play an important role in the representation of finite distributive lattices as congruence lattices of finite lattices.

2.1. Elements and lattices

In a nontrivial finite lattice L, an element a is *join-reducible* if $a = 0$ or if $a = b \vee c$ for some $b < a$ and $c < a$; otherwise, it is *join-irreducible*. Let $\mathrm{J}(L)$ denote the set of all join-irreducible elements of L, regarded as an ordered set under the ordering of L. By definition, $0 \notin \mathrm{J}(L)$. For $a \in L$, set

$$\mathrm{J}(a) = \{\, x \mid x \le a,\ x \in \mathrm{J}(L) \,\} = \mathrm{id}(a) \cap \mathrm{J}(L),$$

that is, $\mathrm{J}(a)$ is $\downarrow a$ formed in $\mathrm{J}(L)$. Note that, by definition, 0 *is not* a join-irreducible element; and similarly, 1 is not a meet-irreducible element.

In a finite lattice, every element is a join of join-irreducible elements (indeed, $a = \bigvee \mathrm{J}(a)$), and similarly for meets.

Dually, we define *meet-reducible* and *meet-irreducible* elements.

An element a is an *atom* if $0 \prec a$ and a *dual atom* if $a \prec 1$. Atoms are join-irreducible.

A lattice L is *atomistic* if every element is a finite join of atoms.

In a bounded lattice L, the element a is a *complement* of the element b iff $a \wedge b = 0$ and $a \vee b = 1$. A *complemented lattice* is a bounded lattice in

© Springer International Publishing Switzerland 2016
G. Grätzer, *The Congruences of a Finite Lattice*,
DOI 10.1007/978-3-319-38798-7_2

which every element has a complement; the lattices of Figure 1.5 are complemented and so are all but one of the lattices of Figure 1.6. The lattice B_n is complemented.

Let $a \in [b, c]$; the element x is a *relative complement of* a *in* $[b, c]$ iff $a \wedge x = b$, and $a \vee x = c$. A *relatively complemented lattice* is a lattice in which every element has a relative complement in any interval containing it. The lattice N_5 of Figure 2.1 is complemented but not relatively complemented.

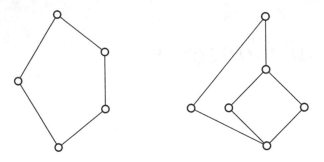

Figure 2.1: The lattices N_5 and N_6.

In a lattice L with zero, let $a \leq b$. A complement of a in $[0, b]$ is called a *sectional complement* of a in b. A lattice L with zero is called *sectionally complemented* if a has a sectional complement in b for all $a \leq b$ in L. The lattice N_6 of Figure 2.1 is sectionally complemented but not relatively complemented.

2.2. Direct and subdirect products

Let L and K be lattices and form the *direct product* $L \times K$ as in Section 1.1.3. Then $L \times K$ is a lattice and $\vee$ and $\wedge$ are computed "componentwise":

$$(a, b) \vee (c, d) = (a \vee c, b \vee d),$$
$$(a, b) \wedge (c, d) = (a \wedge c, b \wedge d).$$

See Figure 2.2 for the example $C_2 \times N_5$.

Obviously, B_n is isomorphic to a direct product of n copies of B_1.

There are two *projection maps* (homomorphisms) associated with this construction:

$$\pi_L \colon L \times K \to L \quad \text{and} \quad \pi_K \colon L \times K \to K,$$

defined by $\pi_L \colon (x, y) \mapsto x$ and by $\pi_K \colon (x, y) \mapsto y$, respectively.

Similarly, we can form the *direct product* $L_1 \times \cdots \times L_n$ with elements $(x_1, \ldots, x_n)$, where $x_i \in L_i$ for $i \leq n$; we denote the projection map

$$(x_1, \ldots, x_i, \ldots, x_n) \mapsto x_i$$

by π_i. If $L_i = L$, for all $i \leq n$, we get the *direct power* L^n. By identifying $x \in L_i$ with $(0, \ldots, 0, x, 0, \ldots, 0)$ (x is the i-th coordinate), we regard L_i as an ideal of $L_1 \times \cdots \times L_n$ for $i \leq n$. The black-filled elements in Figure 2.2 show how we consider C_2 and N_5 ideals of $C_2 \times N_5$.

A very important property of direct products is:

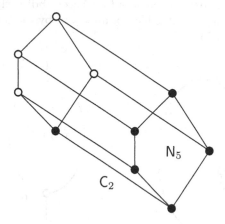

Figure 2.2: $C_2 \times N_5$, a direct product of two lattices.

Theorem 2.1. *Let L and K be lattices, let α be a congruence relation of L, and let β be a congruence relation of K. Define the relation $\alpha \times \beta$ on $L \times K$ by*

$$(a_1, b_1) \equiv (a_2, b_2) \, (\alpha \times \beta) \quad iff \quad a_1 \equiv a_2 \pmod{\alpha} \, and \, b_1 \equiv b_2 \pmod{\beta}.$$

Then $\alpha \times \beta$ is a congruence relation on $L \times K$. Conversely, every congruence relation of $L \times K$ is of this form.

A more general construction is subdirect products. If $L \leq K_1 \times \cdots \times K_n$ and the projection maps π_i are onto maps, for $i \leq n$, then we call L a *subdirect product* of $K_1, \ldots, K_n$.

Trivial examples: L is a subdirect product of L and L if we identify $x \in L$ with $(x, x) \in L^2$ (*diagonal embedding*). In this example, the projection map is an isomorphism. To exclude such trivial cases, let us call a representation of L as a subdirect product of $K_1, \ldots, K_n$ *trivial* if one of the projection maps $\pi_1, \ldots, \pi_n$ is an isomorphism.

A lattice L is called *subdirectly irreducible* iff all representations of L as a subdirect product are trivial. Let L be a subdirect product of K_1 and K_2; then $\ker(\pi_1) \wedge \ker(\pi_2) = 0$. This subdirect product is trivial iff $\ker(\pi_1) = 0$ or $\ker(\pi_2) = 0$. Conversely, if $\alpha_1 \wedge \alpha_2 = 0$ in $\operatorname{Con} K$, then K is a subdirect product of K/α_1 and K/α_2, and this representation is trivial iff $\alpha_1 = 0$ or $\alpha_2 = 0$.

Every simple lattice is subdirectly irreducible. The lattice N_5 is subdirectly irreducible but not simple.

There is a natural correspondence between subdirect representations of a lattice L and sets of congruences $\{\gamma_1, \ldots, \gamma_n\}$ satisfying $\gamma_1 \wedge \cdots \wedge \gamma_n = \mathbf{0}$. This representation is nontrivial iff $\gamma_i \neq \mathbf{0}$ for all $i \leq n$. In this subdirect representation, the factors (the lattices L/γ_i) are subdirectly irreducible iff the congruences γ_i are meet-irreducible, for all $i \leq n$, by the Second Isomorphism Theorem (Theorem 1.5).

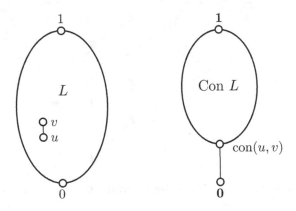

Figure 2.3: A subdirectly irreducible lattice and its congruence lattice.

For a finite lattice L, the lattice $\operatorname{Con} L$ is finite, so we can represent $\mathbf{0}$ as a meet of meet-irreducible congruences, and we obtain:

Theorem 2.2. *Every finite lattice L is a subdirect product of subdirectly irreducible lattices.*

This result (*Birkhoff's Subdirect Representation Theorem*) is true for any algebra in any variety (a class of algebras defined by identities, such as the class of all lattices or the class of all groups).

Finite subdirectly irreducible lattices are easy to recognize. If L is such a lattice, then the meet α of all the $> \mathbf{0}$ elements is $> \mathbf{0}$. Obviously, α is an atom, the unique atom of $\operatorname{Con} L$. Conversely, if $\operatorname{Con} L$ has a unique atom, then all $> \mathbf{0}$ congruences are $\geq \alpha$, so their meet cannot be $\mathbf{0}$. We call α the *base congruence* of L (called *monolith* in many publications).

If $u \neq v$ and $u \equiv v \pmod{\alpha}$, then $\alpha = \operatorname{con}(u, v)$. So

$$\operatorname{Con} L = \{\mathbf{0}\} \cup [\operatorname{con}(u, v), \mathbf{1}],$$

as illustrated in Figure 2.3.

Let L be a finite subdirectly irreducible lattice with $\operatorname{con}(u, v)$ the base congruence, where $u \prec v \in L$. By inserting two elements as shown in Figure 2.4, we embed L into a simple lattice.

Lemma 2.3. *Every finite subdirectly irreducible lattice can be embedded into a simple lattice with at most two extra elements.*

Note that every finite lattice can be embedded into a finite simple lattice; in general, we need more than two elements. For a stronger statement, see Lemma 14.3.

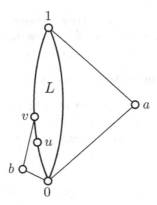

Figure 2.4: Embedding into a simple lattice.

2.3. Terms and identities

From the variables $x_1, \ldots, x_n$, we can form (n-*ary*) *terms* in the usual manner using $\vee, \wedge$, and parentheses. Examples of terms are: x_1, x_3, $x_1 \vee x_1$, $(x_1 \wedge x_2) \vee (x_3 \wedge x_1)$, $(x_3 \wedge x_1) \vee ((x_3 \vee x_2) \wedge (x_1 \vee x_2))$.

An n-ary term p defines a function in n variables (a *term function*, or simply, a *term*) on a lattice L. For example, if

$$p = (x_1 \wedge x_3) \vee (x_3 \vee x_2)$$

and $a, b, c \in L$, then

$$p(a, b, c) = (a \wedge c) \vee (c \vee b) = b \vee c.$$

If p is a *unary* ($n = 1$) lattice term, then $p(a) = a$ for any $a \in L$. If p is *binary*, then $p(a, b) = a$, or $p(a, b) = b$, or $p(a, b) = a \wedge b$, or $p(a, b) = a \vee b$ for all $a, b \in L$.

If $p = p(x_1, \ldots, x_n)$ is an n-ary term and L is a lattice, then by substituting some variables by elements of L, we get a function on L of n-variables. Such functions are called *term functions*. Unary term functions of the form

$$p(x) = p(a_1, \ldots, a_{i-1}, x, a_{i+1}, \ldots, x_n),$$

where $a_1, \ldots, a_{i-1}, a_{i+1}, \ldots, a_n \in L$, play the most important role, see Section 3.1.

A term (function), in fact, any term function, p is *isotone*; that is, if $a_1 \leq b_1, \ldots, a_n \leq b_n$, then $p(a_1, \ldots, a_n) \leq p(b_1, \ldots, b_n)$. Furthermore,

$$a_1 \wedge \cdots \wedge a_n \leq p(a_1, \ldots, a_n) \leq a_1 \vee \cdots \vee a_n.$$

Note that many publications in Lattice Theory and Universal Algebra use *polynomials* and *polynomial functions* for terms and term functions.

Terms have many uses. We briefly discuss three.

(i) The sublattice generated by a set

Lemma 2.4. *Let L be a lattice and let H be a nonempty subset of L. Then $a \in \mathrm{sub}(H)$ (the sublattice generated by H) iff $a = p(h_1, \ldots, h_n)$ for some integer $n \geq 1$, for some n-ary term p, and for some $h_1, \ldots, h_n \in H$.*

(ii) Identities

A *lattice identity* (resp., *lattice inequality*)—also called *equation*—is an expression of the form $p = q$ (resp., $p \leq q$), where p and q are terms. An *identity* $p = q$ (resp., $p \leq q$) *holds in the lattice L* iff $p(a_1, \ldots, a_n) = q(a_1, \ldots, a_n)$ (resp., $p(a_1, \ldots, a_n) \leq q(a_1, \ldots, a_n)$) holds for all $a_1, \ldots, a_n \in L$. The identity $p = q$ is equivalent to the two inequalities $p \leq q$ and $q \leq p$; the inequality $p \leq q$ is equivalent to the identity $p \vee q = q$.

The most important properties of identities are given by

Lemma 2.5. *Identities are preserved under the formation of sublattices, homomorphic images, direct products, and ideal lattices.*

A lattice L is called *distributive* if the identities

$$x \wedge (y \vee z) = (x \wedge y) \vee (x \wedge z),$$
$$x \vee (y \wedge z) = (x \vee y) \wedge (x \vee z)$$

hold in L. In fact, it is enough to assume one of these identities, because the two identities are equivalent.

As we have just noted, the identity $x \wedge (y \vee z) = (x \wedge y) \vee (x \wedge z)$ is equivalent to the two inequalities:

$$x \wedge (y \vee z) \leq (x \wedge y) \vee (x \wedge z),$$
$$(x \wedge y) \vee (x \wedge z) \leq x \wedge (y \vee z).$$

However, the second inequality holds in any lattice. So a lattice is distributive iff the inequality $x \wedge (y \vee z) \leq (x \wedge y) \vee (x \wedge z)$ holds. By duality, we get a similar statement about the second identity defining distributivity.

The class of all distributive lattices will be denoted by **D**. A *boolean lattice* is a distributive complemented lattice. A finite boolean lattice is isomorphic to some B_n.

A lattice is called *modular* if the identity

$$(x \wedge y) \vee (x \wedge z) = x \wedge (y \vee (x \wedge z))$$

holds. Note that this identity is equivalent to the following implication:

$$x \geq z \text{ implies that } (x \wedge y) \vee z = x \wedge (y \vee z).$$

The class of all modular lattices will be denoted by **M**.

Every distributive lattice is modular. The lattice M_3 is modular but not distributive. All the lattices of Figures 1.5 and 1.6 are modular except for Part $\{1, 2, 3, 4\}$ and N_5.

A class of lattices **V** is called a *variety* if it is defined by a set of identities. The classes **D** and **M** are examples of varieties, and so are **L**, the variety of all lattices and **T**, the (trivial) variety of one-element lattices.

(iii) Free lattices

Starting with a set H, we can form the set of all terms over H, collapsing two terms if their equality follows from the lattice axioms. We thus form the "free-est" lattice over H. For instance, if we start with $H = \{a, b\}$, then we obtain the four-element lattice, $F(2)$, of Figure 2.5.

We obtain more interesting examples if we start with an ordered set P, and require that the ordering in P be preserved. For instance, if we start with $P = \{a, b, c\}$ with $a < b$, then we get the corresponding nine-element "free" lattice, $F(P)$, of Figure 2.5.

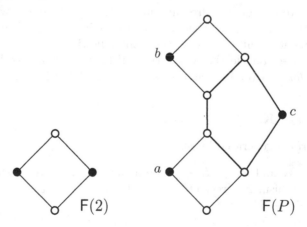

Figure 2.5: Two free lattices.

Sometimes, we need free lattices with respect to some special conditions. The following result illustrates this.

Lemma 2.6. *Let x, y, and z be elements of a lattice L and let $x \vee y$, $y \vee z$, $z \vee x$ be pairwise incomparable. Then* $\operatorname{sub}(\{x \vee y, y \vee z, z \vee x\}) \cong \mathsf{B}_3$.

Lemma 2.6 is illustrated by Figure 2.6.

We will also need "free distributive lattices", obtained by collapsing two terms if their equality follows from the lattice axioms and the distributive identities. Starting with a three-element set $H = \{x, y, z\}$, we then obtain the lattice, $\operatorname{Free}_D(3)$, of Figure 2.7.

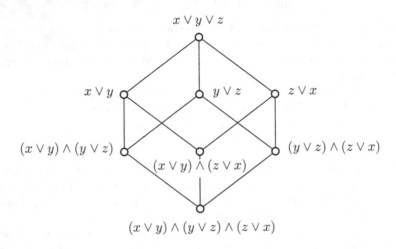

Figure 2.6: A free lattice with special relations.

Similarly, we can define "free modular lattices." Starting with a three-element set $H = \{x, y, z\}$, we then obtain the lattice, $\operatorname{Free}_M(3)$, of Figure 2.8.

An equivalent definition of freeness is the following:

Let H be a set and let $\mathbf{K}$ be a variety of lattices. A lattice $\operatorname{Free}_{\mathbf{K}}(H)$ is called a *free lattice* over $\mathbf{K}$ *generated by* H iff the following three conditions are satisfied:

(i) $\operatorname{Free}_{\mathbf{K}}(H) \in \mathbf{K}$.

(ii) $\operatorname{Free}_{\mathbf{K}}(H)$ is generated by H.

(iii) Let $L \in \mathbf{K}$ and let $\psi \colon H \to L$ be a map; then there exists a (lattice) homomorphism $\varphi \colon \operatorname{Free}_{\mathbf{K}}(H) \to L$ extending ψ (that is, satisfying $\varphi a = \psi a$ for all $a \in H$).

2.4. Gluing

In Section 1.1.3 glued sums of ordered sets, $P \dotplus Q$, applied to an ordered set P with largest element 1_P and an ordered set Q with smallest element 0_Q

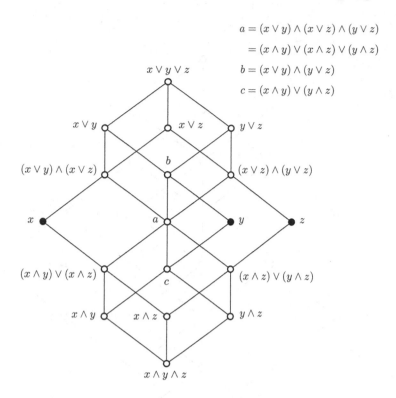

$$a = (x \vee y) \wedge (x \vee z) \wedge (y \vee z)$$
$$= (x \wedge y) \vee (x \wedge z) \vee (y \wedge z)$$
$$b = (x \vee y) \wedge (y \vee z)$$
$$c = (x \wedge y) \vee (y \wedge z)$$

Figure 2.7: The free distributive lattice on three generators, $\mathrm{Free}_{\mathbf{D}}(3)$.

were introduced. This applies to any two lattices K with a unit and L with a zero. A natural generalization of this construction is gluing.

Let K and L be lattices, let F be a filter of K, and let I be an ideal of L. If F is isomorphic to I (with φ the isomorphism), then we can form the lattice G, the *gluing* of K and L over F and I (with respect to φ), defined as follows:

We form the disjoint union $K \cup L$, and identify $a \in F$ with $\varphi a \in I$, for all $a \in F$, to obtain the set G. We order G as follows (see Figure 2.9):

$$a \leq b \quad \text{iff} \quad \begin{cases} a \leq_K b & \text{if } a, b \in K; \\ a \leq_L b & \text{if } a, b \in L; \\ a \leq_K x \text{ and } \varphi x \leq_L b & \text{if } a \in K \text{ and } b \in L; \\ & \text{for some } x \in F. \end{cases}$$

$$u = (x \wedge y) \vee (y \wedge z) \vee (x \wedge z)$$
$$v = (x \vee y) \wedge (y \vee z) \wedge (x \vee z)$$
$$x_1 = (x \wedge v) \vee u$$
$$y_1 = (y \wedge v) \vee u$$
$$z_1 = (z \wedge v) \vee u$$

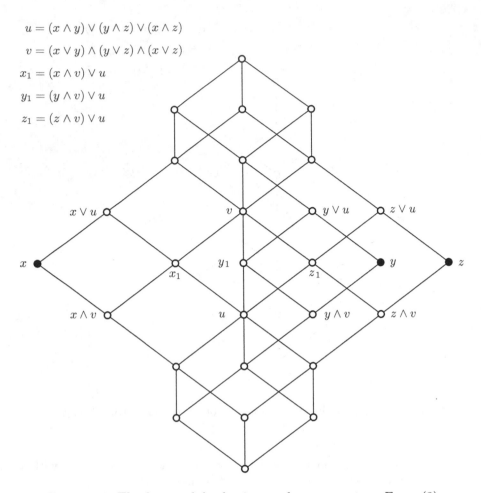

Figure 2.8: The free modular lattice on three generators, $\text{Free}_M(3)$.

Lemma 2.7. *G is an ordered set, in fact, G is a lattice. The join in G is described by*

$$a \vee_G b = \begin{cases} a \vee_K b & \text{if } a, b \in K; \\ a \vee_L b & \text{if } a, b \in L; \\ \varphi(a \vee_K x) \vee_L b & \text{if } a \in K \text{ and } b \in L \text{ for any } b \geq x \in F, \end{cases}$$

and dually for the meet. If L has a zero, 0_L, then the last clause for the join may be rephrased:

$$a \vee_G b = \varphi(a \vee_K 0_L) \vee_L b \quad \text{if } a \in K \text{ and } b \in L.$$

G contains K and L as sublattices; in fact, K is an ideal and L is a filter of G.

An example of gluing is shown in Figure 2.10. There are more sophisticated examples in this book; see, for instance, Chapters 12 and 18.

Lemma 2.8. *Let K, L, F, I, and G be given as above. Let A be a lattice containing K and L as sublattices so that $K \cap L = I = F$. Then $K \cup L$ is a sublattice of A and it is isomorphic to G.*

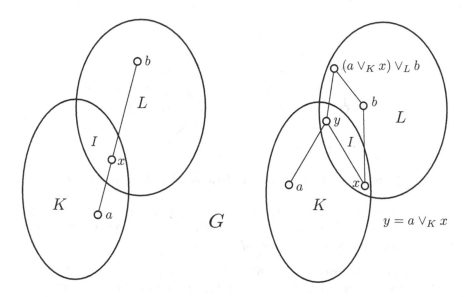

Figure 2.9: Defining gluing.

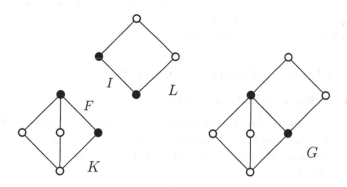

Figure 2.10: An easy gluing example.

Now if α_K is a binary relation on K and α_L is a binary relation on L, we define the *reflexive product* $\alpha_K \overset{r}{\circ} \alpha_L$ as $\alpha_K \cup \alpha_L \cup (\alpha_K \circ \alpha_L)$.

We can easily describe the congruences of G.

Lemma 2.9. *A congruence* α *of* G *can be uniquely written in the form*

$$\alpha = \alpha_K \overset{r}{\circ} \alpha_L,$$

where α_K *is a congruence of* K *and* α_L *is a congruence of* L *satisfying the condition that* α_K *restricted to* F *equals* α_L *restricted to* I *(under the identification of elements by* φ*).*

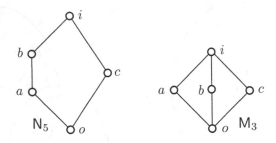

Figure 2.11: The two characteristic nondistributive lattices.

Conversely, if α_K *is a congruence of* K *and* α_L *is a congruence of* L *satisfying the condition that* α_K *restricted to* F *equals* α_L *restricted to* I, *then* $\alpha = \alpha_K \overset{r}{\circ} \alpha_L$ *is a congruence of* G.

Let A and B be lattices, F_A a filter of A, I_B an ideal of B, and F_B a filter of B. Let us assume that F_A, I_B, and F_B are isomorphic. We now define what it means to obtain C by *gluing* B to A *k-times*. For $k = 1$, let C be the gluing of A and B over F_A and I_B with the filter F_B regarded as a filter F_C of C. Now if C_{k-1} with the filter $F_{C_{k-1}}$ is the gluing of B to A $k-1$-times, then we glue C_{k-1} and B over $F_{C_{k-1}}$ and I_B to obtain C the gluing of B to A k-times with the filter F_B regarded as a filter F_C of C.

2.5. Modular and distributive lattices

2.5.1   The characterization theorems

The two typical examples of nondistributive lattices are N_5 and M_3, whose diagrams are given (again) in Figure 2.11. The following characterization theorem follows immediately by inspecting the diagrams of the free lattices in Figures 2.5 and 2.8.

Theorem 2.10.

  (i) *A lattice* L *is modular iff it does not contain* N_5 *as a sublattice.*

  (ii) *A modular lattice* L *is distributive iff it does not contain* M_3 *as a sublattice.*

(iii) *A lattice L is distributive iff L contains neither N_5 nor M_3 as a sublattice.*

Theorem 2.11. *Let L be a modular lattice, let $a \in L$, and let U and V be sublattices with the property $u \wedge v = a$ for all $u \in U$ and $v \in V$. Then $\mathrm{sub}(U \cup V)$ is isomorphic to $U \times V$ under the isomorphism ($u \in U$ and $v \in V$)*

$$u \vee v \mapsto (u, v).$$

Conversely, a lattice L satisfying this property is modular.

Corollary 2.12. *Let L be a modular lattice and let $a, b \in L$. Then*

$$\mathrm{sub}([a \wedge b, a] \cup [a \wedge b, b]),$$

that is, the sublattice of L generated by $[a \wedge b, a] \cup [a \wedge b, b]$, is isomorphic to the direct product

$$[a \wedge b, a] \times [a \wedge b, b].$$

In the distributive case, the sublattice generated by $[a \wedge b, a] \cup [a \wedge b, b]$ is the interval $[a \wedge b, a \vee b]$; this does not hold for modular lattices, as exemplified by M_3.

Let G be the gluing of the lattices K and L over F and I, as in Section 2.4.

Lemma 2.13. *If K and L are modular, so is the gluing G of K and L. If K and L are distributive, so is G.*

The distributive identity easily implies that every n-ary term equals one we get by joining meets of variables. So we get:

Lemma 2.14. *A finitely generated distributive lattice is finite.*

2.5.2 Finite distributive lattices

For a nontrivial finite distributive lattice D, the ordered set $\mathrm{J}(D)$ is "equivalent" to the lattice D in the following sense:

Theorem 2.15. *Let D be a nontrivial finite distributive lattice. Then the map*

$$\varphi \colon a \mapsto \mathrm{J}(a)$$

is an isomorphism between D and $\mathrm{Down}(\mathrm{J}(D))$.

Corollary 2.16. *The correspondence $D \mapsto \mathrm{J}(D)$ makes the class of all nontrivial finite distributive lattices correspond to the class of all finite ordered sets; isomorphic lattices correspond to isomorphic ordered sets, and vice versa.*

In particular, $D \cong \mathrm{Down}(\mathrm{J}(D))$ and $P \cong \mathrm{J}(\mathrm{Down}\,P)$.

Let D and E be nontrivial finite distributive lattices, and let $\varphi\colon D \to E$ be a bounded homomorphism. Then with every $x \in \mathrm{J}(E)$, we can associate the smallest $y \in D$ with $\varphi y \geq x$. It turns out that $y \in \mathrm{J}(D)$, so we obtain an isotone map $\mathrm{J}(\varphi)\colon \mathrm{J}(E) \to \mathrm{J}(D)$.

Let P and Q be ordered sets, and let $\psi\colon P \to Q$ be a isotone map. Then with every $I \in \mathrm{Down}\,Q$, we can associate $\psi^{-1}I$. It turns out that $\psi^{-1} \in \mathrm{Down}\,P$, so we obtain the isotone map $\mathrm{Down}(\psi)\colon \mathrm{Down}\,Q \to \mathrm{Down}\,P$.

Theorem 2.17. *Let D and E be nontrivial finite distributive lattices, and let $\varphi\colon D \to E$ be a bounded homomorphism. Let φ_D and φ_E be the isomorphisms between D and $\mathrm{Down}(\mathrm{J}(D))$ and between E and $\mathrm{Down}(\mathrm{J}(E))$, respectively. Then the diagram*

$$
\begin{array}{ccc}
D & \xrightarrow{\ \varphi_D\ } & \mathrm{Down}(\mathrm{J}(D)) \\[4pt]
{\scriptstyle \varphi}\Big\downarrow & & \Big\downarrow{\scriptstyle \mathrm{Down}(\mathrm{J}(\varphi))} \\[4pt]
E & \xrightarrow{\ \varphi_E\ } & \mathrm{Down}(\mathrm{J}(E))
\end{array}
$$

commutes, that is, $\mathrm{Down}(\mathrm{J}(\varphi))\varphi_D = \varphi_E\varphi$.

Let U be a finite order. If U is an antichain, then $\mathrm{Pow}\,U \cong \mathrm{Down}\,U$, the finite boolean lattice of the power set of H. Since $\mathrm{Down}(\mathrm{J}(D))$ is a sublattice of the finite boolean lattice of all subsets of $\mathrm{J}(D)$, we get

Corollary 2.18. *Every finite distributive lattice D can be embedded into a finite boolean lattice. If D is nontrivial, then it can be embedded into B_n, where $n = |\mathrm{J}(D)|$*

In Theorem 2.15, instead of $\mathrm{J}(D)$ and down sets, we could work with the dual concepts: $\mathrm{M}(D)$ (the ordered set of meet-irreducible elements of D) and up-sets. However, surprisingly, $\mathrm{J}(D)$ and $\mathrm{M}(D)$ are isomorphic. To see this, for $a \in \mathrm{M}(D)$, let $a^\dagger$ denote the smallest element x of D not below a (that is, with $x \not\leq a$). By distributivity, it is easy to see that $a^\dagger \in \mathrm{J}(D)$; in fact, we have the following.

Theorem 2.19. *Let D be a nontrivial finite distributive lattice. Then the map*

$$\varphi\colon a \mapsto a^\dagger$$

is an isomorphism between the ordered sets $\mathrm{M}(D)$ and $\mathrm{J}(D)$.

2.5.3   Finite modular lattices

Take a look at the two positions of the pair of intervals $[a,b]$ and $[c,d]$ in Figure 2.12. In either case, we will write $[a,b] \sim [c,d]$, and say that $[a,b]$ is *perspective* to $[c,d]$. If we want to show whether the perspectivity is "up" or "down", we will write $[a, b] \overset{\text{up}}{\sim} [c, d]$ in the first case and $[a, b] \overset{\text{dn}}{\sim} [c, d]$ in the second case.

If for some natural number n and intervals $[e_i, f_i]$, for $0 \leq i \leq n$,

$$[a, b] = [e_0, f_0] \sim [e_1, f_1] \sim \cdots \sim [e_n, f_n] = [c, d],$$

then we say that $[a, b]$ is *projective* to $[c, d]$ and write $[a, b] \approx [c, d]$.

One of the most important properties of a modular lattice is stated in the following result:

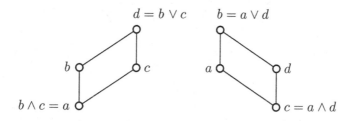

Figure 2.12: $[a, b] \overset{\text{up}}{\sim} [c, d]$ and $[a, b] \overset{\text{dn}}{\sim} [c, d]$.

Theorem 2.20 (Isomorphism Theorem for Modular Lattices). *Let L be a modular lattice and let $[a, b] \overset{\text{up}}{\sim} [c, d]$ in L. Then*

$$\varphi_c \colon x \mapsto x \vee c, \quad x \in [a, b],$$

is an isomorphism of $[a, b]$ and $[c, d]$. The inverse isomorphism is

$$\psi_b \colon y \mapsto y \wedge b, \quad y \in [c, d].$$

(See Figure 2.13.)

Corollary 2.21. *In a modular lattice, projective intervals are isomorphic.*

Corollary 2.22. *In a modular lattice if a prime interval $\mathfrak{p}$ is projective to an interval $\mathfrak{q}$, then $\mathfrak{q}$ is also prime.*

Let us call the finite lattice L *semimodular* or *upper semimodular* if for $a, b, c \in L$, the covering $a \prec b$ implies that $a \vee c \prec b \vee c$ or $a \vee c = b \vee c$. The dual of an upper semimodular lattice is a *lower semimodular* lattice.

Lemma 2.23. *A modular lattice is both upper and lower semimodular. For a finite lattice, the converse also holds: a finite upper and lower semimodular lattice is modular, and conversely.*

The lattice S_8, in Figure 2.14, is an example of a semimodular lattice that is not modular. See Section 10.2 for an interesting use of this lattice.

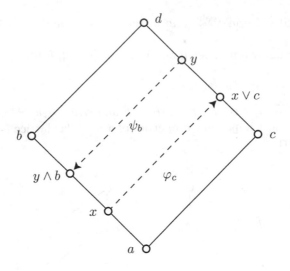

Figure 2.13: The isomorphisms φ_c and ψ_b.

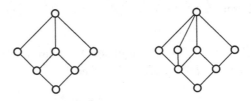

Figure 2.14: The lattices S_7 and S_8.

The following is even more trivial than Lemma 2.13:

Lemma 2.24. *If K and L are finite semimodular lattices, so is the gluing G of K and L.*

A large class of semimodular lattices is provided by

Lemma 2.25. *Let A be a nonempty set. Then* Part A *is semimodular; it is not modular unless $|A| \leq 3$.*

Congruences

3.1. Congruence spreading

Let a, b, c, d be elements of a lattice L. If

$$a \equiv b \pmod{\boldsymbol{\alpha}} \quad \text{implies that} \quad c \equiv d \pmod{\boldsymbol{\alpha}},$$

for any congruence relation $\boldsymbol{\alpha}$ of L, then we can say that $a \equiv b$ *congruence-forces* $c \equiv d$.

In Section 1.3.3 we saw that $a \equiv b \pmod{\boldsymbol{\alpha}}$ iff $a \wedge b \equiv a \vee b \pmod{\boldsymbol{\alpha}}$; therefore, to investigate congruence-forcing, it is enough to deal with comparable pairs, $a \leq b$ and $c \leq d$. Instead of comparable pairs, we will deal with intervals $[a, b]$ and $[c, d]$.

Projectivity (see Section 2.5.3) is sufficient for the study of congruence-forcing (or congruence spreading) in some classes of lattices (for instance, in the class of modular lattices). In general, however, we have to introduce somewhat more general concepts and notation.

As illustrated in Figure 3.1, we say that $[a, b]$ is *up congruence-perspective* onto $[c, d]$ and write $[a, b] \overset{\text{up}}{\twoheadrightarrow} [c, d]$ iff there is an $a_1 \in [a, b]$ with $[a_1, b] \overset{\text{up}}{\sim} [c, d]$; similarly, $[a, b]$ is *down congruence-perspective* onto $[c, d]$ and we shall write $[a, b] \overset{\text{dn}}{\twoheadrightarrow} [c, d]$ iff there is a $b_1 \in [a, b]$ with $[a, b_1] \overset{\text{dn}}{\sim} [c, d]$. If $[a, b] \overset{\text{up}}{\twoheadrightarrow} [c, d]$ or $[a, b] \overset{\text{dn}}{\twoheadrightarrow} [c, d]$, then $[a, b]$ is *congruence-perspective* onto $[c, d]$ and we write $[a, b] \twoheadrightarrow [c, d]$. If for some natural number n and intervals $[e_i, f_i]$, for $0 \leq i \leq n$,

$$[a, b] = [e_0, f_0] \twoheadrightarrow [e_1, f_1] \twoheadrightarrow \cdots \twoheadrightarrow [e_n, f_n] = [c, d],$$

© Springer International Publishing Switzerland 2016
G. Grätzer, *The Congruences of a Finite Lattice*,
DOI 10.1007/978-3-319-38798-7_3

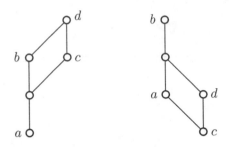

Figure 3.1: $[a,b] \stackrel{\text{up}}{\twoheadrightarrow} [c,d]$ and $[a,b] \stackrel{\text{dn}}{\twoheadrightarrow} [c,d]$.

then we call $[a,b]$ *congruence-projective* onto $[c,d]$, and we write $[a,b] \Rightarrow [c,d]$. If $[a,b] \Rightarrow [c,d]$ and $[c,d] \Rightarrow [a,b]$, then we write $[a,b] \Leftrightarrow [c,d]$.

Also if (see Section 2.5.3) $[a,b] \stackrel{\text{up}}{\sim} [c,d]$, then $[a,b] \stackrel{\text{up}}{\twoheadrightarrow} [c,d]$ and $[c,d] \stackrel{\text{dn}}{\twoheadrightarrow} [a,b]$; if $[a,b] \stackrel{\text{dn}}{\sim} [c,d]$, then $[a,b] \stackrel{\text{dn}}{\twoheadrightarrow} [c,d]$ and $[c,d] \stackrel{\text{up}}{\twoheadrightarrow} [a,b]$; if $[a,b] \sim [c,d]$, then $[a,b] \twoheadrightarrow [c,d]$ and $[c,d] \twoheadrightarrow [a,b]$; if $[a,b] \approx [c,d]$, then $[a,b] \Rightarrow [c,d]$ and $[c,d] \Rightarrow [a,b]$. But while $\sim$, $\approx$, and $\Leftrightarrow$ are symmetric, the relations $\twoheadrightarrow$ and $\Rightarrow$ are not. In particular, if $a \leq c \leq d \leq b$, then $[a,b] \Rightarrow [c,d]$.

If $[a,b] \Rightarrow [c,d]$, then there is a unary term function p with $p(a) = c$ and $p(c) = d$. It is easy to see what special kinds of unary term functions are utilized in the relation $\Rightarrow$.

Intuitively, "$a \equiv b$ congruence-forces $c \equiv d$" iff $[c,d]$ is put together from pieces $[c',d']$ each satisfying $[a,b] \Rightarrow [c',d']$. To state this more precisely, we describe $\mathrm{con}(a,b)$, the smallest congruence relation under which $a \equiv b$ (introduced in Section 1.3.3); see R. P. Dilworth [44]. We use the Technical Lemma to prove this result.

Theorem 3.1. *Let L be a lattice, $a,b,c,d \in L$, $a \leq b$, $c \leq d$. Then $c \equiv d$ (mod $\mathrm{con}(a,b)$) iff, for some sequence*

$$c = e_0 \leq e_1 \leq \cdots \leq e_m = d,$$

we have

$$[a,b] \Rightarrow [e_j, e_{j+1}], \qquad for \quad j = 0, \ldots, m-1.$$

This result can be usefully augmented by the following lemma; for an application see Lemma 23.2.

Lemma 3.2. *Let L be a lattice, $a,b,c,d \in L$ with $a \leq b$ and $c \leq d$. Then $[a,b]$ is congruence-projective to $[c,d]$ iff the following condition is satisfied: There is an integer m and there are elements $e_0, \ldots, e_{m-1} \in L$ such that*

$$p_m(a, e_0, \ldots, e_{m-1}) = c,$$
$$p_m(b, e_0, \ldots, e_{m-1}) = d,$$

where the term p_m is defined by

$$p_m(x, y_0, \ldots, y_{m-1}) = \cdots (((x \vee y_0) \wedge y_1) \vee y_2) \wedge \cdots.$$

Let L be a lattice and $H \subseteq L^2$. To compute $\mathrm{con}(H)$, the smallest congruence relation $\boldsymbol{\alpha}$ under which $a \equiv b \pmod{\boldsymbol{\alpha}}$, for all $(a, b) \in H$, we use the formula

$$\mathrm{con}(H) = \bigvee (\, \mathrm{con}(a, b) \mid (a, b) \in H \,).$$

We also need a formula for joins:

Lemma 3.3. *Let L be a lattice and let $\boldsymbol{\alpha}_i$, $i \in I$, be congruence relations of L. Then $a \equiv b \pmod{\bigvee (\boldsymbol{\alpha}_i \mid i \in I)}$ iff there is a sequence*

$$z_0 = a \wedge b \leq z_1 \leq \cdots \leq z_n = a \vee b$$

such that, for each j with $0 \leq j < n$, there is an $i_j \in I$ satisfying $z_j \equiv z_{j+1}$ $\pmod{\boldsymbol{\alpha}_{i_j}}$.

This is an easy but profoundly important result. For instance, the well-known result of N. Funayama and T. Nakayama [53]—which provides the foundation for this book—immediately follows. (Another typical application is Lemma 3.13.)

Theorem 3.4. *The lattice $\mathrm{Con}\, L$ is distributive for any lattice L.*

Proof. Let $\boldsymbol{\alpha}, \boldsymbol{\beta}, \boldsymbol{\gamma} \in \mathrm{Con}\, L$. As we note on page 24, to verify the identity $\boldsymbol{\alpha} \wedge (\boldsymbol{\beta} \vee \boldsymbol{\gamma}) = (\boldsymbol{\alpha} \wedge \boldsymbol{\beta}) \vee (\boldsymbol{\alpha} \wedge \boldsymbol{\gamma})$, it is sufficient to verify the inequality $\boldsymbol{\alpha} \wedge (\boldsymbol{\beta} \vee \boldsymbol{\gamma}) \leq (\boldsymbol{\alpha} \wedge \boldsymbol{\beta}) \vee (\boldsymbol{\alpha} \wedge \boldsymbol{\gamma})$. So let $a \equiv b \pmod{\boldsymbol{\alpha} \wedge (\boldsymbol{\beta} \vee \boldsymbol{\gamma})}$. Then $a \equiv b$ $\pmod{\boldsymbol{\alpha}}$ and so $a \wedge b \equiv a \vee b \pmod{\boldsymbol{\alpha}}$, and also $a \equiv b \pmod{\boldsymbol{\beta} \vee \boldsymbol{\gamma}}$, so by Lemma 3.3, there is a sequence

$$z_0 = a \wedge b \leq z_1 \leq \cdots \leq z_n = a \vee b$$

such that $z_j \equiv z_{j+1} \pmod{\boldsymbol{\beta}}$ or $z_j \equiv z_{j+1} \pmod{\boldsymbol{\gamma}}$ for each j with $0 \leq j < n$. For each j, we have $z_j \equiv z_{j+1} \pmod{\boldsymbol{\alpha}}$, so either $z_j \equiv z_{j+1} \pmod{\boldsymbol{\alpha} \wedge \boldsymbol{\gamma}}$ or $z_j \equiv z_{j+1} \pmod{\boldsymbol{\beta} \wedge \boldsymbol{\gamma}}$, proving that $a \equiv b \pmod{(\boldsymbol{\alpha} \wedge \boldsymbol{\beta}) \vee (\boldsymbol{\alpha} \wedge \boldsymbol{\gamma})}$. $\square$

By combining Theorem 3.1 and Lemma 3.3, we get:

Corollary 3.5. *Let L be a lattice, let $H \subseteq L^2$, and let $a, b \in L$ with $a \leq b$. Then $a \equiv b \pmod{\mathrm{con}(H)}$ iff, for some integer n, there exists a sequence*

$$a = c_0 \leq c_1 \leq \cdots \leq c_n = b$$

such that, for each i with $0 \leq i < n$, there exists a $(d_i, e_i) \in H$ satisfying

$$[d_i \wedge e_i, d_i \vee e_i] \Rightarrow [c_i, c_{i+1}].$$

There is another congruence structure we can associate with a lattice L: the ordered set of principal congruences, $\operatorname{Princ} L$, a subset of $\operatorname{Con} L$. We dedicate Part VI to the study of this structure.

Lemma 3.6. *For a lattice L, the ordered set $\operatorname{Princ} L$ is a directed ordered set with zero. If L is bounded, so is $\operatorname{Princ} L$.*

Indeed, for $a, b, c, d \in L$, an upper bound of $\operatorname{con}(a, b)$ and $\operatorname{con}(c, d)$ is $\operatorname{con}(a \wedge b \wedge c \wedge d, a \vee b \vee c \vee d)$.

3.2.   Finite lattices and prime intervals

There is a much stronger version of the Technical Lemma for finite lattices, see G. Grätzer [72].

Lemma 3.7. *Let L be a finite lattice. Let δ be an equivalence relation on L with intervals as equivalence classes. Then δ is a congruence relation iff the following condition, $(C_\vee)$, and its dual, $(C_\wedge)$, hold:*
$(C_\vee)$
   If x is covered by $y, z \in L$ and $x \equiv y \pmod{\delta}$, then $z \equiv y \vee z \pmod{\delta}$.

Proof. To prove the join-substitution property: if $x \leq y$ and $x \equiv y \pmod{\delta}$, then

$$(1) \qquad\qquad x \vee z \equiv y \vee z \pmod{\delta}.$$

This is trivial if $y = z$, so we assume that $y \neq z$. Clearly, we can also assume that $x < y$ and $x < z$.

Let $U = [x, y \vee z]$. We induct on $\operatorname{length} U$, the length of U.

Using the fact that the intersection of two convex sublattices is either $\varnothing$ or a convex sublattice, it follows that, for every interval V of L, the classes of $\delta \rceil V$ are intervals. Hence, we can assume that $x = y \wedge z$; indeed, otherwise the induction hypothesis applies to $V = [y \wedge z, y \vee z]$ and $\delta \rceil V$, since $\operatorname{length} V < \operatorname{length} U$, yielding (1).

Note that $\operatorname{length} U \geq 2$. If $\operatorname{length} U = 2$, then (1) is stated in $(C_\vee)$.

So we can also assume that $\operatorname{length} U > 2$. Pick the elements $y_1, z_1 \in L$ so that $x \prec y_1 \leq y$ and $x \prec z_1 \leq z$. The elements y_1 and z_1 are distinct, since $y_1 = z_1$ would contradict that $x = y \wedge z = y_1 \wedge z_1$. Let $w = y_1 \vee z_1$. Since the δ-classes are intervals, $x \equiv y_1 \pmod{\delta}$, therefore, $(C_\vee)$ yields that

$$(2) \qquad\qquad z_1 \equiv w \pmod{\delta}.$$

Let $I = [y_1, y \vee z]$ and $J = [z_1, y \vee z]$. Then $\operatorname{length} I, \operatorname{length} J < \operatorname{length} U$. Hence, the induction hypothesis applies to I and $\delta \rceil I$, and we obtain that $w \equiv y \vee w \pmod{\delta}$. Combining this with (2), by the transitivity of δ, we conclude that

$$(3) \qquad\qquad z_1 \equiv y \vee w \pmod{\delta}.$$

Therefore, applying the induction hypothesis to J and $\delta\rceil J$, we conclude from (3) that

$$x \vee z = z \vee z_1 \equiv z \vee (y \vee w) = y \vee z \quad (\text{mod } \delta),$$

proving (1).

We get the meet-substitution property by duality.   □

This lemma found applications in some recent papers of G. Czédli and myself.

Let $\text{Prime}(L)$ denote the set of prime intervals of a finite lattice L. Let $\mathfrak{p} = [a, b]$ be a prime interval in L (that is, $a \prec b$ in L) and let α be a congruence relation of L.

In a finite lattice L, the formula

$$\alpha = \bigvee(\, \text{con}(\mathfrak{p}) \mid 0_\mathfrak{p} \equiv 1_\mathfrak{p} \quad (\text{mod } \alpha)\,)$$

immediately yields that the congruences in $J(\text{Con } L)$ (to be denoted by $\text{Con}_J L$) are the congruences of the form $\text{con}(\mathfrak{p})$ for some $\mathfrak{p} \in \text{Prime}(L)$. Of course, a join-irreducible congruence α can be expressed, as a rule, in many ways in the form $\text{con}(\mathfrak{p})$.

Since a prime interval cannot contain a three element chain, the following two lemmas easily follow from the Technical Lemma and from Corollary 3.5.

Lemma 3.8. *Let L be a finite lattice and let $\mathfrak{p}$ and $\mathfrak{q}$ be prime intervals in L. Then $\text{con}(\mathfrak{p}) \geq \text{con}(\mathfrak{q})$ iff $\mathfrak{p} \Rightarrow \mathfrak{q}$.*

This condition is easy to visualize using Figure 3.2; the sequence of congruence-perspectivities, as a rule, has to go through nonprime intervals (intervals of arbitrary size) to get from $\mathfrak{p}$ to $\mathfrak{q}$.

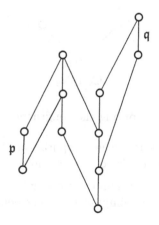

Figure 3.2: Congruence spreading from prime interval to prime interval.

Lemma 3.9. *Let L be a finite lattice, let $\mathfrak{p}$ be a prime interval of L, and let $[a,b]$ be an interval of L. If $\mathfrak{p}$ is collapsed under $\mathrm{con}(a,b)$, then there is a prime interval $\mathfrak{q}$ in $[a,b]$ satisfying $\mathfrak{q} \Rightarrow \mathfrak{p}$.*

In view of Theorem 2.15, we get the following:

Theorem 3.10. *Let L be a finite lattice. The relation $\Rightarrow$ is a preordering on $\mathrm{Prime}(L)$. The equivalence classes under $\Leftrightarrow$ form an ordered set isomorphic to $\mathrm{Con_J}\,L$.*

If the finite lattice L is atomistic, then the join-irreducible congruences are even simpler to find. Indeed if $[a,b]$ is a prime interval, and p is an atom with $p \le b$ and $p \nleq a$, then $[a,b] \overset{\mathrm{dn}}{\sim} [0,p]$, so $\mathrm{con}(a,b) = \mathrm{con}(0,p)$.

Corollary 3.11. *Let L be a finite atomistic lattice. Then every join-irreducible congruence can be represented in the form $\mathrm{con}(0,p)$, where p is an atom. The relation $p \Rightarrow q$, defined as $[0,p] \Rightarrow [0,q]$, introduces a preordering on the set of atoms of L. The equivalence classes under the preordering form an ordered set isomorphic to $\mathrm{Con_J}\,L$.*

We use Figure 3.3 to illustrate how we compute the congruence lattice of N_5 using Theorem 3.10. N_5 has five prime intervals: $[o,a]$, $[a,b]$, $[b,i]$, $[o,c]$, $[c,i]$. The equivalence classes are $\alpha = \{[a,b]\}$, $\beta = \{[o,c],[b,i]\}$, and $\gamma = \{[o,a],[c,i]\}$. The ordering $\alpha < \gamma$ holds because $[c,i] \overset{\mathrm{dn}}{\twoheadrightarrow} [o,b] \overset{\mathrm{up}}{\twoheadrightarrow} [a,b]$. Similarly, $\alpha < \beta$.

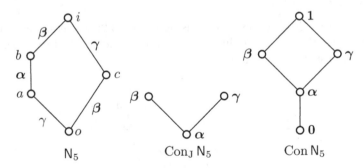

Figure 3.3: Computing the congruence lattice of N_5.

It is important to note that the computation of $\mathfrak{p} \Rightarrow \mathfrak{q}$ may involve non-prime intervals. For instance, let $\mathfrak{p} = [o,c]$ and $\mathfrak{q} = [a,b]$ in N_5. Then $\mathfrak{p} \Rightarrow \mathfrak{q}$, because $\mathfrak{p} \overset{\mathrm{up}}{\twoheadrightarrow} [a,i] \overset{\mathrm{dn}}{\twoheadrightarrow} \mathfrak{q}$, but we cannot get $\mathfrak{p} \Rightarrow \mathfrak{q}$ involving only prime intervals.

As another example, we compute the congruence lattice of S_8; see Figure 3.4.

By Corollary 2.23, in a modular lattice if $\mathfrak{p}$ and $\mathfrak{q}$ are prime intervals, then $\mathfrak{p} \overset{\mathrm{up}}{\twoheadrightarrow} \mathfrak{q}$ implies that $\mathfrak{p} \overset{\mathrm{up}}{\sim} \mathfrak{q}$ and $\mathfrak{p} \overset{\mathrm{dn}}{\twoheadrightarrow} \mathfrak{q}$ implies that $\mathfrak{p} \overset{\mathrm{dn}}{\sim} \mathfrak{q}$; thus $\mathfrak{p} \Rightarrow \mathfrak{q}$ implies

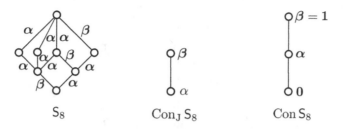

S_8 $\mathrm{Con_J}\, S_8$ $\mathrm{Con}\, S_8$

Figure 3.4: Computing the congruence lattice of S_8.

that $\mathfrak{p} \approx \mathfrak{q}$. Therefore, Theorem 3.10 tells us that $\mathrm{Con_J}\, L$ is an antichain in a finite modular lattice L, so $\mathrm{Con}\, L$ is boolean (as noted in Section 2.5.2).

Corollary 3.12. *The congruence lattice of a finite modular lattice is boolean.*

By a *colored lattice* we will mean a finite lattice (some) of whose prime intervals are labeled so that if the prime intervals $\mathfrak{p}$ and $\mathfrak{q}$ are of the same color, then $\mathrm{con}(\mathfrak{p}) = \mathrm{con}(\mathfrak{q})$. These labels represent (a subset of) the equivalence classes of prime intervals, as stated in Theorem 3.10. In Figure 3.4, every prime interval of S_8 is labeled; in Figure 9.2, only some are labeled. The coloring helps in the intuitive understanding of some constructions.

If the prime interval $\mathfrak{p}$ is of color c, then we define $\mathrm{con}(c)$ as $\mathrm{con}(\mathfrak{p})$.

3.3.   Congruence-preserving extensions and variants

Let L be a lattice and $K \leq L$. How do the congruences of L relate to the congruences of K?

Every congruence $\boldsymbol{\alpha}$ *restricts* to K: the relation $\boldsymbol{\alpha} \cap K^2 = \boldsymbol{\alpha}\rceil K$ on K is a congruence of K. So we get the *restriction map*:

$$\mathrm{re}\colon \mathrm{Con}\, L \to \mathrm{Con}\, K,$$

that maps a congruence $\boldsymbol{\alpha}$ of L to $\boldsymbol{\alpha}\rceil K$.

Lemma 3.13. *Let $K \leq L$ be lattices. Then* $\mathrm{re}\colon \mathrm{Con}\, L \to \mathrm{Con}\, K$ *is a* $\{\wedge, 0, 1\}$-*homomorphism.*

For instance, if $K = \{o, a, i\}$ and $L = \mathsf{M}_3$ (see Figure 2.11), then $\mathrm{Con}\, K$ is isomorphic to B_2, but only $\mathbf{0}$ and $\mathbf{1}$ are restrictions of congruences in L. As another example, take the lattice L of Figure 3.5 and its sublattice K, the black-filled elements; in this case, $\mathrm{Con}\, L \cong \mathrm{Con}\, K \cong \mathsf{B}_2$, but again only $\mathbf{0}$ and $\mathbf{1}$ are restrictions. There is no natural relationship between the congruences of K and L.

If K is an ideal in L (or any convex sublattice), we can say a lot more.

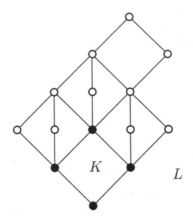

Figure 3.5: Illustrating the map re.

Lemma 3.14. *Let $K \leq L$ be lattices. If K is an ideal of L, then* re: $\mathrm{Con}\, L \to$ $\mathrm{Con}\, K$ *is a bounded homomorphism.*

Proof. By Lemma 3.13, the map re is a $\{\wedge, 0, 1\}$-homomorphism. Let α and β be congruences of L; we have to prove that

$$\alpha{\rceil}K \vee \beta{\rceil}K = (\alpha \vee \beta){\rceil}K.$$

Since $\leq$ is trivial, we prove $\geq$. So let $a, b \in K$, $a \equiv b \pmod{(\alpha \vee \beta){\rceil}K}$; we want to prove that $a \equiv b \pmod{\alpha{\rceil}K \vee \beta{\rceil}K}$. By Lemma 3.3, there is a sequence

$$z_0 = a \wedge b \leq z_1 \leq \cdots \leq z_n = a \vee b$$

such that, for each j with $0 \leq j < n$, either $z_j \equiv z_{j+1} \pmod{\alpha}$ or $z_j \equiv z_{j+1}$ $\pmod{\beta}$ holds in L. Since $a, b \in K$ and K is an ideal, it follows that $z_0, z_1,$ $\ldots, z_n \in K$, so for each j with $0 \leq j < n$, either $z_j \equiv z_{j+1} \pmod{\alpha{\rceil}K}$ or $z_j \equiv z_{j+1} \pmod{\beta{\rceil}K}$ holds, proving that $a \equiv b \pmod{\alpha{\rceil}K \vee \beta{\rceil}K}$. $\square$

Let $K \leq L$ be lattices, and let α be a congruence of K. The congruence $\mathrm{con}_L(\alpha)$ (the congruence $\mathrm{con}(\alpha)$ formed in L) is the smallest congruence γ of L such that $\alpha \leq \gamma{\rceil}K$. Unfortunately, $\mathrm{con}_L(\alpha){\rceil}K$ may be different from α, as in the example of Figure 3.5. We say that the congruence α of K *extends* to L, iff α is the restriction of $\mathrm{con}_L(\alpha)$. Figure 3.6 illustrates this in part. If a congruence α extends, then the congruence classes of α in K extend to congruence classes in L, but there may be congruence classes in L that are not such extensions.

The *extension map*:

$$\mathrm{ext}\colon \mathrm{Con}\, K \to \mathrm{Con}\, L$$

maps a congruence α of K to the congruence $\mathrm{con}_L(\alpha)$ of L. The map ext is a $\{\vee, 0\}$-homomorphism of $\mathrm{Con}\,K$ into $\mathrm{Con}\,L$. In addition, ext preserves the zero, that is, ext is $\{0\}$-*separating*. To summarize:

Lemma 3.15. *Let $K \leq L$ be lattices. Then* ext: $\mathrm{Con}\,K \to \mathrm{Con}\,L$ *is a $\{0\}$-separating join-homomorphism.*

The extension L of K is a *congruence-reflecting extension* (and the sublattice K of L is a *congruence-reflecting sublattice*) if every congruence of K extends to L.

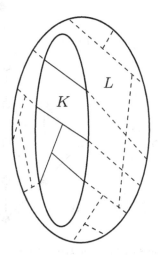

Figure 3.6: A congruence extends.

Utilizing the results of Sections 3.1 and 3.2, we can find many equivalent formulations of the congruence-reflecting property for finite lattices:

Lemma 3.16. *Let L be a finite lattice, and $K \leq L$. Then the following conditions are equivalent:*

(i) *K is a congruence-reflecting sublattice of L.*

(ii) *L is a congruence-reflecting extension of K.*

(iii) *Let $\mathfrak{p}$ and $\mathfrak{q}$ be prime intervals in K; if $\mathfrak{p} \Rightarrow \mathfrak{q}$ in L, then $\mathfrak{p} \Rightarrow \mathfrak{q}$ in K.*

As an example, the reader may want to verify that any sublattice of a distributive lattice is congruence-reflecting.

A much stronger concept—central to this book—is the following. Let K be a lattice. A lattice L is a *congruence-preserving extension* of K (or K is a *congruence-preserving sublattice* of L) if L is an extension and every congruence α of K has *exactly one* extension $\overline{\alpha}$ to L satisfying $\overline{\alpha}\rceil_K = \alpha$.

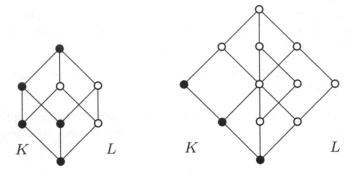

Figure 3.7: Examples of congruence-preserving extensions.

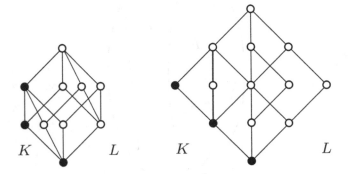

Figure 3.8: Examples of not congruence-preserving extensions.

Of course, $\overline{\alpha} = \mathrm{con}_L(\alpha)$. It follows that $\alpha \mapsto \mathrm{con}_L(\alpha)$ is an isomorphism between $\mathrm{Con}\,K$ and $\mathrm{Con}\,L$.

Two congruence-preserving extensions are shown in Figure 3.7, while Figure 3.8 shows two other extensions that are not congruence-preserving.

We can also obtain congruence-preserving extensions using gluing, based on the following result:

Lemma 3.17. *Let K and L be lattices, let F be a filter of K, and let I be an ideal of L. Let φ be an isomorphism between F and I. Let G be the gluing of K and L over F and I with respect to φ. If L is a congruence-preserving extension of I, then G is a congruence-preserving extension of K.*

If I and L are simple, then L is a congruence-preserving extension of I. So we obtain:

Corollary 3.18. *Let K, L, F, I, and φ be given as above. If I and L are simple lattices, then G is a congruence-preserving extension of K.*

Lemma 3.19. *Let the lattice L be an extension of the lattice K. Then L is a congruence-preserving extension of K iff the following two conditions hold:*

(i) $\operatorname{re}(\operatorname{ext}\alpha) = \alpha$ *for any congruence α of K.*

(ii) $\operatorname{ext}(\operatorname{re}\alpha) = \alpha$ *for any congruence α of L.*

We can say a lot more for finite lattices:

Lemma 3.20. *Let L be a finite lattice, and $K \leq L$. Then L is a congruence-preserving extension of K iff the following two conditions hold:*

(a) *Let $\mathfrak{p}$ and $\mathfrak{q}$ be prime intervals in K; if $\mathfrak{p} \Rightarrow \mathfrak{q}$ in L, then $\mathfrak{p} \Rightarrow \mathfrak{q}$ in K.*

(b) *Let $\mathfrak{p}$ be a prime interval of L. Then there exist a prime interval $\mathfrak{q}$ in K such that $\mathfrak{p} \Leftrightarrow \mathfrak{q}$ in L.*

Lemma 3.19.(ii) is very interesting by itself. It says that every congruence α of L is determined by its restriction to K. In other words,

$$\alpha = \operatorname{con}_L(\alpha\rceil K).$$

We will call such a sublattice K a *congruence-determining sublattice*. We can easily modify Lemma 3.20 to characterize congruence-determining sublattices in finite lattices:

Lemma 3.21. *Let L be a finite lattice L, and $K \leq L$. Then K is a congruence-determining sublattice of L iff for any prime interval $\mathfrak{p}$ in L, there is a prime interval $\mathfrak{q}$ in K satisfying $\mathfrak{p} \Leftrightarrow \mathfrak{q}$ in L.*

Of course, a congruence-preserving sublattice is always congruence-determining. In fact, a sublattice is congruence-preserving iff it is congruence-reflecting and congruence-determining.

Planar Semimodular Lattices

4.1. Basic concepts

Many concepts discussed in this section are diagram dependent. This is discussed in detail in G. Czédli and G. Grätzer [33], a chapter in LTS1.

A planar lattice L has a *left boundary chain*, $C_l(L)$, and a *right boundary chain*, $C_r(L)$. If C and D are maximal chains in the interval $[a, b]$, and there is no element of L between C and D, then we call $C \cup D$ a *cell*. A four-element cell is a *4-cell* or *covering square*, that is, cover-preserving four-element boolean sublattice of L. A diagram of M_3 has exactly two 4-cells and three covering squares.

A planar lattice is called a *4-cell lattice* if all of its cells are 4-cells. For example, M_3 is a 4-cell lattice but N_5 is not.

Planar semimodular lattices can be characterized by cells, see G. Grätzer and E. Knapp [85, Lemmas 4 and 5].

Lemma 4.1. *Let L be a planar lattice.*

(i) *If L is semimodular, then it is a 4-cell lattice. If A, B are 4-cells of L with the same bottom, then these 4-cells have the same top.*

(ii) *If L has a planar 4-cell diagram E in which no two 4-cells with the same bottom have distinct tops, then L is semimodular.*

As defined in G. Grätzer and E. Knapp [85], a semimodular lattice L is *slim* if L contains no covering M_3 sublattice. Let L be a slim semimodular lattice. Two prime intervals of L are *consecutive* if they are opposite sides of a 4-cell.

As in G. Czédli and E. T. Schmidt [38], maximal sequences of consecutive prime intervals form a *trajectory*.So a trajectory is an equivalence class of the transitive reflexive closure of the "consecutive" relation. See Figure 4.1 for two examples.

We denote by Traj L the set of all trajectories of L.

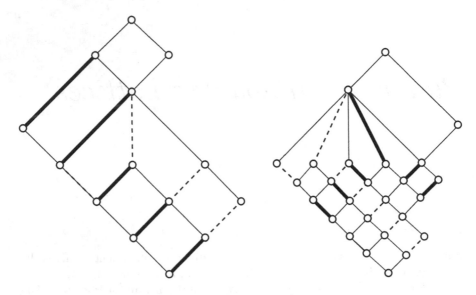

Figure 4.1: Two trajectories.

4.2.   SPS lattices

An *SPS lattice* is a slim, planar, semimodular lattice.

For the following lemma, see G. Grätzer and E. Knapp [85, Lemma 6].

Lemma 4.2. *An SPS lattice L is distributive iff* S_7 *(see Figure 2.14) is not a cover-preserving sublattice of L.*

For an overview of this topic, see G. Czédli and G. Grätzer [33], Chapter 3 of LTS1.

Let us call the elements $u, v, w \in L$ *pairwise disjoint over the element a* provided that $a = u \wedge v = v \wedge w = w \wedge u$.

The first and third statement of the next lemma can be found in the literature (see G. Grätzer and E. Knapp [85]–[88], G. Czédli and E. T. Schmidt [40]–[41]).

Lemma 4.3. *Let L be an SPS lattice.*

(i) *An element of L has at most two covers.*

(ii) *If the elements $u, v, w \in L$ are pairwise disjoint over a, then two of them are comparable.*

(iii) *Let $x \in L$ cover three distinct elements u, v, and w. Then the set $\{u, v, w\}$ generates an S_7 sublattice.*

Lemma 4.3(i) and (ii) state in different ways that there are only two directions "to go up" from an element. The next lemma states this in one more way. This important statement follows from G. Czédli and E. T. Schmidt [38, Lemma 2.8].

Lemma 4.4. *Let L be an SPS lattice. Let $\mathfrak{q}, \mathfrak{q}_1, \mathfrak{q}_2$ be pairwise distinct prime intervals of L satisfying $\mathfrak{q}_1 \overset{\mathrm{dn}}{\sim} \mathfrak{q}$ and $\mathfrak{q}_2 \overset{\mathrm{dn}}{\sim} \mathfrak{q}$. Then $\mathfrak{q}_1 \sim \mathfrak{q}_2$.*

SPS lattices are meet-semidistributive; see LTF for the relevant definitions, especially Section 2.7 by K. Adaricheva. For a current overview of semidistributivity, see the chapters by K. Adaricheva and J. B. Nation in the second volume of LTS1, [153].

4.3. Forks

Our goal is to construct all planar semimodular lattices from planar distributive lattices. This section is based on G. Czédli and E. T. Schmidt [40].

We construct a lattice extension $L[S]$ of L as follows.

Firstly, we replace S by a copy of S_7, the lattice of Figure 2.14, introducing three new elements.

Secondly, we do a series of steps—each step introducing one new element: if there is a chain $u \prec v \prec w$ such that v is a new element but u and w are not, and $T = \{u \wedge x, x, u, w = x \vee u\}$ is a 4-cell in the original lattice L, see Figure 4.2, then we insert a new element y such that $x \wedge u \prec y \prec x \vee u$ and $y \prec v$. Figure 4.2 illustrates the construction.

Let $L[S]$ denote the lattice we obtain when the procedure terminates. We say that $L[S]$ is obtained from L by *inserting a fork* at the 4-cell S. It is easy to see that $L[S]$ is an SPS lattice. For the new elements we shall use the notation of Figure 4.4.

In an SPS lattice L, an element u covering $n \geq 3$ elements together generate, up to isomorphism a unique sublattice, we call this sublattice a multifork. For $n = 3$, it is S_7, for $n = 4$, we get the lattice of Figure 4.3; see G. Czédli [18].

Consider the trajectory containing (determined by) a prime interval $\mathfrak{p}$; the part of the trajectory to the left of $\mathfrak{p}$ (including $\mathfrak{p}$) is called the *left wing* of $\mathfrak{p}$; see G. Czédli and G. Grätzer [32]. The left wing of $[o, a_l]$ is also called the *left wing* of the covering square $S = \{o, a_l, a_r, t\}$. We define the *right wing* symmetrically. Note that trajectories start and end in prime intervals on the boundary. So a left wing starts with a prime interval on the boundary. For a fork, we use the notation of Figure 4.4. This notation will be utilized in Chapter 24.

The direct product of two nontrivial chains is a *grid*.

Now we can state the main result of G. Czédli and E. T. Schmidt [40].

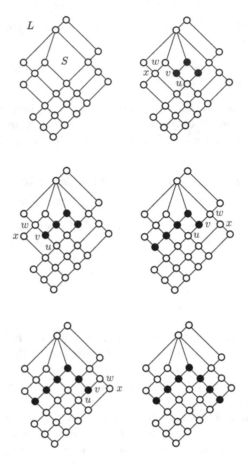

Figure 4.2: Inserting a fork into L at S.

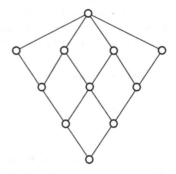

Figure 4.3: A multifork with $n = 4$.

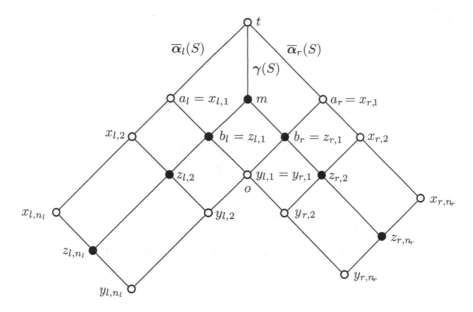

Figure 4.4: Notation for the fork construction.

Theorem 4.5. *An SPS lattice with at least three elements can be constructed from a grid by the following two steps:*

(i) *inserting forks,*

(ii) *removing corners,*

first applying (i) *and then* (ii).

G. Czédli and G. Grätzer [32] (see also Chapter 3 in LTS1) presents a twin of this construction, called *resection*.

There are other constructions that yield all planar semimodular lattices. Planar semimodular lattices can also be described by *Jordan–Hölder permutations*, see G. Czédli and E. T. Schmidt [42], G. Czédli, L. Ozsvárt, and B. Udvari [34], and G. Czédli [21]. These descriptions are generalized to a larger class of lattices, see G. Czédli [19] and K. Adaricheva and G. Czédli [1].

There is a description with matrices, see G. Czédli [16], with smaller diagrams as in G. Czédli [24]; they can also be built from smaller building blocks called rectangular lattices, discussed in the next section.

4.4. Rectangular lattices

Following G. Grätzer and E. Knapp [87], a semimodular lattice L is *rectangular* if the left boundary chain, $C_l(L)$, has exactly one doubly-irreducible

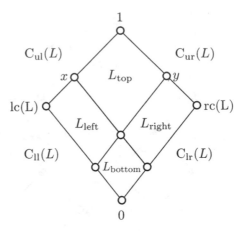

Figure 4.5: Decomposing a slim rectangular lattice.

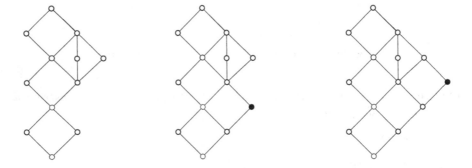

Figure 4.6: Two steps to a congruence-preserving rectangular extension.

element, lc(L) and the right boundary chain, $C_r(D)$, has exactly one doubly-irreducible element, rc(D), and these elements are complementary, that is,

$$lc(D) \vee rc(D) = 1,$$
$$lc(D) \wedge rc(D) = 0.$$

A rectangular lattice L is a *patch* lattice, if the *corners*, lc(L) and rc(L), are dual atoms of L.

Lemma 4.6 (G. Grätzer and E. Knapp [88]). *Let L be a rectangular lattice. Then the intervals $[0, lc(L)]$, $[lc(L), 1]$, $[0, rc(L)]$, and $[rc(L), 1]$ are chains.*

So the chains $C_l(L)$ and $C_r(L)$ are split into two, a lower and an upper part: $C_{ll}(L) = [0, lc(L)]$, $C_{ul}(L) = [lc(L), 1]$, $C_{lr}(L) = [0, rc(L)]$, and $C_{ur}(L) = [rc(L), 1]$ (C_{ll}, C_{ul}, C_{lr}, and C_{ur}, for short).

The structure of rectangular lattices is easily described utilizing Theorem 4.5.

Theorem 4.7 (G. Czédli and E. T. Schmidt [36]). *L is a slim rectangular lattice iff it can be obtained from a grid by inserting forks.*

For a slim rectangular lattice L, let $x \in C_{ul}(L) - \{1, lc(L)\}$ and let $y \in C_{ur}(L) - \{1, rc(L)\}$. We introduce some notation (see Figure 4.5):

$$L_{top}(x, y) = [x \wedge y, 1],$$
$$L_{left}(x, y) = [lc(L) \wedge y, x],$$
$$L_{right}(x, y) = [x \wedge rc(L), y],$$
$$L_{bottom}(x, y) = [0, (lc(L) \wedge y) \vee (x \wedge rc(L))].$$

The following decomposition theorem is from G. Grätzer and E. Knapp [88].

Theorem 4.8. *Let L be a slim rectangular lattice, and let $x \in C_{ul}(L) - \{1, lc(L)\}$, $y \in C_{ur}(L) - \{1, rc(L)\}$. Then L can be decomposed into four slim rectangular lattices $L_{top}(x, y)$, $L_{left}(x, y)$, $L_{right}(x, y)$, $L_{bottom}(x, y)$, and the lattice L can be reconstructed from these by repeated gluing.*

Let L be a nontrivial lattice. If L cannot be obtained as a gluing of two nontrivial lattices, we call L *gluing indecomposable*.

The following result of G. Czédli and E. T. Schmidt [40] is an easy consequence of Theorem 4.8.

Theorem 4.9. *Let L be an SPS lattice with at least four elements. Then L is a patch lattice iff it is gluing indecomposable.*

The next result of G. Czédli and E. T. Schmidt [41] follows easily from Theorem 4.7.

Theorem 4.10. *A patch lattice L can be obtained from the four-element boolean lattice by inserting forks.*

See G. Grätzer [64] for an alternative approach to the results of G. Czédli and E. T. Schmidt [41].

Finally, start with a planar semimodular lattice L. Can we add a corner? Yes, unless L is rectangular. So we get the following result, illustrated by Figure 4.6. Three more steps are needed to get a rectangular extension.

Lemma 4.11. *Let K be a planar semimodular lattice. Then K has a rectangular extension L. In fact, L is a congruence-preserving extension of K.*

Part II

Some Special Techniques

Chapter

5

Chopped Lattices

The first basic technique is the use of a chopped lattice, a finite meet-semi-lattice $(M, \wedge)$ regarded as a *partial algebra* $(M, \wedge, \vee)$, where $\vee$ is a partial operation. It turns out that the ideals of a chopped lattice form a lattice with the same congruence lattice as that of the chopped lattice. So to construct a finite lattice with a given congruence lattice it is enough to construct such a chopped lattice. The problem is how to ensure that the ideal lattice of the chopped lattice has some given properties. As an example, we will look at sectionally complemented lattices.

Chopped lattices were introduced by G. Grätzer and H. Lakser, published first in [56]. They were named in G. Grätzer and E. T. Schmidt [126] and generalized to the infinite case in G. Grätzer and E. T. Schmidt [123].

5.1. Basic definitions

An (n-ary) *partial operation* on a nonempty set A is a map from a subset of A^n to A. For $n = 2$, we call the partial operation *binary*. A *partial algebra* is a nonempty set A with partial operations defined on A.

A finite meet-semilattice $(M, \wedge)$ may be regarded as a partial algebra, $(M, \wedge, \vee)$, called a *chopped lattice*, where $\wedge$ is an operation and $\vee$ is a partial operation: $a \vee b$ is the least upper bound of a and b, provided that it exists.

We can obtain an example of a chopped lattice by taking a finite lattice with unit, 1, and defining $M = L - \{1\}$. The converse also holds: by adding a new unit 1 to a chopped lattice M, we obtain a finite lattice L, and chopping off the unit element, we get M back.

© Springer International Publishing Switzerland 2016
G. Grätzer, *The Congruences of a Finite Lattice*,
DOI 10.1007/978-3-319-38798-7_5

A more useful example is obtained with merging. Let C and D be lattices such that $J = C \cap D$ is an ideal in both C and D. Then, with the natural ordering, $\mathrm{Merge}(C, D) = C \cup D$, called the *merging* of C and D, is a chopped lattice. Note that if $a \vee b = c$ in $\mathrm{Merge}(C, D)$, then either $a, b, c \in C$ and $a \vee b = c$ in C or $a, b, c \in D$ and $a \vee b = c$ in D.

Among finite ordered sets, chopped lattices are easy to spot.

Lemma 5.1. *Let M be a finite ordered set and let* Max *be the set of maximal elements of M. If $\downarrow m$ is a lattice, for each $m \in$ Max, and if $m \wedge n$ exists, for all m, $n \in$ Max, then M is a meet-semilattice.*

Proof. Indeed, for $x, y \in M$,

$$x \wedge y = (x \wedge_m a) \wedge_a (y \wedge_n a),$$

where $m, n \in$ Max, $x \leq m$, $y \leq n$, $a = m \wedge n$, and $\wedge_m$, $\wedge_n$, $\wedge_a$ denotes the meet in the lattice $\downarrow m$, $\downarrow n$, and $\downarrow a$, respectively. It is easy to see that $x \wedge y$ is the greatest lower bound of x and y in M. $\qquad\square$

A meet-subsemilattice A of a chopped lattice M is a *sublattice*, if $c = a \vee b$ in A, then $c = a \vee b$ in M.

We define an equivalence relation $\boldsymbol{\alpha}$ to be a *congruence* of a chopped lattice M as we defined it for lattices in Section 1.3.3: we require that $(\mathrm{SP}_\wedge)$ and $(\mathrm{SP}_\vee)$ hold, the latter with the proviso: whenever $a \vee c$ and $b \vee d$ exist. The set $\mathrm{Con}\, M$ of all congruence relations of M ordered by set inclusion is a lattice.

Lemma 5.2. *Let M be a chopped lattice and let $\boldsymbol{\alpha}$ be an equivalence relation on M satisfying the following two conditions for $x, y, z \in M$:*

(1) *If $x \equiv y \pmod{\boldsymbol{\alpha}}$; then $x \wedge z \equiv y \wedge z \pmod{\boldsymbol{\alpha}}$.*

(2) *If $x \equiv y \pmod{\boldsymbol{\alpha}}$ and $x \vee z$ and $y \vee z$ exist, then $x \vee z \equiv y \vee z \pmod{\boldsymbol{\alpha}}$.*

Then $\boldsymbol{\alpha}$ is a congruence relation on M.

Proof. Condition (1) states that $\boldsymbol{\alpha}$ preserves $\wedge$.

Now let $x, y, u, v \in S$ with $x \equiv y \pmod{\boldsymbol{\alpha}}$ and $u \equiv v \pmod{\boldsymbol{\alpha}}$; let $x \vee u$ and $y \vee v$ exist. Then $x \equiv x \wedge y \equiv y \pmod{\boldsymbol{\alpha}}$ and $(x \wedge y) \vee u$ and $(x \wedge y) \vee v$ exist. Thus, by condition (2),

$$x \vee u \equiv (x \wedge y) \vee u \equiv (x \wedge y) \vee v \equiv y \vee v \pmod{\boldsymbol{\alpha}}. \qquad\square$$

A nonempty subset I of the chopped lattice M is an *ideal* iff it is a down set with the property:

(Id) $a, b \in I$ implies that $a \vee b \in I$, provided that $a \vee b$ exists in M.

The set $\mathrm{Id}\, M$ of all ideals of M ordered by set inclusion is a lattice. For $I, J \in \mathrm{Id}\, M$, the meet is $I \cap J$, but the join is a bit more complicated to describe.

Lemma 5.3. *Let I and J be ideals of the chopped lattice M. Define*

$$U(I, J)_0 = I \cup J,$$
$$U(I, J)_i = \{\, x \mid x \leq u \vee v,\ u,\ v \in U(I, J)_{i-1} \,\} \ \textit{for } 0 < i < \omega.$$

Then

$$I \vee J = \bigcup(U(I, J)_i \mid i < \omega).$$

Proof. Define $U = \bigcup(U(I, J)_i \mid i < \omega)$. If K is an ideal of M, then $I \subseteq K$ and $J \subseteq K$ imply—by induction—that $U \subseteq K$. So it is sufficient to prove that U is an ideal of M.

Obviously, U is a down set. Also, U has property (Id), since if $a, b \in U$ and $a \vee b$ exists in M, then $a, b \in U(I, J)_n$, for some $0 < n < \omega$, and so $a \vee b \in U(I, J)_{n+1} \subseteq U$. $\qquad\square$

Most lattice concepts and notations for them will be used for chopped lattices without further explanation.

In the literature, infinite chopped lattices are also defined, but we will not need them in this book.

5.2. Compatible vectors of elements

Let M be a chopped lattice, and let $\mathrm{Max}(M)$ (Max if M is understood) be the set of maximal elements of M. Then $M = \bigcup(\mathrm{id}(m) \mid m \in \mathrm{Max})$ and each $\mathrm{id}(m)$ is a (finite) lattice. A *vector* (associated with M) is of the form $(i_m \mid m \in \mathrm{Max})$, where $i_m \leq m$ for all $m \in M$. We order the vectors componentwise.

With every ideal I of M, we can associate the vector $(i_m \mid m \in \mathrm{Max})$ defined by $I \cap \mathrm{id}(m) = \mathrm{id}(i_m)$. Clearly, $I = \bigcup(\mathrm{id}(i_m) \mid m \in M)$. Such vectors are easy to characterize. Let us call the vector $(j_m \mid m \in \mathrm{Max})$ *compatible* if $j_m \wedge n = j_n \wedge m$ for all $m, n \in \mathrm{Max}$.

Lemma 5.4. *Let M be a chopped lattice.*

(i) *There is a one-to-one correspondence between ideals and compatible vectors of M.*

(ii) *Given any vector $\mathbf{g} = (g_m \mid m \in \mathrm{Max})$, there is a smallest compatible vector $\overline{\mathbf{g}} = (i_m \mid m \in \mathrm{Max})$ containing $\mathbf{g}$.*

(iii) *Let I and J be ideals of M, with corresponding compatible vectors*

$$(i_m \mid m \in \mathrm{Max}) \textit{ and } (j_m \mid m \in \mathrm{Max}).$$

Then

(a) $I \leq J$ in $\operatorname{Id} M$ iff $i_m \leq j_m$ for all $m \in \text{Max}$.

(b) *The compatible vector corresponding to* $I \wedge J$ *is* $(i_m \wedge j_m \mid m \in \text{Max})$.

(c) *Let* $\mathbf{a} = (i_m \vee j_m \mid m \in \text{Max})$. *Then the compatible vector corresponding to* $I \vee J$ *is* $\overline{\mathbf{a}}$.

Proof.

(i) Let I be an ideal of M. Then $(i_m \mid m \in \text{Max})$ is compatible since $i_m \wedge m \wedge n$ and $i_n \wedge m \wedge n$ both generate the principal ideal $I \cap \operatorname{id}(m) \cap \operatorname{id}(n)$ for all $m, n \in \text{Max}$.

Conversely, let $(j_m \mid m \in \text{Max})$ be compatible, and define

$$I = \bigcup(\operatorname{id}(j_m) \mid m \in \text{Max}).$$

Observe that

$$(1) \qquad\qquad I \cap \operatorname{id}(m) = \operatorname{id}(j_m),$$

for $m \in \text{Max}$. Indeed, if $x \in I \cap \operatorname{id}(m)$ and $x \in \operatorname{id}(j_n)$, for $n \in \text{Max}$, then $x \leq m \wedge j_n = n \wedge j_m$ (since $(j_m \mid m \in \text{Max})$ is compatible), so $x \leq j_m$, that is, $x \in \operatorname{id}(j_m)$. The reverse inclusion is obvious.

I is obviously a down set. To verify property (Id) for I, let $a, b \in I$ and let us assume that $a \vee b$ exists in M. Then $a \vee b \leq m$, for some $m \in \text{Max}$, so $a \leq m$ and $b \leq m$. By (1), we get $a \leq j_m$ and $b \leq j_m$, so $a \vee b \leq j_m \in I$. Since I is a down set, it follows that $a \vee b \in I$, verifying property (Id).

(ii) Obviously, the vector $(m \mid m \in \text{Max})$ contains all other vectors and it is compatible. Since the componentwise meet of compatible vectors is compatible, the statement follows.

(iii) is obvious since, by (ii), we are dealing with a closure system (see Section 1.2.2). $\qquad\square$

5.3. Compatible vectors of congruences

Let M be a chopped lattice. With any congruence $\boldsymbol{\alpha}$ of M, we can associate the *restriction vector* $(\boldsymbol{\alpha}]_m \mid m \in \text{Max})$, where $\boldsymbol{\alpha}]_m$ is the restriction of $\boldsymbol{\alpha}$ to $\operatorname{id}(m)$. The restriction $\boldsymbol{\alpha}]_m$ is a congruence of the lattice $\operatorname{id}(m)$.

Let $\boldsymbol{\gamma}_m$ be a congruence of $\operatorname{id}(m)$ for all $m \in \text{Max}$. The *congruence vector* $(\boldsymbol{\gamma}_m \mid m \in \text{Max})$ is called *compatible* if $\boldsymbol{\gamma}_m$ restricted to $\operatorname{id}(m \wedge n)$ is the same as $\boldsymbol{\gamma}_n$ restricted to $\operatorname{id}(m \wedge n)$ for $m, n \in \text{Max}$. Obviously, a restriction vector is compatible. The converse also holds.

Lemma 5.5. *Let* $(\boldsymbol{\gamma}_m \mid m \in \text{Max})$ *be a compatible congruence vector of a chopped lattice* M. *Then there is a unique congruence* $\boldsymbol{\alpha}$ *of* M *such that the restriction vector of* $\boldsymbol{\alpha}$ *agrees with* $(\boldsymbol{\gamma}_m \mid m \in \text{Max})$.

Proof. Let $(\gamma_m \mid m \in \mathrm{Max})$ be a compatible congruence vector. We define a binary relation α on M as follows:

Let $m, n \in \mathrm{Max}$. For $x \in \mathrm{id}(m)$ and $y \in \mathrm{id}(n)$, let $x \equiv y \pmod{\alpha}$ iff $x \equiv x \wedge y \pmod{\gamma_m}$ and $y \equiv x \wedge y \pmod{\gamma_n}$.

Obviously, α is reflexive and symmetric. To prove transitivity, let $m, n, k \in \mathrm{Max}$, and let $x \in \mathrm{id}(m)$, $y \in \mathrm{id}(n)$, $z \in \mathrm{id}(k)$; let $x \equiv y \pmod{\alpha}$ and $y \equiv z \pmod{\alpha}$, that is,

$$(2) \qquad x \equiv x \wedge y \pmod{\gamma_m},$$
$$(3) \qquad y \equiv x \wedge y \pmod{\gamma_n},$$
$$(4) \qquad y \equiv y \wedge z \pmod{\gamma_n},$$
$$(5) \qquad z \equiv y \wedge z \pmod{\gamma_k}.$$

Then meeting the congruence (4) with x (in the lattice $\mathrm{id}(n)$), we get

$$(6) \qquad x \wedge y \equiv x \wedge y \wedge z \pmod{\gamma_n},$$

and from (3), by meeting with z, we obtain

$$(7) \qquad y \wedge z \equiv x \wedge y \wedge z \pmod{\gamma_n}.$$

Since $x \wedge y$ and $x \wedge y \wedge z \in \mathrm{id}(m)$, by compatibility, (6) implies that

$$(8) \qquad x \wedge y \equiv x \wedge y \wedge z \pmod{\gamma_m}.$$

Now (2) and (8) yield

$$(9) \qquad x \equiv x \wedge y \wedge z \pmod{\gamma_m}.$$

Similarly,

$$(10) \qquad z \equiv x \wedge y \wedge z \pmod{\gamma_k}.$$

γ_m is a lattice congruence on $\mathrm{id}(m)$ and $x \wedge y \wedge z \leq x \wedge z \leq x$, so

$$(11) \qquad x \equiv x \wedge z \pmod{\gamma_m}.$$

Similarly,

$$(12) \qquad z \equiv x \wedge z \pmod{\gamma_k}.$$

Equations (11) and (12) yield that $x \equiv z \pmod{\alpha}$, proving transitivity.

(SP$_\wedge$) is easy: let $x \in \mathrm{id}(m)$, $y \in \mathrm{id}(n)$, $z \in M$; if $x \equiv y \pmod{\alpha}$, then $x \wedge z \equiv y \wedge z \pmod{\alpha}$ because $x \wedge z \equiv x \wedge y \wedge z \pmod{\gamma_m}$ and $y \wedge z \equiv x \wedge y \wedge z \pmod{\gamma_n}$.

Finally, we verify (SP$_\vee$). Let $x \equiv y \pmod{\alpha}$ and $z \in M$, and let us assume that $x \vee z$ and $y \vee z$ exist. Then there are $p, q \in \mathrm{Max}$ such that

$x \vee z \in \mathrm{id}(p)$ and $y \vee z \in \mathrm{id}(q)$. By compatibility, $x \equiv x \wedge y \pmod{\gamma_p}$, so $x \vee z \equiv (x \wedge y) \vee z \pmod{\gamma_p}$. Since $(x \wedge y) \vee z \leq (x \vee z) \wedge (y \vee z) \leq x \vee z$, we also have

$$x \vee z \equiv (x \vee z) \wedge (y \vee z) \pmod{\gamma_p}.$$

Similarly,

$$y \vee z \equiv (x \vee z) \wedge (y \vee z) \pmod{\gamma_q}.$$

The last two displayed equations show that $x \vee z \equiv y \vee z \pmod{\alpha}$. $\square$

5.4.   From the chopped lattice to the ideal lattice

The map $m \mapsto \mathrm{id}(m)$ embeds the chopped lattice M with zero into the lattice $\mathrm{Id}\, M$, so we can regard $\mathrm{Id}\, M$ an extension. It is, in fact, a congruence-preserving extension (G. Grätzer and H. Lakser [89], proof first published in [55]):

Theorem 5.6. *Let M be a chopped lattice. Then $\mathrm{Id}\, M$ is a congruence-preserving extension of M.*

Proof. Let α be a congruence relation of M. If $I, J \in \mathrm{Id}\, M$, define

$$I \equiv J \pmod{\overline{\alpha}} \quad \text{iff} \quad I/\alpha = J/\alpha.$$

Obviously, $\overline{\alpha}$ is an equivalence relation. Let $I \equiv J \pmod{\overline{\alpha}}$, $N \in \mathrm{Id}\, M$, and $x \in I \cap N$. Then $x \equiv y \pmod{\alpha}$, for some $y \in J$, and so $x \equiv x \wedge y \pmod{\alpha}$ and $x \wedge y \in J \cap N$. This shows that $(I \cap N)/\alpha \subseteq (J \cap N)/\alpha$. Similarly, $(J \cap N)/\alpha \subseteq (I \cap N)/\alpha$, so $I \cap N \equiv J \cap N \pmod{\overline{\alpha}}$.

To prove $I \vee N \equiv J \vee N \pmod{\overline{\alpha}}$, by symmetry, it is sufficient to verify that $I \vee N \subseteq (J \vee N)/\alpha$. By Lemma 5.3, this is equivalent to proving that $U_n \subseteq (J \vee N)/\alpha$ for $n < \omega$. This is obvious for $n = 0$.

Now assume that $U_{n-1} \subseteq (J \vee N)/\alpha$ and let $x \in U_n$. Then $x \leq t_1 \vee t_2$ for some $t_1, t_2 \in U_{n-1}$. Thus $t_1 \equiv u_1 \pmod{\alpha}$ and $t_2 \equiv u_2 \pmod{\alpha}$, for some $u_1, u_2 \in J \vee N$, and so $t_1 \equiv t_1 \wedge u_1 \pmod{\alpha}$ and $t_2 \equiv t_2 \wedge u_2 \pmod{\alpha}$. Observe that $t_1 \vee t_2$ is an upper bound for $\{t_2 \wedge u_1, t_2 \wedge u_2\}$; consequently, $(t_1 \wedge u_1) \vee (t_2 \wedge u_2)$ exists. Therefore,

$$t_1 \vee t_2 \equiv (t_1 \wedge u_1) \vee (t_2 \wedge u_2) \pmod{\alpha}.$$

Finally,

$$x \equiv x \wedge (t_1 \vee t_2) = x \wedge ((t_1 \wedge u_1) \vee (t_2 \wedge u_2)) \pmod{\alpha},$$

and

$$x \wedge ((t_1 \wedge u_1) \vee (t_2 \wedge u_2)) \in J \vee N.$$

Thus $x \in (J \vee N)/\alpha$, completing the induction, verifying that $\overline{\alpha}$ is a congruence relation of $\mathrm{Id}\, M$.

If $a \equiv b \pmod{\boldsymbol{\alpha}}$ and $x \in \mathrm{id}(a)$, then $x \equiv x \wedge b \pmod{\boldsymbol{\alpha}}$. Thus $\mathrm{id}(a) \subseteq \mathrm{id}(b)/\boldsymbol{\alpha}$. Similarly, $\mathrm{id}(b) \subseteq \mathrm{id}(a)/\boldsymbol{\alpha}$, and so $\mathrm{id}(a) \equiv \mathrm{id}(b) \pmod{\overline{\boldsymbol{\alpha}}}$. Conversely, if $\mathrm{id}(a) \equiv \mathrm{id}(b) \pmod{\overline{\boldsymbol{\alpha}}}$, then $a \equiv b_1 \pmod{\boldsymbol{\alpha}}$ and $a_1 \equiv b \pmod{\boldsymbol{\alpha}}$ for some $a_1 \le a$ and $b_1 \le b$. Forming the join of these two congruences, we get $a \equiv b \pmod{\boldsymbol{\alpha}}$. Thus $\overline{\boldsymbol{\alpha}}$ has all the properties required by Lemma 5.3.

To show the uniqueness, let $\boldsymbol{\gamma}$ be a congruence relation of $\mathrm{Id}\, M$ satisfying $\mathrm{id}(a) \equiv \mathrm{id}(b) \pmod{\boldsymbol{\gamma}}$ iff $a \equiv b \pmod{\boldsymbol{\alpha}}$. Let $I, J \in \mathrm{Id}\, M$, $I \equiv J \pmod{\boldsymbol{\gamma}}$, and $x \in I$. Then

$$\mathrm{id}(x) \cap I \equiv \mathrm{id}(x) \cap J \pmod{\boldsymbol{\gamma}},$$
$$\mathrm{id}(x) \cap I = \mathrm{id}(x),$$
$$\mathrm{id}(x) \cap J = \mathrm{id}(y)$$

for some $y \in J$. Thus $\mathrm{id}(x) \equiv \mathrm{id}(y) \pmod{\boldsymbol{\gamma}}$, and so $x \equiv y \pmod{\boldsymbol{\alpha}}$, proving that $I \subseteq J/\boldsymbol{\alpha}$. Similarly, $J \subseteq I/\boldsymbol{\alpha}$, and so $I \equiv J \pmod{\overline{\boldsymbol{\alpha}}}$. Conversely, if $I \equiv J \pmod{\overline{\boldsymbol{\alpha}}}$, then take all congruences of the form $x \equiv y \pmod{\boldsymbol{\alpha}}$, $x \in I$, $y \in J$. By our assumption regarding $\boldsymbol{\gamma}$, we get the congruence $\mathrm{id}(x) \equiv \mathrm{id}(y) \pmod{\boldsymbol{\gamma}}$, and by our definition of $\overline{\boldsymbol{\alpha}}$, the join of all these congruences yields $I \equiv J \pmod{\boldsymbol{\alpha}}$. Thus $\boldsymbol{\gamma} = \overline{\boldsymbol{\alpha}}$. $\qquad\square$

This result is very useful. It means that in order to construct a finite lattice L to represent a given finite distributive lattice D as a congruence lattice, it is sufficient to construct a chopped lattice M with $\mathrm{Con}\, M \cong D$, since $\mathrm{Con}\, M \cong \mathrm{Con}(\mathrm{Id}\, M) = \mathrm{Con}\, L$, where $L = \mathrm{Id}\, M$, and L is a finite lattice.

This result also allows us to construct congruence-preserving extensions.

Corollary 5.7. *Let $M = \mathrm{Merge}(A, B)$ be a chopped lattice with $A = \mathrm{id}(a)$ and $B = \mathrm{id}(b)$. If $a \wedge b > 0$ and B is simple, then $\mathrm{Id}\, M$ is a congruence-preserving extension of A.*

Proof. Let $\boldsymbol{\alpha}$ be a congruence of A. Then $(\boldsymbol{\alpha}, \boldsymbol{\gamma})$ is a compatible congruence vector iff

$$\boldsymbol{\gamma} = \begin{cases} \mathbf{0} & \text{if } \boldsymbol{\alpha} \text{ is discrete on } [0, a \wedge b]; \\ \mathbf{1}, & \text{otherwise.} \end{cases}$$

So $\boldsymbol{\gamma}$ is determined by $\boldsymbol{\alpha}$ and the statement follows. $\qquad\square$

5.5. Sectional complementation

We introduce *sectionally complemented chopped lattices* as we did for lattices in Section 2.1.

We illustrate the use of compatible vectors with two results on sectionally complemented chopped lattices. The first result is from G. Grätzer and E. T. Schmidt [129].

Lemma 5.8 (Atom Lemma). *Let M be a chopped lattice with two maximal elements m_1 and m_2. We assume that $\mathrm{id}(m_1)$ and $\mathrm{id}(m_2)$ are sectionally complemented lattices. If $p = m_1 \wedge m_2$ is an atom, then $\mathrm{Id}\, M$ is sectionally complemented.*

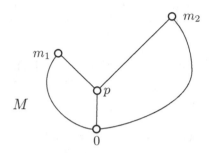

Figure 5.1: Atom Lemma illustrated.

Proof. Figure 5.1 illustrates the setup.

To show that $\mathrm{Id}\, M$ is sectionally complemented, let $I \subseteq J$ be two ideals of M, represented by the compatible vectors (i_1, i_2) and (j_1, j_2), respectively. Let s_1 be the sectional complement of i_1 in j_1 and let s_2 be the sectional complement of i_2 in j_2. If $p \wedge s_1 = p \wedge s_2$, then (s_1, s_2) is a compatible vector, representing an ideal S that is a sectional complement of I in J. Otherwise, without loss of generality, we can assume that $p \wedge s_1 = 0$ and $p \wedge s_2 = p$. Since $\mathrm{id}(m_2)$ is sectionally complemented, there is a sectional complement s_2' of p in $[0, s_2]$. Then (s_1, s_2') satisfies $p \wedge s_1 = p \wedge s_2'\ (= 0)$, and so it is compatible; therefore, (s_1, s_2') represents an ideal S of M. Obviously, $I \wedge S = \{0\}$.

From $p \wedge s_2 = p$, it follows that $p \leq s_2 \leq j_2$. Since J is an ideal and $j_2 \wedge p = p$, it follows that $j_1 \wedge p = p$, that is, $p \leq j_1$. Obviously, $I \vee S \subseteq J$. So to show that $I \vee S = J$, it is sufficient to verify that $j_1, j_2 \in I \vee S$. Evidently, $j_1 = i_1 \vee s_1 \in I \vee S$. Note that $p \leq j_1 = i_1 \vee s_1 \in I \vee S$. Thus $p, s_2', i_2 \in I \vee S$, and therefore

$$p \vee s_2' \vee i_2 = (p \vee s_2') \vee i_2 = s_2 \vee i_2 = j_2 \in I \vee S. \qquad \square$$

The second result (G. Grätzer, H. Lakser, and M. Roddy [101]) shows that the ideal lattice of a sectionally complemented chopped lattice is not always sectionally complemented.

Theorem 5.9. *There is a sectionally complemented chopped lattice M whose ideal lattice $\mathrm{Id}\, M$ is not sectionally complemented.*

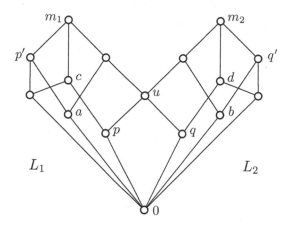

Figure 5.2: The chopped lattice M.

Proof. Let M be the chopped lattice of Figure 5.2, where $L_1 = \mathrm{id}(m_1)$ and $L_2 = \mathrm{id}(m_2)$. Note that p is meet-irreducible in $\mathrm{id}(m_2)$ and q is meet-irreducible in $\mathrm{id}(m_1)$.

The unit element of the ideal lattice of M is the compatible vector (m_1, m_2). We show that the compatible vector (a, b) has no complement in the ideal lattice of M.

Assume, to the contrary, that the compatible vector (s, t) is a complement of (a, b). Since $(a, b) \le (a \vee u, m_2)$, a compatible vector, $(s, t) \not\le (a \vee u, m_2)$, that is,

$$(13) \qquad\qquad s \not\le a \vee u.$$

Similarly, by considering $(m_1, b \vee u)$, we conclude that

$$(14) \qquad\qquad t \not\le b \vee u.$$

Now $(a, b) \le (p', q')$, which is a compatible vector. Thus $(s, t) \not\le (p', q')$, and so either $s \not\le p'$ or $t \not\le q'$. Without loss of generality, we may assume that $s \not\le p'$. It then follows by (13) that s can be only c or m_1. Then, since $s \wedge a = 0$, we conclude that $s = c$. Thus $s \wedge u = p$, and so $t \wedge u = p$. But p is meet-irreducible in L_2. Thus $t = p \le b \vee u$, contradicting (14). $\square$

This result illustrates that the Atom Lemma (Lemma 5.8) cannot be extended to the case where $[0, m_1 \wedge m_2]$ is a four-element boolean lattice.

Boolean Triples

In Part IV, we construct congruence-preserving extensions of finite lattices, extensions with special properties, such as sectionally complemented, semi-modular, and so on.

It is easy to construct a proper congruence-preserving extension of a finite lattice. In the early 1990s, G. Grätzer and E. T. Schmidt raised the question in [126] whether *every* lattice has a proper congruence-preserving extension. (See also G. Grätzer and E. T. Schmidt [123].)

It took almost a decade for the answer to appear in G. Grätzer and F. Wehrung [145]. For infinite lattices, the affirmative answer was provided by the boolean triples construction, which is described in this chapter. It is interesting that boolean triples also provide a very important tool for finite lattices.

6.1. The general construction

In this section, I describe a congruence-preserving extension of a (finite) lattice L, introduced in G. Grätzer and F. Wehrung [145]. We will see that this generalizes a construction of E. T. Schmidt [174] for bounded distributive lattices.

For a lattice L, let us call the triple $(x, y, z) \in L^3$ *boolean* iff

(F)
$$x = (x \vee y) \wedge (x \vee z),$$
$$y = (y \vee x) \wedge (y \vee z),$$
$$z = (z \vee x) \wedge (z \vee y),$$

© Springer International Publishing Switzerland 2016
G. Grätzer, *The Congruences of a Finite Lattice*,
DOI 10.1007/978-3-319-38798-7_6

where F stands for "Fixed point definition".

Note that by Lemma 2.6, if (F) holds then $\mathrm{sub}(\{x, y, z\})$ is boolean.

(F) is a "Fixed point definition" because the triple (x, y, z) satisfies (F) iff $p(x, y, z) = (x, y, z)$, where

$$p = ((x \vee y) \wedge (x \vee z), (y \vee x) \wedge (y \vee z), (z \vee x) \wedge (z \vee y)).$$

We denote by $\mathsf{M}_3[L] \subseteq L^3$ the ordered set of boolean triples of L (ordered as an ordered subset of L^3, that is, componentwise). If we apply the construction to an interval $[a, b]$ of L, we write $\mathsf{M}_3[a, b]$ for $\mathsf{M}_3[[a, b]]$.

Observe that any boolean triple $(x, y, z) \in L^3$ satisfies

(B) $$x \wedge y = y \wedge z = z \wedge x.$$

where B stands for "balanced". Indeed, if (x, y, z) is boolean, then

$$x \wedge y = y \wedge z = z \wedge x = (x \vee y) \wedge (y \vee z) \wedge (z \vee x).$$

We call such triples *balanced.*

Here are some of the basic properties of boolean triples.

Lemma 6.1. *Let L be a lattice.*

(i) $(x, y, z) \in L^3$ *is boolean iff there is a triple $(u, v, w) \in L^3$ satisfying*

(E)
$$x = u \wedge v,$$
$$y = u \wedge w,$$
$$z = v \wedge w.$$

(ii) $\mathsf{M}_3[L]$ *is a closure system in L^3. For $(x, y, z) \in L^3$, the closure is*

$$\overline{(x, y, z)} = ((x \vee y) \wedge (x \vee z), (y \vee x) \wedge (y \vee z), (z \vee x) \wedge (z \vee y)).$$

(iii) *If L has 0, then the ordered subset $\{ (x, 0, 0) \mid x \in L \}$ is a sublattice of $\mathsf{M}_3[L]$ and $\gamma \colon x \mapsto (x, 0, 0)$ is an isomorphism between L and this sublattice.*

(iv) *If L is bounded, then $\mathsf{M}_3[L]$ has a spanning M_3, that is, a $\{0, 1\}$-sublattice isomorphic to M_3, namely,*

$$\{(0, 0, 0), (1, 0, 0), (0, 1, 0), (0, 0, 1), (1, 1, 1)\}.$$

Remark. (E) is the "Existential definition" because (x, y, z) is boolean iff there *exists* a triple (u, v, w) satisfying (E).

Proof.

(i) If (x, y, z) is boolean, then $u = x \vee y$, $v = x \vee z$, and $w = y \vee z$ satisfy (E). Conversely, if there is a triple $(u, v, w) \in L^3$ satisfying (E), then by Lemma 2.6, the sublattice generated by x, y, z is isomorphic to a quotient of B_3 and x, y, z are the images of the three atoms of B_3. Thus $(x \vee y) \wedge (x \vee z) = x$, the first equation in (F). The other two equations are proved similarly.

(ii) $\mathsf{M}_3[L] \neq \varnothing$; for instance, for all $x \in L$, the diagonal element $(x, x, x) \in \mathsf{M}_3[L]$.

For $(x, y, z) \in L^3$, define $u = x \vee y$, $v = x \vee z$, and $w = y \vee z$. Set $x_1 = u \wedge v$, $y_1 = u \wedge w$, and $z_1 = v \wedge w$. Then (x_1, y_1, z_1) is boolean by (i) and $(x, y, z) \leq (x_1, y_1, z_1)$ in L^3. Now if $(x, y, z) \leq (x_2, y_2, z_2)$ in L^3 and (x_2, y_2, z_2) is boolean, then

$$
\begin{aligned}
x_2 &= (x_2 \vee y_2) \wedge (x_2 \vee z_2) && \text{(by (F))} \\
&\geq (x \vee y) \wedge (x \vee z) && \text{(by } (x, y, z) \leq (x_2, y_2, z_2)) \\
&= u \wedge v = x_1,
\end{aligned}
$$

and similarly, $y_2 \geq y_1$, $z_2 \geq z_1$. Thus $(x_2, y_2, z_2) \geq (x_1, y_1, z_1)$, and so (x_1, y_1, z_1) is the smallest boolean triple containing (x, y, z).

(iii) and (iv) are obvious. $\qquad\square$

$\mathsf{M}_3[L]$ is difficult to draw, in general. Figure 6.1 shows the diagram of $\mathsf{M}_3[\mathsf{C}_3]$ with the three-element chain $\mathsf{C}_3 = \{0, a, 1\}$.

If C is an arbitrary bounded chain, with bounds 0 and 1, it is easy to picture $\mathsf{M}_3[C]$, as sketched in Figure 6.1. The element $(x, y, z) \in C^3$ is boolean iff it is of the form (x, y, y), or (y, x, y), or (y, y, x), where $y \leq x$ in C. So the diagram is made up of three isomorphic "flaps" overlapping on the diagonal. Two of the flaps form the "base", a planar lattice: C_3^2, the third one (shaded) comes up out of the plane pointing in the direction of the viewer.

We get some more examples of $\mathsf{M}_3[L]$ from the following observation.

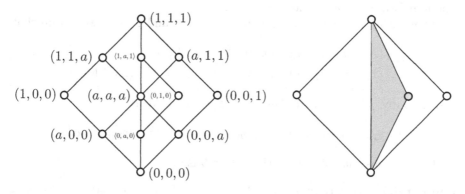

Figure 6.1: The lattice $\mathsf{M}_3[\mathsf{C}_3]$ with a sketch.

Lemma 6.2. *Let the lattice L have a direct product decomposition: $L = L_1 \times L_2$. Then $\mathsf{M}_3[L] \cong \mathsf{M}_3[L_1] \times \mathsf{M}_3[L_2]$.*

Proof. This is obvious, since $((x_1, y_1), (x_2, y_2), (x_3, y_3))$ is a boolean triple iff (x_1, x_2, x_3) and (y_1, y_2, y_3) both are ($x_i \in L_1$ and $y_i \in L_2$ for $i = 1, 2, 3$). □

6.2. The congruence-preserving extension property

Let L be a nontrivial lattice with zero and let

$$\gamma \colon x \mapsto (x, 0, 0) \in \mathsf{M}_3[L]$$

be an embedding of L into $\mathsf{M}_3[L]$.

Here is the main result of G. Grätzer and F. Wehrung [145]:

Theorem 6.3. *$\mathsf{M}_3[L]$ is a congruence-preserving extension of γL.*

The next two lemmas prove this theorem.

For a congruence α of L, let α^3 denote the congruence of L^3 defined componentwise. Let $\mathsf{M}_3[\alpha]$ be the restriction of α^3 to $\mathsf{M}_3[L]$.

Lemma 6.4. *$\mathsf{M}_3[\alpha]$ is a congruence relation of $\mathsf{M}_3[L]$.*

Proof. $\mathsf{M}_3[\alpha]$ is obviously an equivalence relation on $\mathsf{M}_3[L]$. Since $\mathsf{M}_3[L]$ is a meet-subsemilattice of L^3, it is clear that $\mathsf{M}_3[\alpha]$ satisfies $(\mathrm{SP}_\wedge)$. To verify $(\mathrm{SP}_\vee)$ for $\mathsf{M}_3[\alpha]$, let $(x_1, y_1, z_1), (x_2, y_2, z_2) \in \mathsf{M}_3[L]$, let

$$(x_1, y_1, z_1) \equiv (x_2, y_2, z_2) \pmod{\mathsf{M}_3[\alpha]},$$

and let $(u, v, w) \in \mathsf{M}_3[L]$. Set

$$(x'_i, y'_i, z'_i) = (x_i, y_i, z_i) \vee (u, v, w)$$

(the join formed in $\mathsf{M}_3[L]$) for $i = 1, 2$.

Then, using Lemma 6.1.(ii) for $x_1 \vee u$, $y_1 \vee v$, and $z_1 \vee w$, we obtain that

$$x'_1 = (x_1 \vee u \vee y_1 \vee v) \wedge (x_1 \vee u \vee z_1 \vee w)$$
$$\equiv (x_2 \vee u \vee y_2 \vee v) \wedge (x_2 \vee u \vee z_2 \vee w) = x'_2 \pmod{\mathsf{M}_3[\alpha]},$$

and similarly, $y'_1 \equiv y'_2 \pmod{\mathsf{M}_3[\alpha]}$, $z'_1 \equiv z'_2 \pmod{\mathsf{M}_3[\alpha]}$, hence

$$(x_1, y_1, z_1) \vee (u, v, w) \equiv (x_2, y_2, z_2) \vee (u, v, w) \pmod{\mathsf{M}_3[\alpha]}. □$$

It is obvious that $\mathsf{M}_3[\alpha]$ restricted to γL is $\gamma \alpha$.

Lemma 6.5. *Every congruence of $\mathsf{M}_3[L]$ is of the form $\mathsf{M}_3[\alpha]$ for a suitable congruence α of L.*

Proof. Let γ be a congruence of $\mathsf{M}_3[L]$, and let α denote the congruence of L obtained by restricting γ to the sublattice $L' = \{\,(x,0,0) \mid x \in L\,\}$ of $\mathsf{M}_3[L]$, that is, for $x, y \in L$, $x \equiv y \pmod{\alpha}$ iff $(x,0,0) \equiv (y,0,0) \pmod{\gamma}$. We prove that $\gamma = \mathsf{M}_3[\alpha]$.

To show that $\gamma \subseteq \mathsf{M}_3[\alpha]$, let

$$(1) \qquad (x_1, y_1, z_1) \equiv (x_2, y_2, z_2) \pmod{\gamma}.$$

Meeting the congruence (1) with $(1,0,0)$ yields

$$(x_1, 0, 0) \equiv (x_2, 0, 0) \pmod{\gamma},$$

and so

$$(2) \qquad (x_1, 0, 0) \equiv (x_2, 0, 0) \pmod{\mathsf{M}_3[\alpha]}.$$

Meeting the congruence (1) with $(0,1,0)$ yields

$$(3) \qquad (0, y_1, 0) \equiv (0, y_2, 0) \pmod{\gamma}.$$

Since

$$(0, y_1, 0) \vee (0, 0, 1) = \overline{(0, y_1, 1)} = (y_1, y_1, 1),$$

and

$$(y_1, y_1, 1) \wedge (1, 0, 0) = (y_1, 0, 0),$$

and similarly for $(0, y_2, 0)$, joining the congruence (3) with $(0,1,0)$ and then meeting with $(1,0,0)$, yields

$$(y_1, 0, 0) \equiv (y_2, 0, 0) \pmod{\gamma},$$

and so

$$(4) \qquad (0, y_1, 0) \equiv (0, y_2, 0) \pmod{\mathsf{M}_3[\alpha]}.$$

Similarly,

$$(5) \qquad (0, 0, z_1) \equiv (0, 0, z_2) \pmod{\mathsf{M}_3[\alpha]}.$$

Joining the congruences (2), (4), and (5), we obtain

$$(6) \qquad (x_1, y_1, z_1) \equiv (x_2, y_2, z_2) \pmod{\mathsf{M}_3[\alpha]},$$

proving that $\gamma \subseteq \mathsf{M}_3[\alpha]$.

To prove the converse, $\mathsf{M}_3[\alpha] \subseteq \gamma$, take

$$(7) \qquad (x_1, y_1, z_1) \equiv (x_2, y_2, z_2) \pmod{\mathsf{M}_3[\alpha]}$$

in $\mathsf{M}_3[L]$; equivalently,

(8) $$(x_1, 0, 0) \equiv (x_2, 0, 0) \pmod{\gamma},$$
(9) $$(y_1, 0, 0) \equiv (y_2, 0, 0) \pmod{\gamma},$$
(10) $$(z_1, 0, 0) \equiv (z_2, 0, 0) \pmod{\gamma}$$

in $\mathsf{M}_3[L]$.

Joining the congruence (9) with $(0, 0, 1)$ and then meeting the result with $(0, 1, 0)$, we get (as in the computation following (3)):

(11) $$(0, y_1, 0) \equiv (0, y_2, 0) \pmod{\gamma}.$$

Similarly, from (10), we conclude that

(12) $$(0, 0, z_1) \equiv (0, 0, z_2) \pmod{\gamma}.$$

Finally, joining the congruences (8), (11), and (12), we get

(13) $$(x_1, y_1, z_1) \equiv (x_2, y_2, z_2) \pmod{\gamma},$$

that is, $\mathsf{M}_3[\alpha] \subseteq \gamma$. This completes the proof of this lemma.   □

6.3.   The distributive case

Let D be a bounded distributive lattice. In 1974 (25 years before the publication of G. Grätzer and F. Wehrung [145]), E. T. Schmidt [174] defined $\mathsf{M}_3[D]$ as the set of balanced triples $(x, y, z) \in D^3$ (defined in Section 6.1), regarded as an ordered subset of D^3. Schmidt proved the following result:

Theorem 6.6. *Let D be a bounded distributive lattice. Then $\mathsf{M}_3[D]$ (the set of balanced triples $(x, y, z) \in D^3$) is a modular lattice. The map*

$$\gamma \colon x \mapsto (x, 0, 0) \in \mathsf{M}_3[D]$$

is an embedding of D into $\mathsf{M}_3[D]$, and $\mathsf{M}_3[D]$ is a congruence-preserving extension of γD.

Proof. Examining Figure 2.7 (page 27), we immediately see that in a distributive lattice D, conditions (B) and (F) are equivalent, so $\mathsf{M}_3[D]$ is the boolean triple construction. Therefore, this result follows from the results in Sections 6.1 and 6.2, except for the modularity. A direct computation of this is not so easy—although entertaining. However, we can do it without computation. Observe that it is enough to prove modularity for a finite D. Now if D is finite, then by Lemma 6.2 and Corollary 2.18, $\mathsf{M}_3[D]$ can be embedded into $\mathsf{M}_3[\mathsf{B}_n] \cong (\mathsf{M}_3[\mathsf{C}_2])^n \cong \mathsf{M}_3^n$, a modular lattice; hence $\mathsf{M}_3[D]$ is modular.   □

6.4.   Two interesting intervals

Let L be a bounded lattice. For an arbitrary $a \in L$, consider the following interval $\mathsf{M}_3[L, a]$ of $\mathsf{M}_3[L]$ (illustrated in Figure 6.2):

$$\mathsf{M}_3[L, a] = [(0, a, 0), (1, 1, 1)] \subseteq \mathsf{M}_3[L].$$

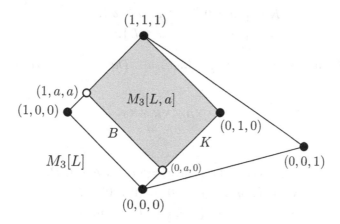

Figure 6.2: The shaded area: the lattice $\mathsf{M}_3[L, a]$.

Define the map (with image $B = \gamma_a L$ in Figure 6.2):

$$\gamma_a : x \mapsto (x, a, x \wedge a).$$

Lemma 6.7. γ_a *is an embedding of* L *into* $\mathsf{M}_3[L, a]$.

Proof. γ_a is obviously one-to-one and meet-preserving. It also preserves the join because

$$\overline{(x \vee y, a, (x \wedge a) \vee (y \wedge a))} = (x \vee y, a, (x \vee y) \wedge a)$$

for $x, y \in L$. $\square$

The following result of G. Grätzer and E. T. Schmidt [131] is a generalization of Theorem 6.3:

Theorem 6.8. $\mathsf{M}_3[L, a]$ *is a congruence-preserving extension of* $\gamma_a L$.

The proof is presented in the following lemmas.

Lemma 6.9. *Let* $(x, y, z) \in \mathsf{M}_3[L, a]$. *Then* $a \leq y$ *and*

(14) $$x \wedge a = z \wedge a.$$

Proof. Since (x, y, z) is boolean, it is balanced, so $x \wedge y = z \wedge y$. Therefore, $x \wedge a = (x \wedge y) \wedge a = (z \wedge y) \wedge a = z \wedge a$, as claimed. $\square$

We need an easy decomposition statement for the elements of $\mathsf{M}_3[L, a]$. Let us use the notation (B and K are marked in Figure 6.2):

$$B = \{\, (x, a, x \wedge a) \mid x \in L \,\}\ (= \gamma_a L),$$
$$K = \{\, (0, x, 0) \mid x \in L,\ x \geq a \,\},$$
$$J = \{\, (x \wedge a, a, x) \mid x \in L \,\}.$$

Lemma 6.10. *Let* $\mathbf{v} = (x, y, z) \in \mathsf{M}_3[L, a]$. *Then* $\mathbf{v}$ *has a decomposition in* $\mathsf{M}_3[L, a]$:

(15) $$\mathbf{v} = \mathbf{v}_B \vee \mathbf{v}_K \vee \mathbf{v}_J,$$

where

(16) $$\mathbf{v}_B = (x, y, z) \wedge (1, a, a) = (x, a, x \wedge a) \in B,$$
(17) $$\mathbf{v}_K = (x, y, z) \wedge (0, 1, 0) = (0, y, 0) \in K,$$
(18) $$\mathbf{v}_J = (x, y, z) \wedge (a, a, 1) = (z \wedge a, a, z) \in J.$$

Proof. (16) follows from (14). By symmetry, (18) follows, and (17) is trivial. Finally, the right side of (15) componentwise joins to the left side—in view of (14). $\square$

We can now describe the congruences of $\mathsf{M}_3[L, a]$.

Lemma 6.11. *Let* γ *be a congruence of* $\mathsf{M}_3[\alpha, a]$ *and let* $\mathbf{v}, \mathbf{w} \in \mathsf{M}_3[L, a]$. *Then*

(19) $$\mathbf{v} \equiv \mathbf{w} \pmod{\gamma},$$

iff

(20) $$\mathbf{v}_B \equiv \mathbf{w}_B \pmod{\gamma},$$
(21) $$\mathbf{v}_K \equiv \mathbf{w}_K \pmod{\gamma},$$
(22) $$\mathbf{v}_J \equiv \mathbf{w}_J \pmod{\gamma}.$$

Proof. (19) implies (20) by (16). Similarly, for (21) and (22). Conversely, (20)–(22) imply (19) by (15). $\square$

Lemma 6.12. *For a congruence* α *of* L, *let* $\mathsf{M}_3[\alpha, a]$ *be the restriction of* α^3 *to* $\mathsf{M}_3[L, a]$. *Then* $\mathsf{M}_3[\alpha, a]$ *is a congruence of* $\mathsf{M}_3[L, a]$, *and every congruence of* $\mathsf{M}_3[L, a]$ *is of the form* $\mathsf{M}_3[\alpha, a]$, *for a unique congruence* α *of* L.

Proof. It follows from Lemma 6.4 that $\mathsf{M}_3[\boldsymbol{\alpha}, a]$ is a congruence of $\mathsf{M}_3[L, a]$. Let $\mathbf{v} = (x, y, z), \mathbf{v}' = (x', y', z') \in \mathsf{M}_3[L, a]$. Let $\boldsymbol{\gamma}$ be a congruence of $\mathsf{M}_3[L, a]$, and let $\boldsymbol{\alpha}$ be the restriction of $\boldsymbol{\gamma}$ to L with respect to the embedding γ_a. By Lemma 6.11,

$$\mathbf{v} \equiv \mathbf{v}' \pmod{\boldsymbol{\gamma}}$$

iff (20)–(22) hold. Note that $\mathbf{v}_B, \mathbf{v}'_B \in \gamma_a L$, so (20) is equivalent to $\mathbf{v}_B \equiv \mathbf{v}'_B \pmod{\boldsymbol{\alpha}}$.

Now consider

$$p(\mathbf{x}) = (\mathbf{x} \vee (0, 1, 0)) \wedge (1, a, a).$$

Then $p((x \wedge a, a, x)) = (x, 1, x) \wedge (1, a, a) = (x, a, x \wedge a)$. So (22) implies that $p(\mathbf{v}_J) \equiv p(\mathbf{v}'_J) \pmod{\boldsymbol{\gamma}}$, and symmetrically. Thus (22) is equivalent to

$$p(\mathbf{v}_J) \equiv p(\mathbf{v}'_J) \pmod{\boldsymbol{\gamma}},$$

that is, to

$$p(\mathbf{v}_J) \equiv p(\mathbf{v}'_J) \pmod{\boldsymbol{\alpha}},$$

since $p(\mathbf{v}_J), p(\mathbf{v}'_J) \in \gamma_a L$.

Now consider

$$q(\mathbf{x}) = (\mathbf{x} \vee (a, a, 1)) \wedge (1, a, a).$$

Then $q((0, x, 0)) = (x, a, x \wedge a)$ for $x \geq a$. So $q(\mathbf{v}_K) \equiv q(\mathbf{v}'_K) \pmod{\boldsymbol{\gamma}}$, that is, $q(\mathbf{v}_K) \equiv q(\mathbf{v}'_K) \pmod{\boldsymbol{\alpha}}$.

Finally, define

$$r(\mathbf{x}) = (\mathbf{x} \vee (a, a, 1)) \wedge (0, 1, 0).$$

Then $q((x, x \wedge a, a)) = (0, x, 0)$. So $q(\mathbf{v}_B) \equiv q(\mathbf{v}'_B) \pmod{\boldsymbol{\gamma}}$. From these it follows that $\mathbf{v}_K \equiv \mathbf{v}'_K \pmod{\boldsymbol{\gamma}}$ is equivalent to $q(\mathbf{v}_K) \equiv q(\mathbf{v}'_K) \pmod{\boldsymbol{\gamma}}$ and $q(\mathbf{v}_K), q(\mathbf{v}'_K) \in \gamma_a L$, so the latter is equivalent to $q(\mathbf{v}_K) \equiv q(\mathbf{v}'_K) \pmod{\boldsymbol{\alpha}}$.

We conclude that the congruence $\mathbf{v} \equiv \mathbf{v}' \pmod{\boldsymbol{\gamma}}$ in $\mathsf{M}_3[L, a]$ is equivalent to the following three congruences in L:

$$\begin{aligned}
\mathbf{v}_B &\equiv \mathbf{v}'_B & \pmod{\boldsymbol{\alpha}}, \\
p(\mathbf{v}_J) &\equiv p(\mathbf{v}'_J) & \pmod{\boldsymbol{\alpha}}, \\
q(\mathbf{v}_K) &\equiv q(\mathbf{v}'_K) & \pmod{\boldsymbol{\alpha}},
\end{aligned}$$

concluding the proof of the lemma. $\square$

In Chapter 18, we need a smaller interval introduced in G. Grätzer and E. T. Schmidt [135]; namely, for $a, b \in L$ with $a < b$, we introduce the interval $\mathsf{M}_3[L, a, b]$ of $\mathsf{M}_3[L]$:

$$\mathsf{M}_3[L, a, b] = [(0, a, 0), (1, b, b)] \subseteq \mathsf{M}_3[L].$$

Again,

$$\gamma_a : x \mapsto (x, a, x \wedge a)$$

is a (convex) embedding of L into $\mathsf{M}_3[L, a, b]$, see Figure 6.3. (Note that if L is bounded, then $\mathsf{M}_3[L, a] = \mathsf{M}_3[L, a, 1]$.)

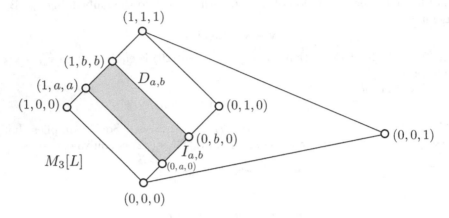

Figure 6.3: The shaded area: the lattice $\mathsf{M}_3[L, a, b]$.

Using the notation (illustrated in Figure 6.4; the black-filled elements form J):

$$B = \{\, (x, a, x \wedge a) \mid x \in L \,\}\ (= \gamma_a L),$$
$$I_{a,b} = [(0, a, 0), (0, b, 0)],$$
$$J = \{\, (x \wedge a, a, x) \mid x \leq b \,\}.$$

We can now generalize Lemmas 6.10–6.12:

Lemma 6.13. *Let* $\mathbf{v} = (x, y, z) \in \mathsf{M}_3[L, a, b]$. *Then* $\mathbf{v}$ *has a decomposition in* $\mathsf{M}_3[L, a, b]$:

$$\mathbf{v} = \mathbf{v}_B \vee \mathbf{v}_{I_{a,b}} \vee \mathbf{v}_J,$$

where

$$\mathbf{v}_B = (x, y, z) \wedge (1, a, a) = (x, a, x \wedge a) \in B,$$
$$\mathbf{v}_{I_{a,b}} = (x, y, z) \wedge (0, b, 0) = (0, y, 0) \in I_{a,b},$$
$$\mathbf{v}_J = (x, y, z) \wedge (a, a, b) = (z \wedge a, a, z) \in J.$$

Lemma 6.14. *Let* γ *be a congruence of* $\mathsf{M}_3[L, a, b]$ *and let* $\mathbf{v}, \mathbf{w} \in \mathsf{M}_3[L, a, b]$. *Then*

$$\mathbf{v} \equiv \mathbf{w} \pmod{\gamma}$$

iff

(23) $$\mathbf{v}_B \equiv \mathbf{w}_B \pmod{\gamma},$$
(24) $$\mathbf{v}_{I_{a,b}} \equiv \mathbf{w}_{I_{a,b}} \pmod{\gamma},$$
(25) $$\mathbf{v}_J \equiv \mathbf{w}_J \pmod{\gamma}.$$

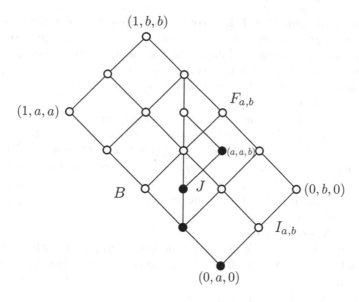

Figure 6.4: The lattice $M_3[L, a, b]$.

Lemma 6.15. *For a congruence α of L, let $M_3[\alpha, a, b]$ be the restriction of α^3 to $M_3[L, a, b]$. Then $M_3[\alpha, a, b]$ is a congruence of $M_3[L, a, b]$, and every congruence of $M_3[L, a, b]$ is of the form $M_3[\alpha, a, b]$ for a unique congruence α of L. It follows that γ_a is a congruence-preserving convex embedding of L into $M_3[L, a, b]$.*

We will use the notation

$$F_{a,b} = [(0, b, 0), (1, b, b)] = \{\, (x, b, x \wedge b) \mid x \in L \,\}.$$

The following two observations are trivial.

Lemma 6.16.

(i) $I_{a,b}$ *is an ideal of* $M_3[L, a, b]$ *and* $I_{a,b}$ *is isomorphic to the interval* $[a, b]$ *of* L.

(ii) $F_{a,b}$ *is a filter of* $M_3[L, a, b]$ *and* $F_{a,b}$ *is isomorphic to* L.

We can say a lot more about $F_{a,b}$. But first we need another lemma.

Lemma 6.17. *Let L be a lattice, let $[u, v]$ and $[u', v']$ be intervals of L, and let $\delta \colon [u, v] \mapsto [u', v']$ be an isomorphism between these two intervals. Let δ and δ^{-1} be term functions in L. Then L is a congruence-preserving extension of $[u, v]$ iff it is a congruence-preserving extension of $[u', v']$.*

Proof. Let us assume that L is a congruence-preserving extension of $[u', v']$. Let α be a congruence relation of $[u, v]$ and let $\delta\alpha$ be the image of α under δ. Since δ is an isomorphism, it follows that $\delta\alpha$ is a congruence of $[u', v']$, and so $\delta\alpha$ has a unique extension to a congruence γ of L.

We claim that γ extends α to L and extends it uniquely.

1. γ extends α.

Let $x \equiv y \pmod{\alpha}$. Then $\delta x \equiv \delta y \pmod{\delta\alpha}$, since δ is an isomorphism. By definition, δ extends $\delta\alpha$, so $\delta x \equiv \delta y \pmod{\alpha}$. Since δ^{-1} is a term function, the last congruence implies that $x \equiv y \pmod{\alpha}$. Conversely, let $x \equiv y \pmod{\gamma}$ and $x, y \in [u, v]$. Then $\delta x, \delta y \in [u', v']$; since γ is a term function, it follows that $\delta x \equiv \delta y \pmod{\gamma}$. Since γ extends $\delta\alpha$, we conclude that $\delta x \equiv \delta y \pmod{\delta\alpha}$. Using that δ^{-1} is an isomorphism, we obtain that $x \equiv y \pmod{\alpha}$, verifying the claim.

2. γ extends α uniquely.

Let γ extend α to L. As in the previous paragraph—*mutatis mutandis*—we conclude that γ extends $\delta\alpha$ to L, hence $\alpha = \gamma$, proving the uniqueness and the claim.

By symmetry, the lemma is proved. □

We have already observed in Lemma 6.12 that $\mathsf{M}_3[L, a, b]$ is a congruence-preserving convex extension of $[(0, a, 0), (1, a, a)]$. Since the isomorphism

$$x \mapsto x \vee (0, b, 0)$$

between $[(0, a, 0), (1, a, a)]$ and $[(0, b, 0), (1, b, b)] = F_{a,b}$ is a term function, and so is the inverse

$$x \mapsto x \wedge (1, a, a),$$

from Lemma 6.17 we conclude the following:

Corollary 6.18. *The lattice $\mathsf{M}_3[L, a, b]$ is a congruence-preserving convex extension of the filter $F_{a,b}$.*

We summarize our results (G. Grätzer and E. T. Schmidt [135]):

Lemma 6.19. *Let L be a bounded lattice, and let $a, b \in L$ with $a < b$. Then there exists a bounded lattice $L_{a,b}$ (with bounds $0_{a,b}$ and $1_{a,b}$) and $u_{a,b}, v_{a,b} \in L_{a,b}$, such that the following conditions are satisfied:*

(i) *$v_{a,b}$ is a complement of $u_{a,b}$.*

(ii) *$F_{a,b} = [v_{a,b}, 1_{a,b}] \cong L$.*

(iii) *$I_{a,b} = [0_{a,b}, v_{a,b}] \cong [a, b]$.*

(iv) *$L_{a,b}$ is a congruence-preserving (convex) extension of $[0_{a,b}, u_{a,b}]$ and of $[v_{a,b}, 1_{a,b}]$.*

(v) *The congruences on* $I_{a,b}$ *and* $F_{a,b}$ *are synchronized, that is, if* α *is a congruence on* L, $\overline{\alpha}$ *is the extension of* α *to* $L_{a,b}$ *(we map* α *to* $F_{a,b}$ *under the isomorphism, and then by* (iv) *we uniquely extend it to* $L_{a,b}$*), and* $x, y \in [a, b]$, *then we can denote by* $x_{F_{a,b}}$, $y_{F_{a,b}} \in F_{a,b}$ *the images of* x, y *in* $F_{a,b}$ *and by* $x_{I_{a,b}}, y_{I_{a,b}} \in I_{a,b}$ *the images of* x, y *in* $I_{a,b}$; *synchronization means that* $x_{F_{a,b}} \equiv y_{F_{a,b}}$ (mod $\overline{\alpha}$) *iff* $x_{I_{a,b}} \equiv y_{I_{a,b}}$ (mod $\overline{\alpha}$).

Proof. Of course, $L_{a,b} = \mathsf{M}_3[L, a, b]$, $0_{a,b} = (0, a, 0)$, $1_{a,b} = (1, b, b)$, $u_{a,b} = (1, a, a)$, and $v_{a,b} = (0, b, 0)$. $\qquad\qquad\qquad\qquad\qquad\qquad$ □

The lattice $L_{a,b}$ is illustrated with $L = \mathsf{C}_5$ in Figure 6.5, the five-element chain, a is the atom and b is the dual atom of S. This figure is the same as Figure 6.4, only the notation is changed.

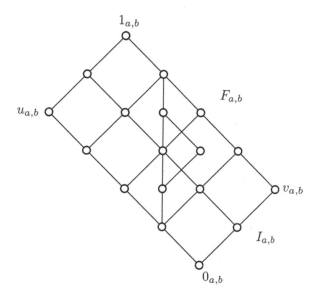

Figure 6.5: The lattice $L_{a,b}$.

Chapter

7

Cubic Extensions

In this chapter, for a finite lattice K, we introduce an extension Cube K with the following properties: (i) The lattice K is a congruence-reflecting sublattice of Cube K. (ii) Con(Cube K) is boolean. (iii) The minimal extension of the meet-irreducible congruences are the dual atoms of Con(Cube K); their ordering is "flattened."

7.1. The construction

Let K be a finite lattice. Following G. Grätzer and E. T. Schmidt [129], for every meet-irreducible congruence γ of K (in formula, $\gamma \in \mathrm{Con_M}\, K$), we form the quotient lattice K/γ, and extend it to a finite *simple* lattice Simp K_γ with zero 0_γ and unit 1_γ, using Lemma 2.3.

Let Cube$_{\mathrm{Simp}}\, K$ be the direct product of the lattices Simp K_γ for $\gamma \in \mathrm{Con_M}\, K$:

$$\mathrm{Cube_{Simp}}\, K = \prod(\mathrm{Simp}\, K_\gamma \mid \gamma \in \mathrm{Con_M}\, K).$$

If the function Simp is understood, we write Cube K for Cube$_{\mathrm{Simp}}\, K$. We regard Simp K_γ, for $\gamma \in \mathrm{Con_M}\, K$, an ideal of Cube K, as in Section 2.2.

For $a \in K$, define Diag$(a) \in$ Cube K as follows:

$$\mathrm{Diag}(a) = (a/\gamma \mid \gamma \in \mathrm{Con_M}\, K).$$

K has a natural (diagonal) embedding into Cube K by

$$\gamma\colon a \mapsto \mathrm{Diag}(a) \quad \text{for } a \in K.$$

© Springer International Publishing Switzerland 2016
G. Grätzer, *The Congruences of a Finite Lattice*,
DOI 10.1007/978-3-319-38798-7_7

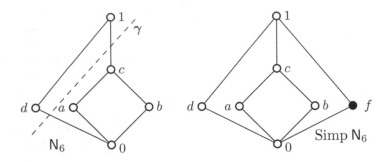

Figure 7.1: The lattices N_6 and Simp N_6.

Let $\mathrm{Diag}(K)$ be the sublattice $\{\,\mathrm{Diag}(a) \mid a \in K\,\}$ of Cube K, and for a congruence $\boldsymbol{\alpha}$ of K, let $\mathrm{Diag}(\boldsymbol{\alpha})$ denote the corresponding congruence of $\mathrm{Diag}(K)$. By identifying a with $\mathrm{Diag}(a)$, for $a \in K$, we can view Cube K as an extension of K; we call Cube K a *cubic extension* of K.

Cubic extensions are hard to draw because they are direct products. Here is a small example. We take the lattice N_6; see Figure 7.1. The lattice N_6 is subdirectly irreducible; it has two meet-irreducible congruences, $\mathbf{0}$ and $\boldsymbol{\gamma}$. Since $N_6/\boldsymbol{\gamma} = C_2$, it is simple, so we can choose Simp $N_6/\boldsymbol{\gamma} = C_2$. The other lattice quotient is $N_6/\mathbf{0} = N_6$ and for this we choose a simple extension Simp N_6, by adding an element; see Figure 7.1. The lattice Cube N_6 is shown in Figure 7.2 along with the embedding Diag of N_6 into Cube N_6. The images of elements of N_6 under Diag are black-filled.

As an alternative, you may draw the chopped lattice M whose ideal lattice is Cube N_6, see Figure 7.3. Unfortunately, the embedding Diag is not easy to

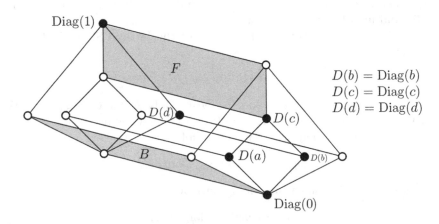

$$D(b) = \mathrm{Diag}(b)$$
$$D(c) = \mathrm{Diag}(c)$$
$$D(d) = \mathrm{Diag}(d)$$

Figure 7.2: The embedding of N_6 into Cube N_6.

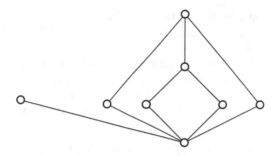

Figure 7.3: The chopped lattice M whose ideals represent Cube N_6.

see with this representation.

7.2. The basic property

The following crucial property of cubic extensions was stated in G. Grätzer and E. T. Schmidt [129].

Theorem 7.1. *Let K be a finite lattice and let* Cube K *be a cubic extension of K. Then K ($=\mathrm{Diag}(K)$) is congruence-reflecting in* Cube K.

Proof. For $\kappa \in \mathrm{Con}\,K$ and $\gamma \in \mathrm{Con}_M K$, define the congruence $\mathrm{Cube}(\kappa,\gamma)$ on the lattice $\mathrm{Simp}\,K_\gamma$ as follows:

$$\mathrm{Cube}(\kappa,\gamma) = \begin{cases} \mathbf{0} & \text{if } \kappa \leq \gamma, \\ \mathbf{1} & \text{if } \kappa \nleq \gamma; \end{cases}$$

and define

$$\mathrm{Cube}(\kappa) = \prod(\,\mathrm{Cube}(\kappa,\gamma) \mid \gamma \in \mathrm{Con}_M K\,),$$

a congruence of the lattice Cube K.

Observe that $\mathrm{Cube}(\mathbf{0}_K) = \mathbf{0}_{\mathrm{Cube}\,K}$ and $\mathrm{Cube}(\mathbf{1}_K) = \mathbf{1}_{\mathrm{Cube}\,K}$, that is, $\mathrm{Cube}(\mathbf{0}) = \mathbf{0}$ and $\mathrm{Cube}(\mathbf{1}) = \mathbf{1}$.

We show that $\mathrm{Diag}(\kappa)$ is the restriction of $\mathrm{Cube}(\kappa)$ to K. First, assume that $\mathrm{Diag}(u) \equiv \mathrm{Diag}(v) \pmod{\mathrm{Diag}(\kappa)}$ in $\mathrm{Diag}(K)$. Then $u \equiv v \pmod{\kappa}$ in K, by the definition of D. The congruence $u \equiv v \pmod{\kappa}$ implies that $u \equiv v \pmod{\gamma}$, for all $\gamma \in \mathrm{Con}_M K$ satisfying $\kappa \leq \gamma$, that is, $u/\gamma = v/\gamma$, for all $\gamma \in \mathrm{Con}_M K$ satisfying $\kappa \leq \gamma$. This, in turn, can be written as

$$u/\gamma \equiv v/\gamma \pmod{\mathrm{Cube}(\kappa,\gamma)}, \text{ for all } \gamma \in \mathrm{Con}_M K \text{ satisfying } \kappa \leq \gamma,$$

since, by definition, $\kappa_\gamma = \kappa$ for $\kappa \leq \gamma$.

Again, by definition, $\kappa_\gamma = 1$ for $\kappa \nleq \gamma$. Therefore, the congruence $u/\gamma \equiv v/\gamma \pmod{\mathrm{Cube}(\kappa,\gamma)}$ always holds.

We conclude that $u/\gamma \equiv v/\gamma$ (mod Cube(κ, γ)) for all $\gamma \in \mathrm{Con}_M K$. This congruence is equivalent to $\mathrm{Diag}(u) \equiv \mathrm{Diag}(v)$ (mod Cube(κ)) in Cube K, which was to be proved.

Second, assume that $\mathrm{Diag}(u) \equiv \mathrm{Diag}(v)$ (mod Cube(κ)) in Cube K. Then $u/\gamma \equiv v/\gamma$ (mod Cube(κ, γ)), for all $\gamma \in \mathrm{Con}_M K$; in particular, for all $\gamma \in \mathrm{Con}_M K$ satisfying $\kappa \leq \gamma$. Thus $u/\gamma = v/\gamma$, for all $\gamma \in \mathrm{Con}_M K$ satisfying $\kappa \leq \gamma$, that is, $u \equiv v$ (mod γ) for all $\gamma \in \mathrm{Con}_M K$ satisfying $\kappa \leq \gamma$. Therefore,

$$u \equiv v \quad (\mathrm{mod} \ \bigwedge(\gamma \in \mathrm{Con}_M K \mid \kappa \leq \gamma)).$$

The lattice Con K is finite, so every congruence is a meet of meet-irreducible congruences; therefore,

$$\kappa = \bigwedge(\gamma \in \mathrm{Con}_M K \mid \kappa \leq \gamma),$$

and so $u \equiv v$ (mod κ) in K, that is, $\mathrm{Diag}(u) \equiv \mathrm{Diag}(v)$ (mod $\mathrm{Diag}(\kappa)$) in $\mathrm{Diag}(K)$. $\qquad \square$

For $\kappa \in \mathrm{Con} \, K$, the set

$$\Delta_\kappa = \{\gamma \in \mathrm{Con}_M K \mid \kappa \nleq \gamma\}$$

is a down set of $\mathrm{Con}_M K$, and every down set of $\mathrm{Con}_M K$ is of the form Δ_κ, for a unique $\kappa \in \mathrm{Con} \, K$. The down set Δ_κ of $\mathrm{Con}_M K$ describes Cube(κ), and conversely.

In the example of Section 7.1 (see Figures 7.1–7.3), the congruence lattice Con $\mathsf{N}_6 = \{0, \gamma, 1\}$. Since Cube$(0) = 0$ and Cube$(1) = 1$, we only have to compute Cube(γ). Clearly, Cube$(\gamma) = 0 \times 1$ (that is, Cube$(\gamma) = 0_{\mathsf{C}_2} \times 1_{\mathrm{Simp} \, \mathsf{N}_6}$); it splits Cube N_6 into two parts as shown by the dashed line in Figure 7.2.

We now summarize the properties of the lattice Cube K.

Theorem 7.2. *Let K be a finite lattice with a cubic extension* Cube K. *Then*

(i) Cube K *is finite.*

(ii) *Choose an atom $s_\gamma \in \mathrm{Simp} \, K_\gamma$, for each $\gamma \in \mathrm{Con}_M K$, and let B be the boolean ideal of* Cube K *generated by these atoms. Then B is a congruence-determining ideal of* Cube K.

(iii) *There are one-to-one correspondences among the subsets of $\mathrm{Con}_M K$, the sets of atoms of B, and the congruences κ of* Cube K; *the subset of $\mathrm{Con}_M K$ corresponding to the congruence κ of* Cube K *is*

$$\delta_\kappa = \{\gamma \in \mathrm{Con}_M K \mid \kappa \nleq \gamma\},$$

which, in turn corresponds to the set

$$S_\kappa = \{\, s_\gamma \mid s_\gamma \equiv 0 \pmod{\kappa} \,\}$$

of atoms of B. Hence, the congruence lattice of Cube K *is a finite boolean lattice.*

(iv) *Every congruence κ of K has an extension* Cube(κ) *to a congruence of* Cube K *corresponding to the down set Δ_κ of* $\mathrm{Con_M}\,K$.

In the example of Section 7.1, for B we can choose the shaded ideal of Figure 7.2.

Corollary 7.3. *Choose a dual atom $t_\gamma \in \mathrm{Simp}\,K_\gamma$, for each $\gamma \in \mathrm{Con_M}\,K$, and define $\bar{t}_\gamma$ as the element of* Cube K *whose γ-component is t_γ and all other components are 1. The element $\bar{t}_\gamma$ is a dual atom of* Cube K. *Let F be the boolean filter of* Cube K *generated by these dual atoms. Then F is a congruence-determining filter of* Cube K.

There are one-to-one correspondences among the subsets of $\mathrm{Con_M}\,K$, *the sets of dual atoms of F, and the congruences κ of* Cube K; *the subset of* $\mathrm{Con_M}\,K$ *corresponding to the congruence κ of* Cube K *is*

$$D_\kappa = \{\, \gamma \in \mathrm{Con_M}\,K \mid \kappa \nleq \gamma \,\},$$

which, in turn, corresponds to the set

$$T_\kappa = \{\, s_\gamma \mid s_\gamma \equiv 1 \pmod{\mathbf{0}} \,\}$$

of dual atoms of F. Also, $|F| = |B|$.

In the example of Section 7.1, we can choose for F the shaded filter of Figure 7.2.

So a cubic extension Cube K of K (i) has a "cubic" congruence lattice (the boolean lattice B_n, an "n-dimensional cube"), and (ii) K and its cubic extension Cube K have the same number of meet-irreducible congruences, and (iii) the congruences κ of K extend to Cube K (but, as a rule, the cubic extension has many more congruences than the Cube(κ)-s). See Sections 14.2 and 15.2 for other small examples, in particular; see Figures 14.3 and 15.2.

Part III

Congruence Lattices
of Finite Lattices

The Dilworth Theorem

In Parts III–V, we discuss a subfield of Lattice Theory that started with the following result—a converse of Theorem 3.4, the Funayama-Nakayama result, [53].

Theorem 8.1 (Dilworth Theorem). *Every finite distributive lattice D can be represented as the congruence lattice of a finite lattice L.*

Our presentation is based on G. Grätzer and E. T. Schmidt [117], where the first proof appeared. In his book (P. Crawley and R. P. Dilworth [13]), Dilworth reproduces the proof from [117]. It is clear from his recollections in [11] that our thinking was very close to his.

In this chapter, we follow G. Grätzer and H. Lakser [89] (published in [59]), and prove this result based on the discussion of chopped lattices in Chapter 5, a simpler proof than the one in [117]. We will also prove that L can be constructed as a sectionally complemented lattice, as stated in [117].

8.1. The representation theorem

By Theorem 5.6, to prove the Dilworth Theorem, it is sufficient to verify the following:

Theorem 8.2. *Let D be a finite distributive lattice. Then there exists a chopped lattice M such that $\operatorname{Con} M$ is isomorphic to D.*

© Springer International Publishing Switzerland 2016
G. Grätzer, *The Congruences of a Finite Lattice*,
DOI 10.1007/978-3-319-38798-7_8

Using the equivalence of nontrivial finite distributive lattices and finite ordered sets (see Section 2.5.2) and using the notation $\mathrm{Con_J}\, M$ (see Section 3.2) for the ordered set of join-irreducible congruences, we can rephrase Theorem 8.2 as follows:

Theorem 8.3. *Let P be a finite order. Then there exists a chopped lattice M such that $\mathrm{Con_J}\, M$ is isomorphic to P.*

We are going to prove the Dilworth Theorem in this form.

8.2. *Proof-by-Picture*

The basic gadget for the construction is the lattice $\mathsf{N_6} = N(p, q)$ of Figure 8.1. The lattice $N(p, q)$ has three congruence relations, namely, **0**, **1**, and $\boldsymbol{\alpha}$, where $\boldsymbol{\alpha}$ is the congruence relation with congruence classes $\{0, q_1, q_2, q\}$ and $\{p_1, p(q)\}$, indicated by the dashed line. Thus $\mathrm{con}(p_1, 0) = \mathbf{1}$. In other words, $p_1 \equiv 0$ "implies" that $q_1 \equiv 0$, but $q_1 \equiv 0$ "does not imply" that $p_1 \equiv 0$. We will use the "gadget" $\mathsf{N_6} = N(p, q)$ to achieve such congruence-forcing.

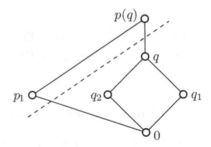

Figure 8.1: The lattice $\mathsf{N_6} = N(p, q)$ and the congruence $\boldsymbol{\alpha}$.

To convey the idea how to prove Theorem 8.3, we present three small examples in which we construct the chopped lattice M from copies of $N(p, q)$.

Example 1: The three-element chain $C = \mathsf{C_3}$. Let $P = \{a, b, c\}$ with $c \prec b \prec a$. We take two copies of the gadget, $N(a, b)$ and $N(b, c)$; they share the ideal $I = \{0, b_1\}$; see Figure 8.2. So we can merge them (in the sense of Section 5.1) and form the chopped lattice

$$M = \mathrm{Merge}(N(a, b), N(b, c))$$

as shown in Figure 8.2.

The congruences of M are easy to find. The isomorphism $P \cong \mathrm{Con_J}\, M$ is given by $x_1 \mapsto \mathrm{con}(0, x)$ for $x \in P$. See Figure 8.1 for the definition of x_1.

The congruences of M can be described by a compatible congruence vector $(\boldsymbol{\alpha}_{a,b}, \boldsymbol{\alpha}_{b,c})$ (see Section 5.3), where $\boldsymbol{\alpha}_{a,b}$ is a congruence of the lattice $N(a, b)$

and $\alpha_{b,c}$ is a congruence of the lattice $N(b,c)$, subject to the condition that $\alpha_{a,b}$ and $\alpha_{b,c}$ agree on I. Looking at Figure 8.1, we see that if the shared congruence on I is $\mathbf{0}$ $(= \mathbf{0}_I)$, then we must have $\alpha_{a,b} = \mathbf{0}$ $(= \mathbf{0}_{N(a,b)})$ and $\alpha_{b,c} = \mathbf{0}$ $(= \mathbf{0}_{N(b,c)})$ or $\alpha_{b,c} = \alpha$ on $N(b,c)$. If the shared congruence on I is $\mathbf{1}$ $(= \mathbf{1}_I)$, then we must have $\alpha_{a,b} = \alpha$ or $\alpha_{a,b} = \mathbf{1}$ $(= \mathbf{1}_{N(a,b)})$ on $N(a,b)$ and

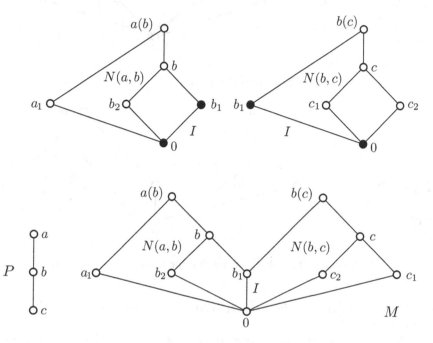

Figure 8.2: The chopped lattice M for $C = \mathsf{C}_3$.

$\alpha_{b,c} = \mathbf{1}$ $(= \mathbf{1}_{N(b,c)})$ on $N(b,c)$. So there are three congruences distinct from $\mathbf{0}$: $(\mathbf{0},\alpha)$, $(\alpha,\mathbf{1})$, $(\mathbf{1},\mathbf{1})$. Therefore, the join-irreducible congruences form the three-element chain.

Example 2: The three-element ordered set V of Figure 8.3, the "ordered set V". We take two copies of the gadget, $N(b,a)$ and $N(c,a)$; they share the ideal $J = \{0, a_1, a_2, a\}$; we merge them to form the chopped lattice

$$M_V = \mathrm{Merge}(N(b,a), N(c,a)),$$

see Figure 8.3. Again, the isomorphism $V \cong \mathrm{Con}_J M_V$ is given by $x_1 \mapsto \mathrm{con}(0, x)$ for $x \in V$.

Example 3: The three-element ordered set H of Figure 8.4, the "ordered set hat". We take two copies of the gadget, $N(a,b)$ and $N(a,c)$; they share the ideal $J = \{0, a_1\}$; we merge them to form the chopped lattice

$$M_V = \mathrm{Merge}(N(a,b), N(a,c)),$$

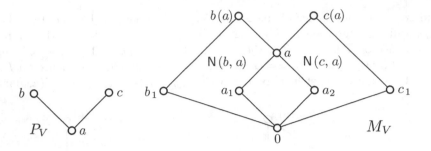

Figure 8.3: The chopped lattice for the ordered set V.

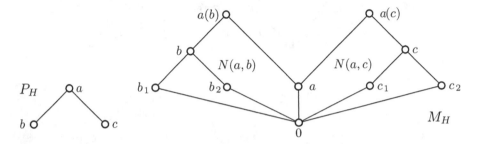

Figure 8.4: The chopped lattice for the ordered set hat.

see Figure 8.4. Again, the isomorphism $H \cong \operatorname{Con_J} M_H$ is given by $x_1 \mapsto \operatorname{con}(0, x)$ for $x \in V$.

The reader should now be able to picture the general proof: instead of the few atoms in these examples, we start with enough atoms to reflect the structure of P, see Figure 8.5. Whenever $b \prec a$ in P, we insert a copy of $N(a, b)$, see Figure 8.6.

8.3. Computing

For a finite ordered set P, let Max be the set of maximal elements in P. We form the set

$$M_0 = \{0\} \cup \{ p_1 \mid p \in \operatorname{Max} \} \cup \bigcup (\{a_1, a_2\} \mid a \in P - \operatorname{Max})$$

consisting of 0, the maximal elements of P indexed by 1, and two copies of the nonmaximal elements of P, indexed by 1 and 2. We make M_0 a meet-semilattice by defining $\inf\{x, y\} = 0$ if $x \neq y$, as illustrated in Figure 8.5. Note that $x \equiv y \pmod{\alpha}$ and $x \neq y$ imply that $x \equiv 0 \pmod{\alpha}$ and $y \equiv 0$

(mod α) in M_0; therefore, the congruence relations of M_0 are in one-to-one correspondence with subsets of P. Thus $\operatorname{Con} M_0$ is a boolean lattice whose atoms are associated with atoms of M_0; the congruence γ_x associated with the atom x has only one nontrivial block $\{0, x\}$.

We construct an extension M of M_0 as follows:

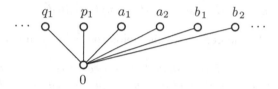

Figure 8.5: The chopped lattice M_0.

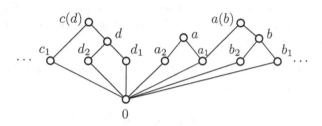

Figure 8.6: The chopped lattice M.

The chopped lattice M consists of four kinds of elements: (i) the zero, 0; (ii) for all maximal elements p of P, the element p_1; (iii) for any nonmaximal element p of P, three elements: p, p_1, p_2; (iv) for each pair $p, q \in P$ with $p \succ q$, a new element, $p(q)$. For $p, q \in P$ with $p \succ q$, we set

$$\mathsf{N}_6 = N(p, q) = \{0, p_1, q, q_1, q_2, p(q)\}.$$

For $x, y \in M$, let us define $x \le y$ to mean that, for some $p, q \in P$ with $p \succ q$, we have $x, y \in N(p, q)$ and $x \le y$ in the lattice $N(p, q)$. It is easily seen that $x \le y$ does not depend on the choice of p and q, and that $\le$ is an ordering. Since, under this ordering, all $N(p, q)$ and $N(p, q) \cap N(p', q')$ ($p \succ q$ and $p' \succ q'$ in P) are lattices and $x, y \in M$, $x \in N(p, q)$, and $y \le x$ imply that $y \in N(p, q)$, we conclude that M is a chopped lattice; in fact, it is a union of the ideals $N(p, q)$ with $p \succ q$ in P, and two such distinct ideals intersect in a one-, two-, or four-element ideal.

Since the chopped lattice M is atomistic, Corollary 3.11 applies. If $p_i \Rightarrow q_j$ in M, for $p, q \in P$ and $i, j \in \{1, 2\}$, then $p \ge q$ in P, and conversely. So the equivalence classes of the atoms under the preordering $\Rightarrow$ form an ordered set isomorphic to $\operatorname{Down} P$. This completes the verification that $\operatorname{Con}_J M \cong P$, and therefore, of Theorem 8.3.

8.4.   Sectionally complemented lattices

Let M be the chopped lattice described in Section 8.3. Since all $N(p,q)$-s are sectionally complemented, so is M.

In G. Grätzer and E. T. Schmidt [117] the following result is proved.

Theorem 8.4. $L = \mathrm{Id}\,M$ *is sectionally complemented.*

Combined with Theorem 8.3, this yields the main result of this chapter.

Theorem 8.5. *Every finite distributive lattice D can be represented as the congruence lattice of a finite sectionally complemented lattice L.*

To prove Theorem 8.4, we will view $L = \mathrm{Id}\,M$ as a closure system.

For an ideal U of M, let $\mathrm{Atom}(U)$ be the set of atoms of M in U. Note that the atoms of M are the $\{p_i\}$, where $p \in P$ and $i \in \{1,2\}$. We start with the following two trivial statements (to simplify the notation, we compute with the indices modulo 2):

Lemma 8.6. *For a set $A \subseteq \mathrm{Atom}(M)$, there is an ideal U with $\mathrm{Atom}(U) = A$ iff A satisfies the condition:*

(Cl)   *For $p \succ q$ in P, if $p_1, q_i \in A$, then $q_{i+1} \in A$.*

Let us call a subset A of $\mathrm{Atom}(M)$ *closed* if it satisfies (Cl). It is obvious that every subset A of $\mathrm{Atom}(M)$ has a closure $\overline{A}$.

Lemma 8.7. *The assignment $I \mapsto \mathrm{Atom}(I)$ is a bijection between the ideals of M and closed subsets of $\mathrm{Atom}(M)$, and*

$$\mathrm{Atom}(I \wedge J) = \mathrm{Atom}(I) \cap \mathrm{Atom}(J),$$
$$\mathrm{Atom}(I \vee J) = \overline{\mathrm{Atom}(I) \cup \mathrm{Atom}(J)}$$

for $I, J \in \mathrm{Id}\,M$. The inverse map assigns to a closed set X of atoms, the ideal $\mathrm{id}(X)$ of M generated by X.

Lemma 8.7 allows us to regard L as the lattice of closed sets in $\mathrm{Atom}(M)$, so $I \in L$ will mean that I is a closed subset of $\mathrm{Atom}(M)$. Thus $I \wedge J = I \cap J$ and $I \vee J = \overline{I \cup J}$ for $I, J \in L$.

Let $I \subseteq J \in L$. Let us say, that $q \in P$ *splits over* (I, J) if there exists a $p \succ q$ in P, with $p_1, q_i \in J - I$ and $q_{i+1} \in I$. If there is a $q \in P$ that splits over (I, J), then $J - I$ is not closed. Let $X = X(I, J)$ be the set of all elements q_i in $J - I$ such that q splits over (I, J). Let $S(I, J) = (J - I) - X$, that is, $S(I, J)$ is the set of all elements q_i in $J - I$ such that q does not split over (I, J).

Lemma 8.8. $S(I, J) \in L$.

Proof. We have to prove that $S = S(I, J)$ is closed.

Let $u \succ v$ in $P, u_1 \in S$, and $v_i \in S$. Since, by the definition of S, the element v does not split over (I, J) and $u_1, v_i \in J - I$, it follows that $v_{i+1} \notin I$. Since $u_1 \in J$ and $v_i \in J$ and J is closed, we obtain that $v_{i+1} \in J$. Thus $v_{i+1} \in J - I$. Since v does not split over (I, J), we get that $v_{i+1} \in S$ by the definition of S. Thus S is closed. $\qquad\square$

We claim that $S = S(I, J)$ is the sectional complement of I in J.
By Lemma 8.7, we have to prove that

(1) $$I \cap S = \varnothing,$$

(2) $$\overline{I \cup S} = J.$$

(1) is obvious from the definition of S.

Since $I \subseteq J$ and $S \subseteq J$, to verify (2), it is sufficient to show that

(3) $$\overline{I \cup S} \supseteq J.$$

Assume, to the contrary, that there is a $q \in P$ and $i \in \{1, 2\}$ such that

(4) $$q_i \in J - \overline{I \cup S}.$$

We can choose q so that it is maximal with respect to this property, that is, if $p > q$ and $p_j \in J$ for some $j \in \{1, 2\}$, then $p_j \in \overline{I \cup S}$.

It follows from (4) that $q_i \in J - (I \cup S) = X$. So, by the definition of X, there is $p \succ q$ in P with $p_1 \in J - I$ and $q_{i+1} \in I$. Since $p_1 \in J$ and $p \succ q$, by the maximality of q, we have that $p_1 \in \overline{I \cup S}$. Also, $q_{i+1} \in I \subseteq \overline{I \cup S}$. So $q_i \in \overline{I \cup S}$ by the definition of closure, contradicting (4). This completes the proof of the claim and of Theorem 8.4.

8.5.   Discussion

The Dilworth Theorem is the foundation of all the results in Parts III–V. Moreover, the results of Part III derive from the results of Parts IV *via* the Dilworth Theorem. For instance, the result of Chapter 15 is: *Every finite lattice K has a congruence-preserving embedding into a finite semimodular lattice L.* Combining this with the Dilworth Theorem, we obtain the main result of Chapter 10: *Every finite distributive lattice D can be represented as the congruence lattice of a finite semimodular lattice L.* Of course, this representation theorem is a relatively easy result compared to the congruence-preserving extension result. Nevertheless, the congruence-preserving extension theorem does not directly imply the representation theorem. Note, additionally, that the proof of the result of Chapter 15 utilizes the result of Chapter 10.

An addendum

The lattice L we construct for Theorem 8.1 has a very interesting congruence structure: every join-irreducible congruence γ is of the form $\mathrm{con}(0, p_\gamma)$, where p_γ is an atom and the p_γ-s generate a boolean ideal. We state this formally.

Theorem 8.9 (Full Dilworth Theorem). *Every finite distributive lattice D can be represented as the congruence lattice of a finite sectionally complemented lattice L with the following property:*

(B) *There is a boolean ideal B in L such that there is a bijection $\gamma \mapsto p_\gamma$ between $\mathrm{Con_J}\, L$ and the atoms of B satisfying $\gamma = \mathrm{con}(0, p_\gamma)$. There is a bijection between the congruences of L and "down sets" of the atoms of B, that is, sets H of atoms satisfying the condition:*

$$p_\gamma \in H \text{ and } \delta \leq \gamma \text{ imply that } p_\delta \in H \text{ for } \gamma, \delta \in \mathrm{Con_J}\, L.$$

Proof. Let $P = \mathrm{J}(D)$ and take the sublattice B generated by $\{\, p_1 \mid p \in P \,\}$. It is easy to see that B is an ideal. The statement about the congruences is implicit in the discussion of Section 8.3. $\square$

The ideal B is, of course, a congruence-determining sublattice of L.

Sectionally complemented chopped lattices

The 1960 result of G. Grätzer and E. T. Schmidt [117]: *Every finite distributive lattice D can be represented as the congruence lattice of a finite sectionally complemented lattice L* constructs a finite sectionally complemented chopped lattice M representing D, and proves that the ideal lattice of M is again sectionally complemented. We have tried with E. T. Schmidt for decades to prove a more general result of this type. It was only in this century that in G. Grätzer, H. Lakser, and M. Roddy [101] an example of a finite sectionally complemented chopped lattice was found whose ideal lattice is not sectionally complemented, as stated in Theorem 5.9; see Figure 5.2.

Problem 8.1. Let M be a finite sectionally complemented chopped lattice. Find reasonable sufficient conditions under which $\mathrm{Id}\, M$ is sectionally complemented.

The reader is referred to G. Grätzer and H. Lakser [96] and [97], G. Grätzer, H. Lakser, and M. Roddy [101], and G. Grätzer and M. Roddy [115], for relevant results.

There are a number of problems that arise from Problem 8.1. We will describe a few; the reader should be able to easily add another dozen.

As in G. Grätzer and E. T. Schmidt [123] and [126], we generalize chopped lattices from the finite to the general case.

Let M be a meet-semilattice satisfying the following condition:

(Ch) $\sup\{a,b\}$ exists for any $a,b \in M$ having a common upper bound in M.

We define in M:
$$a \vee b = \sup\{a,b\},$$
whenever $\sup\{a,b\}$ exists in M. This makes M into a partial lattice called a *chopped lattice*.

For a finite M, this is equivalent to the definition in Section 5.1. We define ideals, congruences, and merging in the general case as we did in the finite case. Unfortunately, the fundamental result, Theorem 5.6 of G. Grätzer and H. Lakser [89], fails for the general case.

Problem 8.2. Investigate generalizations of Theorem 5.6 to the infinite case.

See G. Grätzer and E. T. Schmidt [123] for some related results.

As we note in Section 5.2, every finite chopped lattice M decomposes into lattices: $M = \bigcup(\,\mathrm{id}(m) \mid m \in \mathrm{Max}\,)$.

Problem 8.3. Can chopped lattices, in general, be usefully decomposed into lattices? Could this be utilized in Problem 8.2 by assuming that the chopped lattices decompose into *finitely many* lattices or into lattices with nice properties?

The mcr function

For a natural number n and a class $\mathbf{V}$ of lattices, define $\mathbf{mcr}(n, \mathbf{V})$ (minimal congruence representation) as the smallest integer such that, for any distributive lattice D with n join-irreducible elements, there exists a *finite* lattice $L \in \mathbf{V}$ satisfying $\mathrm{Con}\, L \cong D$ and $|L| \leq \mathbf{mcr}(n, \mathbf{V})$.

We will investigate $\mathbf{mcr}(n, \mathbf{L}) = \mathbf{mcr}(n)$ in the next chapter. In this chapter we proved that $\mathbf{mcr}(n, \mathbf{SecComp}) = O(2^{2n})$, where $\mathbf{SecComp}$ is the class of sectionally complemented lattices.

Problem 8.4. Is $\mathbf{mcr}(n, \mathbf{SecComp}) = O(2^{2n})$ the best possible result?

For any class $\mathbf{S}$ of lattices if the Dilworth Theorem holds for $\mathbf{S}$, then theoretically, the function $\mathbf{mcr}(n, \mathbf{S})$ exists, although it may be difficult to compute.

Of course, for any class $\mathbf{S}$ of lattices for which the Dilworth Theorem holds, we can raise the question what is $\mathbf{mcr}(n, \mathbf{S})$, and chances are that we get an interesting problem. In many cases, however, the Dilworth Theorem fails for $\mathbf{S}$. Here are some nontrivial examples.

(i) The congruence lattice of a finite relatively complemented lattice is boolean. So $\mathbf{SecComp}$ cannot be narrowed to the class of relatively complemented lattices.

(ii) The class **S** = **SecComp** ∩ **DuallySecComp** would also be a logical candidate. Note, however, that by Theorem 11 of G. Grätzer and E. T. Schmidt [117] and Lemma 4.16 of M. F. Janowitz [161], every finite lattice in **SecComp** ∩ **DuallySecComp** has a boolean congruence lattice (see also [59], Theorem II.4.9).

(iii) Similarly, one can ask whether **SecComp** can be narrowed to the class of semimodular sectionally complemented lattices. The discussion in Section IV.3 of [59] (in particular, the top paragraph of p. 240) shows that this cannot be done either. Every finite, semimodular, sectionally complemented lattice has a boolean congruence lattice.

Congruence class sizes

Spectra

The basic gadget in this chapter is the lattice N_6; Figure 8.1 shows N_6 and its only nontrivial congruence α. We can associate with the congruence α the pair $(4, 2)$ measuring the size of the two congruence classes. Which pairs (t_1, t_2) can substitute for $(4, 2)$? In other words, for which pairs of integers (t_1, t_2) is there a finite lattice L such that

(1) L is sectionally complemented;

(2) L has exactly one nontrivial congruence α;

(3) α has exactly two congruence classes: a prime ideal P and a prime filter Q satisfying that $|P| = t_1$ and $|Q| = t_2$.

This question is answered as follows in G. Grätzer and E. T. Schmidt [137]:

Theorem 8.10. *Let* (t_1, t_2) *be a pair of natural numbers. Then there is a finite lattice* L *with properties* (1)–(3) *iff* (t_1, t_2) *satisfies the following three conditions:*

(P_1) $2 \le t_1$ *and* $t_1 \ne 3$.

(P_2) $2 \le t_2$ *and* $t_2 \ne 3$.

(P_3) $t_1 > t_2$.

What can we say about the cardinalities of the congruence classes of a nontrivial congruence in a finite sectionally complemented lattice?

Let L be a finite lattice, and let α be a congruence of L. We denote by Spec α the *spectrum* of α, that is, the family of cardinalities of the congruence classes of α.

In G. Grätzer and E. T. Schmidt [137], spectra are characterized for finite sectionally complemented lattices.

Theorem 8.11. *Let* $S = (m_j \mid j < n)$ *be a family of natural numbers,* $n \geq 1$. *Then there is a nontrivial finite sectionally complemented lattice* L *and a nontrivial congruence* $\boldsymbol{\alpha}$ *of* L *such that* S *is the spectrum of* $\boldsymbol{\alpha}$ *iff* S *satisfies the following conditions:*

(S_1) $2 \leq n$ *and* $n \neq 3$.

(S_2) $2 \leq m_j$ *and* $m_j \neq 3$ *for all* $j < n$.

Corollary 8.12. *Let* $S = (m_j \mid j < n)$ *be a family of natural numbers,* $n > 1$. *Then there is a nontrivial finite sectionally complemented lattice* L *with a unique nontrivial congruence* $\boldsymbol{\alpha}$ *of* L *such that* S *is the spectrum of* $\boldsymbol{\alpha}$ *iff* S *satisfies* (S_1) *and* (S_2), *and additionally:*

(S_3) S *is not constant, that is, there are* $j, j' < n$ *satisfying that* $m_j \neq m_{j'}$.

(S_4) $n \neq 4$.

We only know that the congruence lattice Con L of the lattice L we construct in Theorem 8.11 has three or more elements. Can we prescribe its structure?

Problem 8.5. Let D be a finite distributive lattice. Can we construct the lattice L of Theorem 8.11 that represents $S = (m_j \mid j < n)$ as the spectrum of a nontrivial congruence $\boldsymbol{\alpha}$ so that L also satisfies $D \cong \mathrm{Con}\, L$?

Valuation

There is a more sophisticated way of looking at spectra. Let K be a finite sectionally complemented lattice. Let us represent K in the form $L/\boldsymbol{\alpha}$, where L is a finite sectionally complemented lattice and $\boldsymbol{\alpha}$ is a congruence of L. Then there is a natural map $v \colon K \to \mathbb{N}$ (where $\mathbb{N}$ is the set of natural numbers) defined as follows: Let $a \in K$; then a is represented by a congruence class A of $\boldsymbol{\alpha}$, so we can define $v(a) = |A|$. We call v a *valuation* on K.

Theorem 8.13 (G. Grätzer and E. T. Schmidt [137]). *Let* K *be a nontrivial finite sectionally complemented lattice, and let* $v \colon K \to \mathbb{N}$. *Then there exists a finite sectionally complemented lattice* L *and a nontrivial congruence* $\boldsymbol{\alpha}$ *of* L, *such that there is an isomorphism* $\mathbf{g} \colon K \to L/\boldsymbol{\alpha}$ *satisfying*

$$v(a) = |\mathbf{g}(a)|, \quad \text{for all } a \in K,$$

iff v *satisfies the following two conditions:*

(V_1) v *is antitone.*

(V_2) $2 \leq v(a)$ *and* $v(a) \neq 3$ *for all* $a \in K$.

Corollary 8.14. *Let K be a nontrivial finite sectionally complemented lattice, and let $v\colon K \to \mathbb{N}$. Then there exists a finite sectionally complemented lattice L and a* unique *nontrivial congruence α of L, such that there is an isomorphism $\gamma\colon K \to L/\alpha$ satisfying*

$$v(a) = |\gamma a|, \quad \text{for all } a \in K,$$

iff v satisfies the conditions (V_1) and (V_2), and additionally, v satisfies the following two conditions:

(V_3) *v is not a constant function.*

(V_4) *K is simple.*

We can state Problems 8.5 also for valuations.

Problem 8.6. Let D be a finite distributive lattice.   Can we prove Theorem 8.13 with the additional condition: $\mathrm{Con}\,K \cong D$?

Here is a congruence-preserving extension variant of the valuation problem.

Problem 8.7. Let L be a finite lattice and let α be a nontrivial congruence of L with spectrum v on $K = L/\alpha$. Let $v'\colon K \to \mathbb{N}$ satisfy (V_1) and (V_2). If $v \le v'$ (that is, $v(a) \le v'(a)$, for all $a \in K$), then does there exist a finite congruence-preserving extension L' of L such that the spectrum on L'/α' is v'?

Minimal Representations

9.1. The results

In the proof of the Dilworth Theorem (Theorem 8.1), we construct—for a distributive lattice D with $n \geq 1$ join-irreducible elements—a lattice L satisfying $\operatorname{Con} L \cong D$. The size of this lattice is $O(2^{2n})$.

Can we do better? Using the notation of Section 8.5, what is $\mathbf{mcr}(n)$?

$O(2^{2n})$ was improved to $O(n^3)$ in G. Grätzer and H. Lakser [90]—a substantial improvement from exponential to polynomial size—where it was conjectured that $O(n^3)$ can be improved to $O(n^2)$ and that $O(n^2)$ is best possible. Indeed, $O(n^2)$ is possible, as proved in G. Grätzer, H. Lakser, and E. T. Schmidt [102].

Theorem 9.1. *Let D be a finite distributive lattice with $n \geq 1$ join-irreducible elements. Then there exists a lattice L of $O(n^2)$ elements with $\operatorname{Con} L \cong D$. In fact, there is such a* planar *lattice L.*

Or, equivalently,

Theorem 9.1′. *Let P be a finite ordered set with n elements. Then there exists a lattice L of $O(n^2)$ elements with $\operatorname{Con}_J L \cong P$. In fact, there is such a* planar *lattice L.*

The second part of the conjecture was verified in G. Grätzer, I. Rival, and N. Zaguia [114].

Theorem 9.2. *Let α be a real number satisfying the following condition: Every distributive lattice D with n join-irreducible elements can be represented as the congruence lattice of a lattice L with $O(n^\alpha)$ elements. Then $\alpha \geq 2$.*

© Springer International Publishing Switzerland 2016
G. Grätzer, *The Congruences of a Finite Lattice*,
DOI 10.1007/978-3-319-38798-7_9

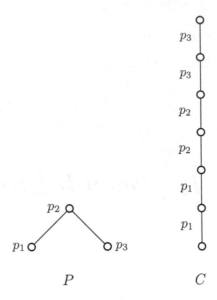

Figure 9.1: P and C.

9.2. *Proof-by-Picture* for minimal construction

We present now the *Proof-by-Picture* of the minimal construction (Theorem 9.1′) with the ordered set P (the hat) of Figure 9.1. We form the chain C from P, as shown by the same diagram. The chain C is of length $2|P| = 6$, it is colored (in the sense of Section 3.2) with elements of P, as illustrated.

We form C^2, and color the prime intervals in the obvious way: a prime interval of C^2 is of the form $[(a,c),(b,c)]$ or $[(c,a),(c,b)]$, where $a \prec b$ in C, and we color both intervals the color of $[a,b]$ in C.

Figure 9.2 shows the two gadgets (colored lattices) we use: M_3 and $\mathsf{N}_{5,5}$. We proceed as follows: To construct L, we take C^2 with the coloring we have just introduced. If both lower edges of a covering square in C^2 have the same

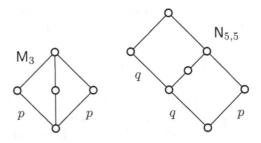

Figure 9.2: The two gadgets $(p < q)$.

color, we add an element to make it a covering M_3, the first gadget. If in C^2 we have a covering $C_3 \times C_2$, where the C_2 is colored by p, the C_3 is colored by q twice, where $p < q$ in P, then we add an element to make it an $N_{5,5}$, the second gadget.

What do these gadgets do? If $\mathfrak{p}$ and $\mathfrak{q}$ are any two prime intervals of M_3, then $\mathfrak{p} \Leftrightarrow \mathfrak{q}$ (using the notation of Section 3.2). If $\mathfrak{p}$ and $\mathfrak{q}$ are any two prime intervals of $N_{5,5}$ of color p and q with $p < q$, then $\mathfrak{q} \Rightarrow \mathfrak{p}$ (but $\mathfrak{p} \Rightarrow \mathfrak{q}$ fails). The lattice L we obtain is shown in Figure 9.3. (The copies of $N_{5,5}$ in the diagram are marked by black-filled elements.)

Every prime interval of C^2 is projective to a prime interval in one of the two copies of C in C^2. The first gadget makes sure that a prime interval in the first copy of C is projective to a prime interval of the same color in the second copy of C. The second gadget orders the prime intervals: if $\mathfrak{p}$ is of color p and $\mathfrak{q}$ is of color q, then $p > q$ implies that $\mathfrak{p} \Rightarrow \mathfrak{q}$. Now Theorem 3.10 shows that the equivalence classes of prime intervals of L form an ordered set isomorphic to P, that is, $\mathrm{Con}_J L \cong P$.

In particular, we make both copies of C congruence-determining sublattices of L.

What is the size of L? The size of C^2 is $(2n+1)^2$. For each color $p \in P$, we add the first gadget four times, adding $4n$ elements. For all $p < q \in P$, there is only one pair of adjacent prime intervals of color q in the first copy of C, and these can be paired with two prime intervals of color p in the second copy of C, yielding two more elements. Since there are at most $n^2/2$ such pairs, we get at most n^2 new elements. So $|L| \leq (2n+1)^2 + 4n + n^2$, so $|L|$ is obviously $O(n^2)$.

This completes the *Proof-by-Picture*.

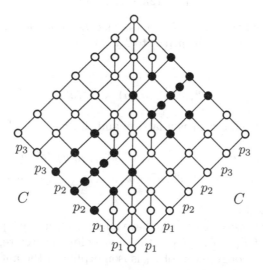

Figure 9.3: The minimal lattice L.

9.3.   The formal construction

We now formally prove Theorem 9.1′. Let P be a finite ordered set of n elements.

Let $P_0 = \{a_0, a_1, \ldots, a_{k-1}\}$ be the set of nonminimal elements of P. If $k = 0$, then P is an antichain, and we can choose L as a chain of length n. Henceforth, we assume that $k > 0$.

Let C_0 be a chain of length $2k$. We color the prime intervals of C_0 as follows: we color the lowermost two prime intervals of C_0 with a_0, the next two with a_1, and so on. Thus, for each $a \in P_0$, there are in C_0 *two prime intervals of color* a (and they are successive). So for each $a \in P_0$, there is a unique subchain $a^b \prec a^m \prec a^t$ such that the prime intervals $[a^b, a^m]$ and $[a^m, a^t]$ have color a, and no other prime interval of C_0 has color a. Observe that, for each a distinct from a_0, there is a $c \in P_0$ with $a^b = c^t$, and, similarly, for each a distinct from a_{k-1}, there is a $c \in P_0$ with $a^t = c^b$. The elements a^m, however, are labeled uniquely.

Let C_1 be a chain of length $n = |P|$. We color the prime intervals of C_1 by an arbitrary bijection with P. Thus, for each $a \in P$, there is in C_1 *exactly one prime interval* of color a; we denote it by $[a^o, a^i]$.

We set $L_0 = C_0 \times C_1$. We will regard C_0 and C_1 as sublattices of L_0 in the usual manner.

As discussed in Section 3.2, we can preorder the prime intervals of L_0 by the preorder $\Rightarrow$ and we can choose the prime intervals in $C_0 \cup C_1$ as representatives of the equivalence classes. Hence, the order of join-irreducible congruences of L_0 is an antichain of cardinality $2k + n$.

Note that

$$|L_0| = (2k+1)(n+1).$$

We next extend the lattice L_0 to a lattice L_1. For each $a \in P_0$, we adjoin two new elements $m_0(a)$ and $m_1(a)$ to L_0; we set

$$(a^b, a^o) \prec m_0(a) \prec (a^m, a^i),$$
$$(a^m, a^o) \prec m_1(a) \prec (a^t, a^i).$$

The resulting ordered set L_1 is a lattice, and, for each $a \in P_0$, the intervals

$$[(a^b, a^o), (a^m, a^i)] = \{(a^b, a^o), (a^m, a^o), m_0(a), (a^b, a^i), (a^m, a^i)\},$$
$$[(a^m, a^o), (a^t, a^i)] = \{(a^m, a^o), (a^t, a^o), m_1(a), (a^m, a^i), (a^t, a^i)\}$$

are isomorphic to M_3.

By adjoining these elements, we have made congruence equivalent any two prime intervals of L_1 in $C_0 \cup C_1$ of the same color. Therefore, the ordered set of join-irreducible congruences of L_1 is isomorphic to the antichain P.

Note that both C_0 and C_1 are congruence-preserving sublattices of L_1.

Observe that

$$|L_1| = (2k+1)(n+1) + 2k.$$

We finally further extend L_1 so as to induce the correct order on the join-irreducible congruences. For each pair $a \succ c$ in P (whereby $a \in P_0$, necessarily), we add a new element $n(a,c)$ to L_1, setting

$$(a^m, c^o) \prec n(a,c) \prec (a^m, c^i).$$

The resulting ordered set L is a lattice. In L, the interval $[a^m, a^t]$ is projective with $[n(a,c), (a^m, c^i)]$, which in turn is projective with $[c^o, c^i]$. So the ordered set of join-irreducible congruences of L is isomorphic to P.

Note that L is a planar lattice, and that in going from L_1 to L, we adjoin no more than kn elements. Thus

$$|L| \leq (2k+1)(n+1) + 2k + kn < 3(n+1)^2.$$

9.4. *Proof-by-Picture* for minimality

We present a *Proof-by-Picture* of Theorem 9.2 by contradiction. Let $\alpha < 2$ be a real number such that every order P with n elements can be represented as $\mathrm{Con}_J L$, for a lattice L with $O(n^\alpha)$ elements, that is, $|L| \leq Cn^\alpha$ for some constant C.

Let $\mathfrak{p} = [a,b]$ and $\mathfrak{q} = [c,d]$ be prime intervals of L such that $\mathrm{con}(\mathfrak{p}) \succ \mathrm{con}(\mathfrak{q})$ in $\mathrm{Con}_J L$. Then by Lemma 3.8, $\mathfrak{p} \Rightarrow \mathfrak{q}$, for instance,

$$\mathfrak{p} = [a,b] \overset{\mathrm{up}}{\sim} [u_1, v_1] \overset{\mathrm{dn}}{\twoheadrightarrow} [u_2, v_2] \overset{\mathrm{up}}{\twoheadrightarrow} [u_3, v_3] \overset{\mathrm{dn}}{\sim} \mathfrak{q} = [c,d],$$

as illustrated in Figure 9.4. Note the elements $h_i \in [u_i, v_i]$, $i = 1, 2, 3$, satisfying $[u_1, h_1] \overset{\mathrm{dn}}{\sim} [u_2, v_2]$, $[h_2, v_2] \overset{\mathrm{up}}{\sim} [u_3, v_3]$, and $[h_3, v_3] \overset{\mathrm{dn}}{\sim} \mathfrak{q}$.

Since $\mathrm{con}(\mathfrak{p}) > \mathrm{con}(\mathfrak{q})$ and

$$\mathrm{con}(\mathfrak{p}) \geq \mathrm{con}(u_1, v_1) \geq \mathrm{con}(u_2, v_2) \geq \mathrm{con}(u_3, v_3) \geq \mathrm{con}(\mathfrak{q}),$$

we must have a $>$ in the last formula; assume, for instance, that

$$\mathrm{con}(\mathfrak{p}) = \mathrm{con}(u_1, v_1) > \mathrm{con}(u_2, v_2).$$

Then $\mathrm{con}(u_1, h_1) = \mathrm{con}(u_2, v_2) < \mathrm{con}(\mathfrak{p})$, so $\mathrm{con}(h_1, v_1) = \mathrm{con}(\mathfrak{p})$. Since $\mathfrak{q}$ is collapsed by $\mathrm{con}(u_1, h_1)$, by Lemma 3.9, there is a prime interval $[e_1, x]$ in $[u_1, h_1]$ so that $\mathfrak{q}$ is collapsed by $\mathrm{con}(e_1, x)$. Now $\mathrm{con}(\mathfrak{p}) > \mathrm{con}(e_1, x) \geq \mathrm{con}(\mathfrak{q})$, so by the assumption $\mathrm{con}(\mathfrak{p}) \succ \mathrm{con}(\mathfrak{q})$, it follows that $\mathrm{con}(e_1, x) = \mathrm{con}(\mathfrak{q})$.So (with $h = h_1$ and $e_2 = v_1$) *we found a three-element chain* $e_1 < h < e_2$ *with* $\mathrm{con}(e_1, h) = \mathrm{con}(\mathfrak{q})$ *and* $\mathrm{con}(h, e_2) = \mathrm{con}(\mathfrak{p})$. Other choices lead to the same result, with maybe $\mathrm{con}(\mathfrak{q})$ on the top.

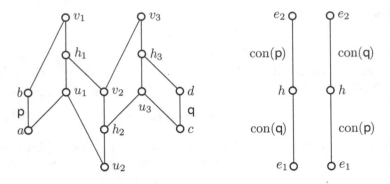

Figure 9.4: $\mathfrak{p} \Rightarrow \mathfrak{q}$ illustrated for $\mathrm{con}(\mathfrak{p}) \succ \mathrm{con}(\mathfrak{q})$ in $\mathrm{Con}_J L$.

The rest is easy combinatorics. Let

$$P_{2n} = \{k_1, \ldots, k_n, m_1, \ldots, m_n\} \ (= B_n)$$

be the bipartite graph with minimal elements k_1, ..., k_n, maximal elements m_1, ..., m_n, so $k_i \prec m_j$ in P, for $1 \le i, j \le n$, and $\{k_1, \ldots, k_n\}$, $\{m_1, \ldots, m_n\}$ are antichains. Figure 9.5 shows B_3.

Let L_{2n} be a lattice with $\mathrm{Con}_J L_{2n} = B_n$ and satisfying $|L_{2n}| \le C(2n)^\alpha = C'n^\alpha$ for some constant C and $C' = 4C$. For $1 \le i$, $j \le n$, we get in L_{2n} a chain $C(i, j)$: $e_1(i, j) < h(i, j) < e_2(i, j)$ in L. Since there are n^2 elements of the form $h(i, j)$ and there are at most $C'n^\alpha$ elements in L_{2n}, therefore, there is an element h occurring as the middle element in $n^2/(C'n^\alpha)$ (that is, $\frac{1}{C'}n^{2-\alpha}$) of these chains. For $h(i, j)$, either $\mathrm{con}(h(i, j), e_1(i, j))$ or $\mathrm{con}(h(i, j), e_2(i, j))$ is a minimal element of B_n. So there are $\frac{1}{2C'}n^{2-\alpha}$ of these elements for which these choices are consistent, say, all $\mathrm{con}(h(i, j), e_1(i, j))$ are maximal. Let A be the set of these $e_1(i, j)$-s.

We get the situation depicted in Figure 9.6: for any $x \in A$, the congruence $\mathrm{con}(h, x)$ corresponds to a maximal element of B_n. Therefore, A is an antichain. Since a finite join of a subset of an antichain of join-irreducible elements cannot contain an element of the antichain not in the subset, we conclude that A under join generates $2^{|A|} - 1$ distinct elements. From $|A| \ge \frac{1}{8C}n^{2-\alpha}$, it now follows that

$$2^{|A|} - 1 = C_1 2^{n^{2-\alpha}},$$

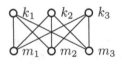

Figure 9.5: The ordered set B_3.

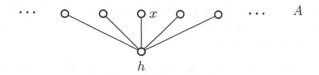

Figure 9.6: A configuration too large in L.

for some constant C_1. So we conclude that L_{2n} has at least $C_1 2^{n^{2-\alpha}}$ elements, contradicting that L_{2n} has at most $C'n^\alpha$ elements, completing the *Proof-by-Picture*.

9.5. Computing minimality

We start by formalizing the situation depicted in Figure 9.4.

Lemma 9.3. *Let L be a finite lattice, and let $v_i, u_i \in L$ satisfy $v_i \prec u_i$ for $i = 1, 2$. Let $\gamma_i = \mathrm{con}(v_i, u_i)$ for $i = 1, 2$. If $\gamma_1 \prec \gamma_2$ in $\mathrm{Con}_J L$, then there is a three-element chain $\{e_1, h, e_2\}$ in L such that $\gamma_i = \alpha(h, e_i)$, for $i = 1, 2$, and $e_1 < h < e_2$ or $e_2 < h < e_1$.*

Proof. Since $v_1 \equiv u_1 \,(\mathrm{con}(v_2, u_2))$ and $v_1 \prec u_1$, by Theorem 3.1, there is a sequence of congruence-perspectivities

$$[v_2, u_2] = [x_1, y_1] \overset{\mathrm{up}}{\twoheadrightarrow} [x_2, y_2] \overset{\mathrm{dn}}{\twoheadrightarrow} [x_3, y_3] \overset{\mathrm{up}}{\twoheadrightarrow} \ldots \overset{\mathrm{dn}}{\twoheadrightarrow} [x_n, y_n] = [v_1, u_1],$$

for some natural number $n > 1$. Obviously,

$$\gamma_2 = \mathrm{con}(x_1, y_1) = \mathrm{con}(x_2, y_2) \geq \mathrm{con}(x_3, y_3) \geq \cdots \geq \mathrm{con}(x_n, y_n) = \gamma_1.$$

Since $\gamma_2 > \gamma_1$, there is a smallest i satisfying $\mathrm{con}(x_i, y_i) < \gamma_2$; obviously, $3 \leq i \leq n$.

Let i be odd, and let $z = x_{i-1} \vee y_i$. Then

$$\mathrm{con}(x_{i-1}, z) = \mathrm{con}(x_i, y_i) < \gamma_2,$$

but $\mathrm{con}(x_{i-1}, y_{i-1}) = \gamma_2$, so $\mathrm{con}(z, y_{i-1}) = \gamma_2$. Since

$$u_1 \equiv v_1 \pmod{\mathrm{con}(x_{i-1}, z)}$$

and $u_1 \prec v_1$, there are $u, v \in [x_{i-1}, z]$, such that $v \prec u$ and $\mathrm{con}(u_1, v_1) \leq \mathrm{con}(u, v)$. So

$$\gamma_1 = \mathrm{con}(u_1, v_1) \leq \mathrm{con}(u, v) < \gamma_2,$$

and $\mathrm{con}(u, v)$ is a join-irreducible congruence, hence by the assumption on γ_1 and γ_2, it follows that $\gamma_1 = \mathrm{con}(u, v)$. Hence we can choose $e_2 = x_{i-1}$, $h = z$, and $e_1 = y_{i-1}$.

If i is even, then we proceed dually. $\square$

The second lemma deals with join-independence. A set A in a finite lattice L is *join-independent* if for any $a \in A$ and subset $A_1 \subseteq A$, the inequality $a \leq \bigvee A_1$ implies that $a \in A_1$.

Lemma 9.4. *Let L be a finite lattice, let $A \subseteq L$, and let $b \in L$ be a lower bound of A in L. If $\{\operatorname{con}(b,x) \mid x \in A\}$ is join-independent in $\operatorname{Con} L$, then A is join-independent in L.*

Proof. Indeed, if $a \leq \bigvee A_1$ for some $A_1 \subseteq A$, then

$$\operatorname{con}(b,a) \leq \bigvee\{\operatorname{con}(b,x) \mid x \in A_1\},$$

a contradiction. □

Finally, observe that in a finite distributive lattice, a set A of join-irreducible elements is join-independent iff the elements are pairwise incomparable. So we obtain the following.

Corollary 9.5. *Let L be a finite lattice, let $A \subseteq L$, and let $b \in L$ be a lower bound of A in L. If $\{\operatorname{con}(b,x) \mid x \in A\}$ is a set of pairwise incomparable join-irreducible congruences, then A is join-independent in L.*

Now we prove Theorem 9.2 with the combinatorial argument of Section 9.4, assisted by Corollary 9.5.

9.6.   Discussion

History

In the early research, R. P. Dilworth, G. Grätzer, and E. T. Schmidt used only the N_6 gadget for constructing lattices with given congruence lattices. Thus the size $O(2^{2n})$.

About 30 years later, I tackled the old problem of G. Birkhoff [8], characterizing congruence lattices of infinitary algebras (see [58] and [59]); in fact, in the stronger form conjectured by R. Wille (see K. Reuter and R. Wille [171]): *Every complete lattice can be represented as the lattice of complete congruences of a suitable complete lattice.* The finite case was earlier done in S.-K. Teo [183].

With H. Lakser (see [92]), we found a substantially simpler proof of my result, using colored chains; see also G. Grätzer and E. T. Schmidt [122]. Next year, we realized (see [93], result announced in G. Grätzer and H. Lakser [91]) that this technique can be utilized to prove that *Every finite distributive lattice D can be represented as the congruence lattice of a finite lattice L of size $O(n^3)$, where n is the number of join-irreducible elements of D.* Soon thereafter, using similar techniques, G. Grätzer, H. Lakser, and E. T. Schmidt [102] found the $O(n^2)$ result.

The third field, not mentioned

The papers flowing from the Dilworth Theorem split into three fields: the finite case, the infinite case, and the complete case: complete congruence lattices of complete lattices. This field started as described in the history section above. Here are some additional papers from this field, in chronological order: G. Grätzer, H. Lakser, and Wolk [109], R. Freese, G. Grätzer, and E. T. Schmidt [49], G. Grätzer and H. Lakser [94], G. Grätzer and E. T. Schmidt [119], [120], [124], [125], [127], [133].

The best result in this field is from G. Grätzer and E. T. Schmidt [124].

Theorem 9.6. *Every complete lattice can be represented as the complete congruence lattice of a complete distributive lattice.*

For a detailed discussion of the complete case, see G. Grätzer [68], Chapter 10 of LTS1.

The class **D** of distributive lattices is the minimal nontrivial variety, so this result cannot be improved by replacing **D** by a smaller variety. However, there are special classes of complete distributive lattices.

The two best known infinitary identities are the *Join-Infinite Distributive Identity*:

(JID) $$a \wedge \bigvee X = \bigvee(a \wedge x \mid x \in X),$$

and its dual, the *Meet-Infinite Distributive Identity*:

(MID) $$a \vee \bigwedge X = \bigwedge(a \vee x \mid x \in X).$$

We will denote by ($\text{JID}_\mathfrak{m}$) the condition that (JID) holds for sets X satisfying $|X| < \mathfrak{m}$, where $\mathfrak{m}$ is a regular cardinal with $\mathfrak{m} > \aleph_0$. We define ($\text{MID}_\mathfrak{m}$) dually.

The lattice we construct for Theorem 9.6 in G. Grätzer and E. T. Schmidt [124] fails both ($\text{JID}_\mathfrak{m}$) and ($\text{MID}_\mathfrak{m}$). In G. Grätzer and E. T. Schmidt [127], we prove the following result.

Theorem 9.7. *Let L be a complete lattice with more than two elements and with a meet-irreducible zero. Then L cannot be represented as the lattice of complete congruence relations of a complete distributive lattice K satisfying* (JID) *and* (MID).

So we can raise the following:

Problem 9.1. Characterize the complete congruence lattices of complete distributive lattices satisfying (JID) and/or (MID).

Or more generally:

Problem 9.2. Characterize the lattices of $\mathfrak{m}$-complete congruences of $\mathfrak{m}$-complete distributive lattices satisfying ($\text{JID}_\mathfrak{m}$) and/or ($\text{MID}_\mathfrak{m}$).

Improved bounds

A somewhat sharper form of Theorem 9.1 is the following:

Theorem 9.8. *For any integer $n \geq 2$,*

$$\frac{1}{16} \frac{n^2}{\log_2 n} < \mathrm{mcr}(n) < 3(n+1)^2.$$

The upper bound was proved in G. Grätzer, H. Lakser, and E. T. Schmidt [102] and the lower bound in G. Grätzer and D. Wang [143].

Y. Zhang [195] noticed that the proof of this inequality can be improved to obtain the following result: for $n \geq 64$,

$$\frac{1}{64} \frac{n^2}{(\log_2 n)^2} < \mathrm{mcr}(n).$$

Different approaches to minimality

A different kind of lower bound is obtained in R. Freese [48]; it is shown that if $\mathrm{Con}_{\mathrm{J}} L$ has e edges ($e > 2$), then

$$\frac{e}{2 \log_2 e} \leq |L|.$$

R. Freese also proves that $\mathrm{Con}_{\mathrm{J}} L$ can be computed in time $O(|L|^2 \log_2 |L|)$.

Consider the optimal *length* of L. E. T. Schmidt [175] constructs a finite lattice L of length $5m$, where m is the number of dual atoms of D (for finite chains, this was done in J. Berman [7]); S.-K. Teo [183] proves that this result is best possible.

Problem 9.3. In Theorem 9.2, is the construction "best" in some sharper sense?

In other words, is there a function $f(n)$ with the properties: (i) Theorem 9.1 holds with $O(f(n))$ in place of $O(n^2)$; (ii) $f(n) < n^2$; (iii) $n^2 = O(f(n))$ fails?

Yet another approach to minimality starts with the elegant inequality for a finite lattice L:

$$|\mathrm{J}(L)| \geq |\mathrm{Con}_{\mathrm{J}} L|$$

of R. Freese, J. Ježek, and J. B. Nation [50]. So if we define

$$\mathrm{je}(L) = |\mathrm{J}(L)| - |\mathrm{Con}_{\mathrm{J}} L|,$$

then $\mathrm{je}(L) \geq 0$ and $\mathrm{je}(L)$ is one measure of the efficiency of the representation of $D = \mathrm{Con}\, L$ as a congruence lattice. For a finite distributive lattice D, define

$$\mathrm{JE}(D) = \min(\mathrm{je}(L) \mid \mathrm{Con}\, L \cong D),$$

where L ranges over finite lattices. Then from this point of view, the best representation of a finite distributive lattice D as a congruence lattice of a finite lattice L is obtained when $\mathrm{je}(L) = \mathrm{JE}(\mathrm{Con}\,L)$. This is the approach we take in G. Grätzer and F. Wehrung [151].

In G. Grätzer and F. Wehrung [151], we compute the *exact* value of $\mathrm{JE}(D)$, yielding the inequality

$$0 \le \mathrm{JE}(D) \le \frac{2}{3}n,$$

and the constant $2/3$ in this estimate is best possible.

Problem 9.4. Determine the least constant k such that for every finite distributive lattice D, there exists a finite sectionally complemented lattice L such that $\mathrm{Con}\,L \cong D$ and

$$\mathrm{J}(L) \le k|\mathrm{J}(D)|.$$

By G. Grätzer and E. T. Schmidt [117], this constant $k \le 2$. The value of the constant defined similarly for the class of atomistic lattices (or the class of all lattices as well) equals $5/3$, by Corollary 5.4 of G. Grätzer and F. Wehrung [151].

Problem 9.5. Let $\mathcal{V}$ be a variety of lattices. If D is a finite distributive lattice representable by a finite lattice in $\mathcal{V}$, compute the least possible value of $\mathrm{J}(L)$, for a finite lattice L in $\mathcal{V}$ such that $\mathrm{Con}\,L \cong D$.

Semimodular Lattices

10.1. The representation theorem

Should we try to get a representation theorem for semimodular lattices? For a class of lattices to be a candidate, it would have to satisfy the following two criteria.

First, there ought to be very many finite simple lattices in the class. For semimodular lattices, the finite partition lattices (Section 1.2.1) are all simple semimodular lattices by O. Ore [167], so we have lots of finite simple semimodular lattices.

Second, there ought to be finite lattices in the class with a three-element congruence lattice. For semimodular lattices, the lattice S_8 of Figure 10.1 is such a lattice.

These criteria satisfied, we looked for a representation theorem for semimodular lattices.

In fact, in G. Grätzer, H. Lakser, and E. T. Schmidt [105], we found a very good representation theorem for semimodular lattices, also obtaining—to our great surprise—planarity and a small size:

Theorem 10.1. *Every finite distributive lattice D can be represented as the congruence lattice of a finite semimodular lattice L. In fact, if D has $n \geq 1$ join-irreducible elements, then L can be constructed as a planar lattice of size $O(n^3)$.*

By Theorem 9.1, the optimal size for any lattice is $O(n^2)$, so the size we obtain in this result seems very good.

© Springer International Publishing Switzerland 2016   113
G. Grätzer, *The Congruences of a Finite Lattice*,
DOI 10.1007/978-3-319-38798-7_10

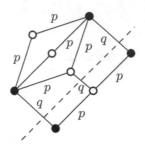

Figure 10.1: The semimodular gadget S_8, colored with $p < q$.

10.2.   *Proof-by-Picture*

The proof of this result is very similar to the proof of Theorem 9.1. The gadget we use now is the semimodular (colored—in the sense of Section 3.2) lattice S_8 of Figure 10.1.

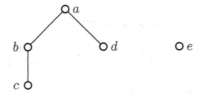

Figure 10.2: The ordered set P.

To represent the ordered set P of Figure 10.2 as $\mathrm{Con_J}\, L$, for a finite planar semimodular lattice L, we start out—see Figure 10.3—by constructing the

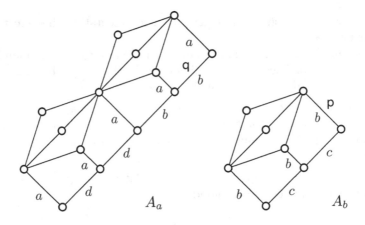

Figure 10.3: The lattices A_a and A_b.

lattices A_a and A_b. We obtain A_a by gluing two copies of S_8 together; and color it with $\{a, b, d\}$ so that $\mathrm{con}(a) > \mathrm{con}(b)$ is accomplished in the top S_8 of A_a and $\mathrm{con}(a) > \mathrm{con}(d)$ is accomplished in the bottom S_8 of A_a. The lattice A_b is S_8 colored by $\{b, c\}$ so that $\mathrm{con}(b) > \mathrm{con}(c)$ in A_b.

Observe that the lattice A_a takes care *of all* $a \succ x$ orderings; in the example, there are only two. We could do three coverings by gluing three copies of S_8 together, and so on.

Form the glued sum S of A_a and A_b; all the covers of P are taken care of in S. There is only one problem: S is not a colored lattice; in this example if $\mathfrak{p}$ is a prime interval of color b in S_b (as in Figure 10.3) and $\mathfrak{q}$ is a prime interval of color b in S_b (as in Figure 10.3), then in S we have $\mathrm{con}(\mathfrak{p}) \wedge \mathrm{con}(\mathfrak{q}) = \mathbf{0}$. Of course, we should have $\mathrm{con}(\mathfrak{p}) = \mathrm{con}(\mathfrak{q}) = \mathrm{con}(b)$. We accomplish this by extending S to the lattice L of Figure 10.4. In L, the black-filled elements form the sublattice S.

As you see, we extend S by adding to it a distributive "grid." The right corner is C_5^2 colored by $\{a, b, c, d\}$; each of the four covering squares colored by the same color twice are made into a covering M_3. This makes the coloring behave properly in the right corner. In the rest of the lattice we do the same: we look for a covering square colored by the same color twice, and make it into a covering M_3. This makes L into a colored lattice: any two prime intervals of the same color generate the same congruence. For instance, if $\mathfrak{p}$ and $\mathfrak{q}$ are prime intervals as in the previous paragraph (see now Figure 10.4), then we find in L the prime intervals $\mathfrak{r}_1, \mathfrak{r}_2, \mathfrak{r}_3, \mathfrak{r}_4$ (see Figure 10.4), so that

$$\mathfrak{p} \overset{\mathrm{up}}{\twoheadrightarrow} \mathfrak{r}_1 \overset{\mathrm{dn}}{\twoheadrightarrow} \mathfrak{r}_2 \overset{\mathrm{up}}{\twoheadrightarrow} \mathfrak{r}_3 \overset{\mathrm{dn}}{\twoheadrightarrow} \mathfrak{r}_4 \overset{\mathrm{up}}{\twoheadrightarrow} \mathfrak{q};$$

therefore, $\mathrm{con}_L(\mathfrak{p}) = \mathrm{con}_L(\mathfrak{q})$.

Finally, we remember the element $e \in P$. We add a "tail" to the lattice and color it e.

It is easy to see that the resulting lattice L is planar and semimodular and that $\mathrm{Con_J}\, L$ is isomorphic to P with the isomorphism $x \mapsto \mathrm{con}(x)$ for $x \in \{a, b, c, d, e\}$.

10.3. Construction and proof

We construct the semimodular lattice L of Theorem 10.1 in several steps.

Take the lattice S_8 with the notation of Figure 10.5. The lattice S_8 has an ideal, I_8, and a filter, F_8, both isomorphic to C_2; we will utilize these for repeated gluings—defined in Section 2.4. The elements of I_8 and F_8 are black-filled in Figure 10.5.

Let D be a finite distributive lattice, and let $P = \mathrm{J}(D)$ be the ordered set of its join-irreducible elements, $n = |P|$. We enumerate

$$p^1,\ p^2,\ \ldots,\ p^m$$

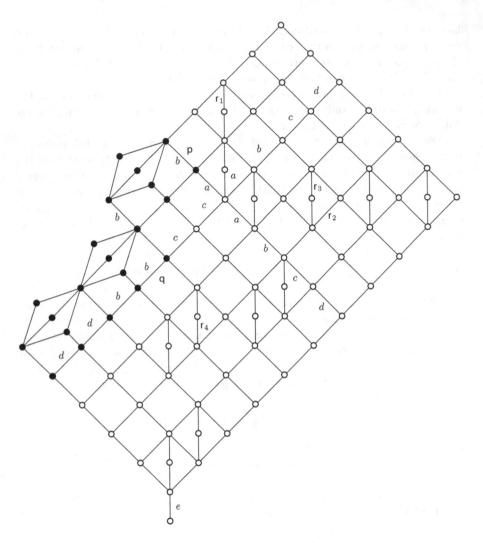

Figure 10.4: The lattice L.

the non-minimal elements of P. For every $p^i, i = 1, 2, \ldots, m$, let

$$\mathrm{cov}(p_i) = \{p_1^i, p_2^i, \ldots, p_{k_i}^i\}$$

denote the set of all lower covers of p^i in P; since p^i is non-minimal, it follows that $k_i > 0$. Let r_1, r_2, $\ldots$, r_t enumerate all elements of P that are incomparable with all other elements.

In the example of Section 10.2, we have, say, $a = p^1$, $b = p^2$, $k_1 = 2$, $k_2 = 1$, $b = p_1^1$, $d = p_2^1$, $c = p_1^2$, $\mathrm{cov}(a) = \{b, d\}$, and $\mathrm{cov}(b) = \{c\}$.

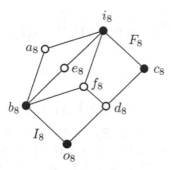

Figure 10.5: The semimodular gadget S_8, with notation.

Step 1: For every i, with $1 \le i \le m$, we construct a lattice A_i with an ideal I_i and a filter F_i, where I_i is a chain of length $2(k_i + \cdots + k_m)$ and F_i is a chain of length $2(k_{i+1} + \cdots + k_m)$.

For the ordered set P of Section 10.2, we construct the lattices A_1 and A_2; see Figure 10.6—the elements of the ideals I_1 and I_2, and of the filters F_1 and F_2 are black-filled.

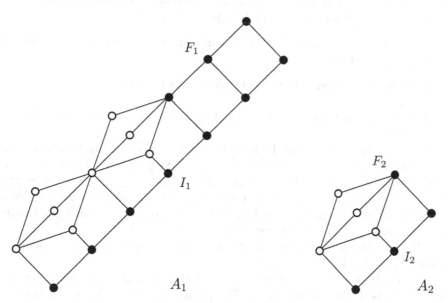

Figure 10.6: The lattices A_1 and A_2.

Now we will use twice the construction, *gluing k-times*, described in Section 2.4. To form A_i, glue S_8 to itself $(k_i - 1)$-times with the ideal I_8 and the filter F_8, to obtain the lattice A_i^1 with a filter $F_{A_i^1}$.

Now take

$$B_2 = \{(0,0),(0,1),(1,0),(1,1)\}$$

with the ideal

$$I_{B_2} = \{(0,0),(1,0)\}$$

and the filter

$$F_{B_2} = \{(0,1),(1,1)\},$$

and glue $2(k_{i+1} + \cdots + k_m)$-times B_2 to A_i^1. The ideal I_i is generated by the element $(0,1)$ of the top B_2, while F_i is generated by the unit element of A_i^1.

We define a coloring μ_i of A_i as follows. On any copy of S_8, $\mu_i[o_8, b_8] = p^i$ and on the j-th copy of S_8, $\mu_i[o_8, d_8] = \mu_i[d_8, c_8] = p_j^i$; on the first two copies of B_2, $\mu_i[(0,1),(1,1)] = p_1^{i+1}$, on the next two copies, $\mu_i[(0,1),(1,1)] = p_2^{i+1}$, after k_{i+1} pairs, the next two satisfies $\mu_i[(0,1),(1,1)] = p_1^{i+2}$, and so on.

Lemma 10.2. μ_i *is a coloring of* A_i. *The join-irreducible congruences of* A_i *are generated by prime intervals of* I_i *and by* $[o_8, b_8]$ *of the bottom* S_8 *in* A_i. *If* $\mathfrak{p}$ *and* $\mathfrak{q}$ *are* $[o_8, b_8]$ *or a prime interval* $[o_8, d_8]$ *or* $[d_8, c_8]$ *of a copy of* S_8 *in* A_i, *then* $\alpha(\mathfrak{p}) \geq \alpha(\mathfrak{q})$ *iff* $\mu_i(\mathfrak{p}) \geq \mu_i(\mathfrak{q})$. *In particular,* $\mathrm{con}(o_8, b_8) \succ \mathrm{con}(o_8, d_8)$ *in* $\mathrm{Con}_J A_i$, *where* o_8, b_8, d_8 *are in a copy of* S_8 *in* A_i. *If* $\mathfrak{p}$ *is a prime interval* $[(0,1),(1,1)]$ *in a copy of* B_2, *then* $\mathrm{con}(\mathfrak{p})$ *is incomparable to any* $\mathrm{con}(\mathfrak{q})$, *where* $\mathfrak{q}$ *is* $[o_8, b_8]$ *or a prime interval of* I_i *different from* $\mathfrak{p}$.

Proof. This is trivial since every prime interval of S_8 is projective to one of $[o_8, b_8], [o_8, d_8], [d_8, c_8]$.   □

Step 2: We define the lattice A by gluing together the (colored) lattices A_i for $1 \leq i \leq m$.

For the ordered set P of Section 10.2, we construct the lattice A; see Figure 10.7.

For $1 \leq i \leq m$, we define, by induction, the lattice $\overline{A}_i$, which contains A_i, and, therefore, F_i, as a filter. Let $\overline{A}_1 = A_1$. Assume that $\overline{A}_i$ with F_i as a filter has been defined. Observe that both F_i and I_{i+1} are chains of length $2(k_{i+1} + \cdots + k_m)$, and so they are isomorphic; in fact, this isomorphism preserves colors. We glue $\overline{A}_i$ to A_{i+1} over F_i and I_{i+1} to obtain $\overline{A}_{i+1}$. Define $A = \overline{A}_m$ and $I_A = I_1$ (see Figure 10.7).

Observe that μ_i on F_i agrees with μ_{i+1} on I_{i+1}; therefore, the μ_i define a coloring μ_A of A for $1 \leq i \leq m$.

Let F_A be the filter of A generated by the element $(0,1)$ of the top B_2 in A_1. F_A is a chain of length m. The prime interval $[o_8, b_8]$ in the bottom S_8 in A_i $(1 \leq i \leq m)$ is projective to a unique prime interval $\mathfrak{p}$ of F_A; define $\mu_A(\mathfrak{p}) = \mu_A[o_8, b_8]$.

Lemma 10.3. μ_A *is a coloring of* A. *The join-irreducible congruences of* A *are generated by prime intervals of* I_A *and* F_A. *Let* $\mathfrak{p}$ *and* $\mathfrak{q}$ *be prime intervals in* I_A *and* F_A.

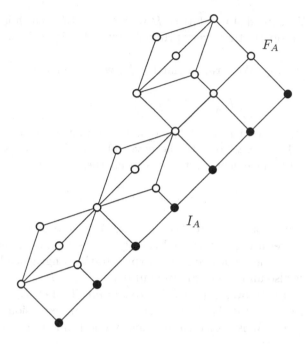

Figure 10.7: The lattice A.

(i) *If $\mathfrak{p}$ and $\mathfrak{q}$ are prime intervals of F_A, then $\mathrm{con}(\mathfrak{p})$ and $\mathrm{con}(\mathfrak{q})$ are incomparable.*

(ii) *If $\mathfrak{p}$ is a prime interval of F_A and $\mathfrak{q}$ is a prime interval of I_A, then $\mathrm{con}(\mathfrak{p})$ and $\mathrm{con}(\mathfrak{q})$ are comparable iff $\mathfrak{p} \subseteq A_i$, for some $1 \leq i \leq m$, $\mathfrak{q}$ is perspective to some $[o_8, d_8]$ or $[d_8, c_8]$ in some S_8 in A_i; in which case, $\mathrm{con}(\mathfrak{p}) \succ \mathrm{con}(\mathfrak{q})$ in $\mathrm{Con_J}\, A$.*

(iii) *If $\mathfrak{p}$ and $\mathfrak{q}$ are prime intervals of I_A, then $\mathrm{con}(\mathfrak{p}) \geq \mathrm{con}(\mathfrak{q})$ iff $\mathfrak{p}$ and $\mathfrak{q}$ are perspective to prime intervals $\mathfrak{p}'$ and $\mathfrak{q}'$ in some A_i, respectively, for some $1 \leq i \leq m$, and $\mathfrak{p}'$ and $\mathfrak{q}'$ are adjacent edges of some S_8 in A_i, in which case, $\mathrm{con}(\mathfrak{p}) = \mathrm{con}(\mathfrak{q})$.*

Proof. This is obvious from Lemma 2.9. $\qquad\square$

Observe that the congruence lattice of A is still quite different from D in two ways, exactly as we discussed this for S in Section 10.2: the congruences that correspond to the r_i-s are still missing; prime intervals in $I_A \cup F_A$ of the same color generate incomparable congruences with one exception: they are adjacent intervals in I_A, perspective to the two prime intervals of some S_8 in some A_i.

Step 3: We extend A to a lattice B with an ideal I_B which is a chain and which has the property that every prime interval of B is projective to a prime interval of I_B.

This step is easy. We form the lattice F_A^2 with the ideal

$$I_{F_A^2} = \{\, (x, 0_{F_A}) \mid x \in F_A \,\},$$

where 0_{F_A} is the zero of F_A. Let 1_{F_A} denote the unit element of F_A and, for $x \in F_A$, $x < 1_{F_A}$, let x^* denote the cover of x in F_A. For every $x \in F_A$, $x < 1_{F_A}$, we add an element x_m to F_A^2 so that the elements

$$(x, x), (x, x^*), (x^*, x), x_m, (x^*, x^*)$$

form a sublattice isomorphic to M_3 with (x, x) as zero and (x^*, x^*) as unit. Let M be the resulting lattice. Obviously, M is a finite planar modular lattice whose congruence lattice is isomorphic to the congruence lattice of F_A. Clearly, $I_{F_A^2}$ is also an ideal of M; we will denote it by I_M.

We glue A to M over F_A and I_M to obtain B. Let I_B be defined as the ideal generated by $(0, 1_{F_A})$. We define μ_B as an extension of μ_A; every prime interval $\mathfrak{p}$ of M is projective to exactly one prime interval $\bar{\mathfrak{p}}$ of I_M, we define $\mu_B(\mathfrak{p}) = \mu_A(\bar{\mathfrak{p}})$.

Lemma 10.4. μ_B is a coloring of B. The join-irreducible congruences of B are generated by prime intervals of I_B. Let $\mathfrak{p}$ and $\mathfrak{q}$ be prime intervals in I_B.

(i) *If $\mathfrak{p}$ and $\mathfrak{q}$ are prime intervals of M, then $\mathrm{con}(\mathfrak{p})$ and $\mathrm{con}(\mathfrak{q})$ are incomparable.*

(ii) *If $\mathfrak{p}$ is a prime interval of M and $\mathfrak{q}$ is a prime interval of I_A, then $\mathrm{con}(\mathfrak{p})$ and $\mathrm{con}(\mathfrak{q})$ are related as $\mathrm{con}_A(\bar{\mathfrak{p}})$ and $\mathrm{con}_A(\mathfrak{q})$ are related in A.*

(iii) *If $\mathfrak{p}$ and $\mathfrak{q}$ are prime intervals of I_A, then $\mathrm{con}(\mathfrak{p})$ and $\mathrm{con}(\mathfrak{q})$ are related exactly as $\mathrm{con}_A(\mathfrak{p})$ and $\mathrm{con}_A(\mathfrak{q})$ are related in A.*

Proof. This is obvious from the congruence structure of M. □

Step 4: We extend B to the lattice L of the Theorem 10.1, as sketched in Figure 10.8.

This is also an easy step. We take a chain C of length n and we color C over P so that the coloring is a bijection. We form the lattice $C \times I_B$. For every pair of prime intervals, $\mathfrak{p} = [a, b]$ of C and $\mathfrak{q} = [c, d]$ of I_B if $\mathfrak{p}$ and $\mathfrak{q}$ have the same color, then we add an element $m(\mathfrak{p}, \mathfrak{q})$ to C over P so that the elements

$$(a, c), (b, c), (a, d), m(\mathfrak{p}, \mathfrak{q}), (b, d)$$

form a sublattice isomorphic to M_3. Let N denote the resulting lattice. N is obviously modular and planar. Set

$$I_N = \{ (x, 0_{I_B}) \mid x \in C \},$$
$$F_N = \{ (1_C, x) \mid x \in I_B \},$$

where 0_{I_B} is the zero of I_B and 1_C is the unit of C. Then I_N is the ideal of N (isomorphic to C) and F_N is a filter of N (isomorphic to I_B). Every prime interval of N is projective to a prime interval of I_N, so we have a natural coloring μ_N on N. Note that this coloring agrees with the coloring μ_B on F_N under the isomorphism with I_B.

We glue N to B over F_N and I_B to obtain L with the coloring μ_L. Set $I_L = I_N$.

It is clear from the construction and from the lemmas that every prime interval of L is projective to a prime interval of I_L and that distinct prime intervals of I_L generate distinct join-irreducible congruences of L.

It remains to see that if $\mathfrak{p}$ and $\mathfrak{q}$ are distinct prime intervals, then $\mathrm{con}(\mathfrak{p}) \geq \mathrm{con}(\mathfrak{q})$ iff $\mu_L(\mathfrak{p}) \geq \mu_L(\mathfrak{q})$. Since P is finite, it is sufficient to prove that $\mathrm{con}(\mathfrak{p}) \succ \mathrm{con}(\mathfrak{q})$ in $\mathrm{Con}_J L$ iff $\mu_L(\mathfrak{p}) \succ \mu_L(\mathfrak{q})$ in $J(D)$. But this is clear since if $\mu_L(\mathfrak{p}) \succ \mu_L(\mathfrak{q})$ in $J(D)$, then $\mu_L(p) = p_i$, for some $1 \leq i \leq m$, and $\mu_L(q) = p_j^i$, for some $1 \leq j \leq k_i$, so $\mathrm{con}(\mathfrak{p}) \succ \mathrm{con}(\mathfrak{q})$ was guaranteed in A_i.

To establish that the size of L is $O(n^3)$, we give a very crude upper bound for $|L|$. Since $2n^2 + 1$ is an upper bound for $|I_i|$, $1 \leq i \leq m$, it follows that $3(2n^2 + 1)$ is an upper bound for $|A_i|$ and $3(2n^2 + 1)n$ is an upper bound

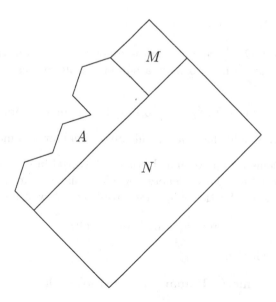

Figure 10.8: Sketching the lattice L.

for $|A|$. Since $|F_A| \leq n+1$, we get the upper bound $(n+1)^2 + n + 1$ for $|M|$. Finally, $|I_B| \leq 2n^2 + 1 + n + 1 = 2n^2 + n + 2$, so $|N| \leq 2(2n^2 + n + 2)(n+1)$. Therefore,

$$3(2n^2 + 1)n + (n+1)^2 + n + 1 + 2(2n^2 + n + 2)(n+1)$$

is an upper bound for L and it is a cubic polynomial in n. This completes the proof of the Theorem 10.1.

It is not difficult to find better upper bounds for $|L|$; for instance,

$$|L| \leq 3n^3 + 2n^2 - 7n + 4.$$

10.4. Discussion

An addendum

In Chapter 15, we need a more detailed version of Theorem 10.1.

Theorem 10.5. *Let P be a finite order. Then there exists a finite* semimodular *lattice L with the following properties:*

(i) $\mathrm{Con}_J L$ *is isomorphic to P.*

(ii) $|L|$ *is $O(|P|^3)$.*

(iii) L *has an congruence-determining sublattice C, which is an ideal and a chain.*

Problems

In Chapter 9, the $O(n^2)$ construction is followed by a proof that this size is optimal. It would be nice to have a similar result for planar semimodular lattices.

Problem 10.1. Is the $O(n^3)$ result optimal for planar semimodular lattices?

Problem 10.2. Would the $O(n^3)$ result be optimal for semimodular lattices?

These problems for rectangular lattices are solved in the next chapter.

Let **Planar** + **SemiMod** denote the class planar semimodular lattices. Using the notation of Section 8.5, these problems ask whether

$$\mathbf{mcr}(n, \mathbf{SemiMod}) = O(n^3)$$

is optimal, and whether

$$\mathbf{mcr}(n, \mathbf{Planar} + \mathbf{SemiMod}) = O(n^3)$$

is optimal, where **SemiMod** is the class of semimodular lattices.

Rectangular Lattices

11.1. Results

In Chapter 10, we proved that every finite distributive lattice D can be represented as the congruence lattice of a finite semimodular lattice L of size $O(n^3)$. In this chapter we discuss what can we say about rectangular lattices, which form a very small subclass of semimodular lattices.

So the first question is, can every finite distributive lattice D be represented as the congruence lattice of a rectangular lattice L?

The answer is surprisingly easy in view of Lemma 4.11 and Theorem 10.1.

Theorem 11.1. *Every finite distributive lattice D can be represented as the congruence lattice of a rectangular lattice L.*

However, we want more, the analogue of Theorem 10.1: a rectangular lattice L of size $O(n^3)$, where n is the number of join-irreducible elements of D. Unfortunately, applying Theorem 10.1 to a finite distributive lattice D with n join-irreducible elements and then adding corners to obtain a rectangular lattice L, we may get a lattice not of size $O(n^3)$ but of size $O(n^4)$, see G. Grätzer and E. Knapp [88].

Nevertheless, G. Grätzer and E. Knapp [87] proved the analogue of Theorem 10.1 for rectangular lattices.

Theorem 11.2. *Every finite distributive lattice D can be represented as the congruence lattice of a rectangular lattice L. In fact, if D has $n \geq 1$ join-irreducible elements, then L can be constructed as a planar lattice of size $O(n^3)$.*

To prove Theorem 11.2, we need a new construction. The *Proof-by-Picture* of this is presented in Section 11.2.

The main result of this field, G. Grätzer and E. Knapp [88], proves that $O(n^3)$ is the best possible in Theorem 11.2.

Theorem 11.3.

(i) *Let P be a finite ordered set with $n \geq 1$ elements. Then P can be represented as the ordered set of join-irreducible congruences of a* rectangular *lattice L satisfying*

$$|L| \leq \frac{2}{3}n^3 + 2n^2 + \frac{4}{3}n + 1.$$

(ii) *Let L_n be a rectangular lattice whose ordered set of join-irreducible congruences is a balanced bipartite ordered set on n elements. Then, for some constant $k > 0$, the inequality $|L_n| \geq kn^3$ holds.*

Apart from the *Proof-by-Picture* section, we do not go into the proofs of these results. We refer the reader to G. Grätzer and E. Knapp [88] and to G. Grätzer [66], Chapter 4 in LTS1.

11.2. *Proof-by-Picture*

Let $P = \{p_1, p_2, p_3\}$ be the ordered set of Figure 11.1, where the enumeration of the elements of P is an extension of the order, that is, if $p_i < p_j$, then $i < j$. We construct a rectangular lattice $L = L_3$ representing P as the ordered set of join-irreducible congruences of L, that is, L satisfies that $\operatorname{Con_J} L \cong P_3$.

Let $P_i = \{p_1, \ldots, p_i\}$ for $i \leq 3$. We inductively construct the lattices L_i, for $1 \leq i \leq 3$, so that the lattice L_i is rectangular and it represents P_i as $\operatorname{Con_J} L_i$.

We define the rectangular lattice L_1 as M_3. Note that $\operatorname{Con_J} L_1 \cong P_1$.

Let $i = 2$. We have the ordered set $P_2 = \{p_1, p_2\}$ with $p_1 \parallel p_2$. Let A_2 be a 0-stacked S_7^+ and $B_2 = \mathsf{C}_2 \times \mathsf{C}_2$ (as shown in Figure 11.2). We glue A_2 and B_2 together to form the lattice K_2. We glue the lattice L_1 to $D_2 = \mathsf{C}_2 \times \mathsf{C}_2$ to obtain the lattice K_2' and then glue the K_2' to K_2, to get L_2—the dashed line in the diagram of L_2 indicates how L_2 was glued together. In the resulting lattice, the top left and top right boundaries are colored with the elements p_1 and p_2, so $\operatorname{Con_J} L_2 \cong P_2$.

Finally, let $i = 3$. The ordered set $P = P_3 = \{p_1, p_2, p_3\}$ has the two coverings $p_1 \prec p_3$ and $p_2 \prec p_3$, so we take A_3 as a 2-stacked S_7^+. Consider the lattice $\mathsf{C}_3 \times \mathsf{C}_4$. We add an element each to the interiors of the intervals $[(0, 1), (1, 2)]$ and $[(1, 2), (2, 3)]$ to form the lattice B_3 and assign a coloring to A_3 and B_3 as indicated in Figure 11.3. Glue A_3 and B_3 to obtain the lattice K_3 (the dashed line indicates the gluing). Set $D_3 = \mathsf{C}_4 \times \mathsf{C}_3$ (not shown in

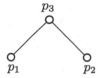

Figure 11.1: The ordered set P.

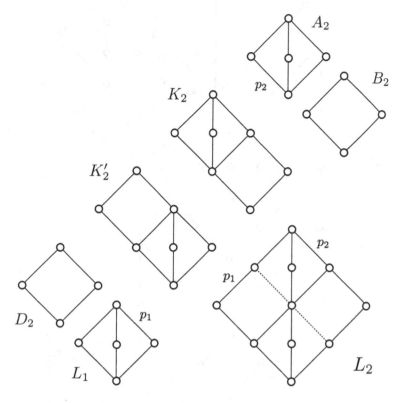

Figure 11.2: The lattices for $i = 2$.

Figure 11.3). Glue L_2 to D_3 and then the result (also not shown in Figure 11.3) to K_3 to form the lattice $L = L_3$ (the dashed line indicates the gluing). Note the coloring of L; it verifies that $\mathrm{Con_J}\, L \cong P$.

11.3. Discussion

The two major open problems are Problems 10.1 and 10.2 stated in Section 10.4.

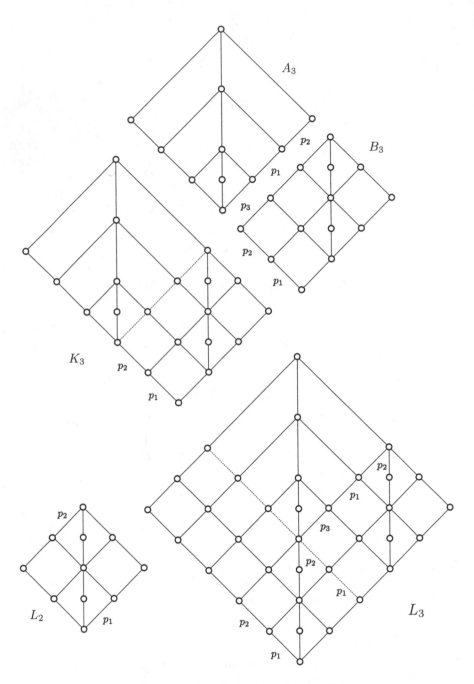

Figure 11.3: The lattices for $i = 3$.

Modular Lattices

12.1. The representation theorem

Corollary 3.12 notes that the congruence lattice of a finite modular lattice is boolean; however, a finite distributive lattice has a representation as the congruence lattice of an *infinite* modular lattice by E. T. Schmidt [174] (see also E. T. Schmidt [180]):

Theorem 12.1. *Every finite distributive lattice D can be represented as the congruence lattice of a modular lattice L.*

We are going to prove this result in the following stronger form provided in G. Grätzer and E. T. Schmidt [134].

Theorem 12.2. *Let P be a finite order. Then there exists a lattice L with the following properties:*

(i) *L is a modular lattice.*

(ii) *$\operatorname{Con} L$ is finite and $\operatorname{Con_J} L$ is isomorphic to P.*

(iii) *L is countably infinite.*

(iv) *L is weakly atomic.*

(v) *L has a unique complemented pair $\{a, a'\} \neq \{0, 1\}$ and $\operatorname{id}(a)$, $\operatorname{id}(a')$ are chains, in fact, successor ordinals.*

(vi) *L is rigid.*

© Springer International Publishing Switzerland 2016
G. Grätzer, *The Congruences of a Finite Lattice*,
DOI 10.1007/978-3-319-38798-7_12

In this result, a lattice L is *weakly atomic* if every nontrivial interval contains a prime interval.

We are constructing a rigid modular lattice L, laying the foundation for the Independence Theorem for Modular Lattices; see Section 12.4. The lattice L we construct is infinite, but "small." As an order, it can be embedded in C^3, where C is a countable ordinal; it is also weakly atomic.

12.2. *Proof-by-Picture*

The gadgets in previous constructions were lattices whose congruence lattices are isomorphic to the three-element chain. The gadget in the modular case is the lattice $M_3[C]$ of Chapter 6 for a chain C. We will use the sketch of this lattice as in Figure 6.1.

Take the three-element ordered set $P = \{p_1, p_2, p_3\}$ with the only relation: $p_1 < p_3$, see Figure 12.1. We present a *Proof-by-Picture* of Theorem 12.2 for the order P.

We start by constructing the lattice L_1 representing $P_1 = \{p_1\}$, see Figure 12.2. The lattice L_1, clearly, is simple and has no automorphism. The congruence representing p_1 is $\mathrm{con}(b_1^1, c_1^1)$.

Figure 12.1: The ordered set P.

From L_1, we obtain the lattice L_2 to represent $\{p_1, p_2\}$, the two-element antichain; see Figure 12.3. In L_2, we designated the elements $b_1^2 < c_1^2 < b_2^2 < c_2^2$ in the chain $\mathrm{fil}(a_2)$, so that p_1 is represented by $\mathrm{con}(b_1^2, c_1^2)$ and p_2 is represented by $\mathrm{con}(b_2^2, c_2^2)$.

Finally, we want to add p_3, so that it is over p_1 but incomparable with p_2. This is accomplished in L_3; see Figure 12.4, where L_2 is only partially drawn.

The lattice L_3 is glued together from four parts; the boundaries are thick lines. The bottom part is L_2 (only partially drawn and shaded), where on the upper left edge (the chain $E = \mathrm{fil}(a_2)$ in L_2) we mark the elements

$$b_1^2 < c_1^2 < b_2^2 < c_2^2,$$

as described above.

We take a chain C of type $\omega + 2$, which we obtain by forming glued sums (introduced in Section 1.1.3)

$$C = [b_1^2, c_1^2] + [b_1^2, c_1^2] + \cdots$$

ω-times and adding $n_0 \prec n_1$ to the top.

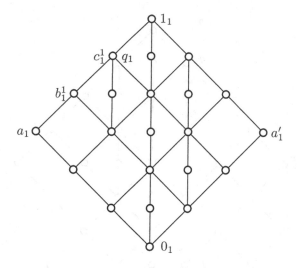

Figure 12.2: The lattice L_1.

The left part of L_3, call it Left, and the right part of L_3, call it Right, are both $C \times E$ and the top part, call it Top, is $M_3[C]$.

Finally, in Left, each square $[b_1^2, c_1^2] \times [b_1^2, c_1^2]$ is made into $M_3[b_1^2, c_1^2]$, and the top square of Left into a covering M_3.

What are the congruences of L_3? Think of L_3 as the base B, a direct product of two chains, $B = \mathrm{id}(a_3) \times \mathrm{id}(a_3')$, with the flaps of the $M_3[X]$-es sticking out (each X is a chain); then a congruence α of L_3 is a congruence of B with the property that it acts the same way on the left edge (a chain) of an $M_3[X]$ as on the right edge (an isomorphic chain).

Therefore, every congruence of L_2 extends, and extends uniquely to L_3. The extension of $\mathrm{con}(b_2^2, c_2^2)$ (which corresponds to p_2) is easy to see: in Left, for $c, d \in C$, $(c, x) \equiv (d, y)$ iff $c = d$ and $x \equiv y$ in L_2. Similarly, in Right. No two distinct element are congruent in Top.

The extension of $\mathrm{con}(b_1^2, c_1^2)$ (which corresponds to p_1) is much larger; since it collapses b_1^2 and c_1^2, it collapses all of C; the quotient lattice is finite, as shown in Figure 12.5.

Collapsing two distinct elements of $\{n_0, n_1\} \times E$ is equivalent to collapsing the two corresponding elements of E. If we collapse two elements of Left with the same C-component, this is obviously equivalent to collapsing the two E-components. If we collapse two elements of Left with the same E-component, this is obviously equivalent to collapsing two elements of the left side of an $M_3[b_1, c_1]$, which makes it equivalent to the collapse of the corresponding two elements in the lower right edge, which is the last case. So there is exactly one new join-irreducible congruence, $\mathrm{con}((a_2, 0_C), (n_1, 0_C))$, which contains $\mathrm{con}(b_1^2, c_1^2)$ but not $\mathrm{con}(b_2^2, c_2^2)$.

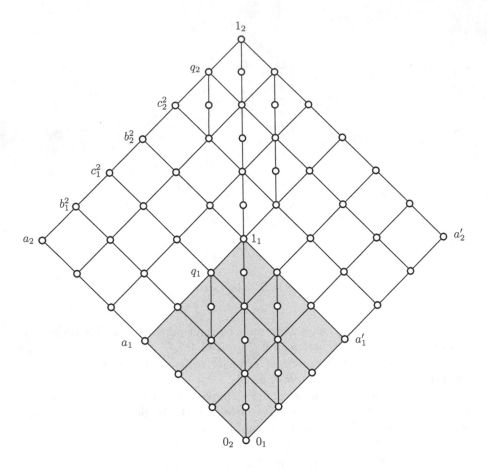

Figure 12.3: The lattice L_2.

In L_3, we marked the elements $b_1^3 < c_1^3 < b_2^3 < c_2^3 < b_3^3 < c_3^3$ in the chain fil(a_3), so that p_1 is represented by con(b_1^3, c_1^3), p_2 is represented by con(b_2^3, c_2^3), and p_3 is represented by con(b_3^3, c_3^3) in L_3 (Figure 12.4).

An automorphism α either fixes a_3 or takes it into a_3'. But if $\alpha a_3 = a_3'$, then α would define an automorphism of L_2 that takes a_2 into a_2', a contradiction. So α fixes a_2 and a_2'. Therefore, α fixes id(a_2) and id(a_2'), and so id(a_2)×id(a_2'). Since every element of L_3 belongs to this direct product, or is on the third flap of an $\mathsf{M}_3[X]$, whose base is in this direct product, we conclude that the base, and therefore, the whole $\mathsf{M}_3[X]$ is fixed by α, so α is the identity map, that is, L_3 is rigid.

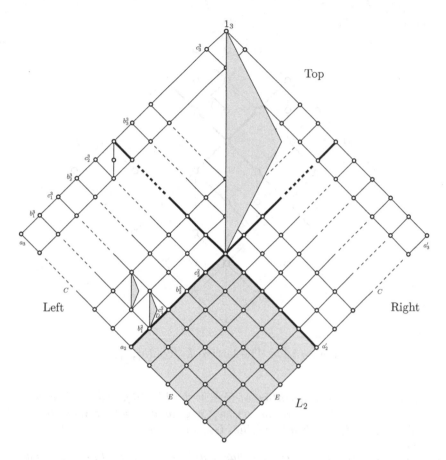

Figure 12.4: The lattice L_3.

12.3. Construction and proof

We begin by constructing the basic building block of the construction, see Figure 12.6.

Lemma 12.3. *Let U be a chain with zero, 0_U, and unit, 1_U. Let $0_U \leq u < v \leq 1_U$, and let V be the interval $[u, v]$ of U. Construct the lattice L as follows:*

(i) *Form the direct product $U \times V$.*

(ii) *Glue the interval $[(0_U, u), (u, v)]$ of $U \times V$ with the filter $[(u, u), (u, v)]$ ($\cong V$) to $\mathsf{M}_3[V]$ with the ideal $\{ (0, 0, v) \mid v \in V \}$ ($\cong V$).*

(iii) *Glue the lattice we obtained with the filter $\{ (1, x, x) \mid x \in V \}$ ($\cong V$) to the interval $[(v, u), (1_U, v)]$ with the ideal $[(v, u), (v, v)]$ ($\cong V$); the resulting lattice is denoted by $U \circledast V$.*

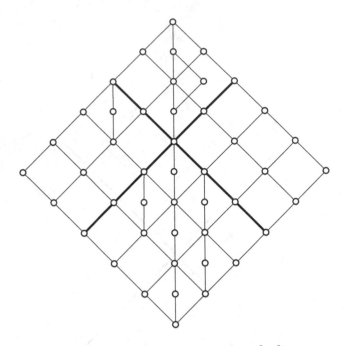

Figure 12.5: The lattice $L_3/\mathrm{con}(b_1^2, c_1^2)$.

(iv) *Consider U as a sublattice of $U \circledast V$ by identifying $x \in U$ with (x, u) if $x \leq u$ or if $v \leq x$; otherwise, $x \in [u, v]$, and we identify x with $(x, 0, 0) \in \mathsf{M}_3[V]$. This identifies U with a principal ideal of $U \circledast V$.*

Then every congruence $\boldsymbol{\alpha}$ of U has a unique extension to a congruence $\overline{\boldsymbol{\alpha}}$ of $U \circledast V$; therefore, $\mathrm{Con}\, U \cong \mathrm{Con}(U \circledast V)$.

Proof. By Lemmas 2.13, 2.9, and 6.5. $\qquad\qquad\square$

We use induction on $n = |P|$, the size of P, to construct a rigid modular lattice R_P with the following properties:

(I_1) R_P is a rigid modular lattice with zero, 0_P, and unit, 1_P.

(I_2) Every principal congruence of R_P is a join of join-irreducible congruences and the join-irreducible congruences of R_P form an ordered set isomorphic to P.

(I_3) R_P has an element a_P with a unique complement a'_P such that the ideals $\mathrm{id}(a_P)$ and $\mathrm{id}(a'_P)$ are isomorphic to a countable successor cardinal α_P, and R_P has a dual atom $q_P \geq a_P$.

(I_4) Every congruence of R_P is determined by its restriction to $\mathrm{fil}(a_P)$.

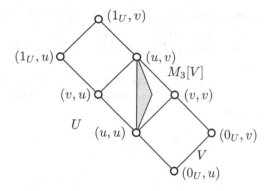

Figure 12.6: The lattice $U \circledast V$.

Observe that (I_4) is equivalent to the following condition:

(I_4') R_P contains a chain

$$a_P \leq b_1^P < c_1^P \leq b_2^P < c_2^P \leq \cdots \leq b_n^P < c_n^P \leq 1_P$$

such that the join-irreducible congruences of R_P are exactly the principal congruences $\mathrm{con}(b_1^P, c_1^P)$, $\mathrm{con}(b_2^P, c_2^P)$, ..., $\mathrm{con}(b_n^P, c_n^P)$.

If $|P| = 1$, then the lattice of Figure 12.2 satisfies these requirements with $\alpha_P = 4$; this lattice is also simple and rigid.

In general, if P is an antichain, we proceed as in the case $n = 2$, see Figure 12.3. We form the glued sum $K = L_1 + L_1 + \cdots + L_1$ with n copies of L_1. Then we add elements (completing covering B_2-s to covering M_3-s) to the lattice C^2, where C is a chain of length $3n$, so that we can embed K into this lattice. It is clear that the lattice we obtain is a planar lattice satisfying (I_1)–(I_4).

Now let P be a finite order, not an antichain, and let us assume that R_S has been constructed for all finite ordered sets S satisfying $|S| < |P|$.

Since P is not an antichain, we can choose a maximal element p of P that is not minimal. Let $q_1, q, \ldots, q_r$ be the elements of P covered by p, with $r \geq 1$.

For the ordered set $Q = P - \{p\}$, there exists a rigid modular lattice R_Q such that (I_1)–(I_4) hold; in particular, there is a chain

$$a_Q \leq b_1^Q < c_1^Q \leq b_2^Q < c_2^Q \leq \cdots \leq b_{n-1}^Q < c_{n-1}^Q \leq 1_Q,$$

or simply,

$$a_Q \leq b_1 < c_1 \leq b_2 < c_2 \leq \cdots \leq b_{n-1} < c_{n-1} \leq 1_Q,$$

such that the join-irreducible congruences of R_Q are the principal congruences $\mathrm{con}(b_1, c_1), \mathrm{con}(b_2, c_2), \ldots, \mathrm{con}(b_{n-1}, c_{n-1})$. Let $C = \mathrm{fil}(a)$ in Q; by (I3) (observing that, by modularity, $\mathrm{id}(a') \cong \mathrm{fil}(a)$), C is a well-ordered chain; it is isomorphic to α_Q.

Let $\mathrm{con}(b_{k_1}, c_{k_1}), \mathrm{con}(b_{k_2}, c_{k_2}), \ldots, \mathrm{con}(b_{k_r}, c_{k_r})$ be the join-irreducible congruences corresponding to $q_1, q_2, \ldots, q_r$, respectively. Without loss of generality, we can assume that

$$b_{k_1} < c_{k_1} \leq b_{k_2} < c_{k_2} \leq \cdots \leq b_{k_r} < c_{k_r}.$$

Let

$$F = [b_{k_1}, c_{k_1}] + [b_{k_2}, c_{k_2}] + \cdots + [b_{k_r}, c_{k_r}].$$

For every natural number i, let F_i be a chain isomorphic to F, and let x^i denote the image of x in F_i under this isomorphism. Finally, we consider the glued sum C of these chains, and the chain $\mathcal{C}$ that is C with two elements, $n_0 \prec n_1$, adjoined,

$$\mathcal{C} = F_1 \dotplus F_2 \dotplus \cdots \dotplus \{n_0, n_1\} = C \dotplus \{n_0, n_1\}.$$

We apply Lemma 12.3 first to $U_1 = E$ and $V_1 = [b_{k_1}, c_{k_1}]$, to obtain the lattice $E \circledast [b_{k_1}, c_{k_1}]$ and filter $\{c_{k_1}\} \times E$ ($\cong E$), and second to $U_2 = E$ and $V_2 = [b_{k_2}, c_{k_2}]$, in order to obtain the lattice $E \circledast [b_{k_2}, c_{k_2}]$ and ideal $\{b_{k_2}\} \times E$ ($\cong E$); and we glue these two lattices together over the given ideal and filter. We proceed similarly and glue $E \circledast [b_{k_3}, c_{k_3}]$ to the resulting lattice, and so on. In r steps, we obtain a lattice L with an ideal and a filter both isomorphic to E.

Now for each $i < \omega$, we take a copy L_i of L, and glue L_2 to L_1, L_3 to the resulting lattice, and so on; in the last step, we glue L_i. Call the lattice we obtained $\overline{L}_i$. Obviously, $\overline{L}_1 \subseteq \overline{L}_2 \subseteq \cdots \subseteq \overline{L}_n$, so we can take $\bigcup(\overline{L}_i \mid i < \omega)$. We adjoin to this lattice $\{n_0, n_1\} \times E$ (where $\{n_0, n_1\}$ is regarded as the two-element chain with $n_0 < n_1$) so that

$$(u, c) \wedge (v, d) = (u \wedge v, c \wedge d),$$
$$(u, c) \vee (v, d) = (u \vee v, c \vee d),$$

for u and/or $v \in \{n_0, n_1\}$ and $c, d \in E$.

We add one more element. E has a dual atom q_C by (I3); the lattice $\{n_0, n_1\} \times \{q_C, 1_Q\}$ is a cover-preserving four-element boolean sublattice of the lattice we have constructed. We add an element w so that $\{n_0, n_1\} \times \{q_C, 1_Q\}$ with w form a cover-preserving M_3. Let Left denote the lattice we have just obtained.

Finally, let us glue R_Q and Left together over the filter E and ideal $\{b_{k_1}\} \times E$ ($\cong E$) in L_0; call the resulting lattice Left$^+$.

To investigate the lattice Left, let us present a more intuitive description. Form the direct product of $\mathcal{C}$ and E; we will call this the *base* of Left. Consider the interval $[(b_{k_j})^i, (c_{k_j})^i]$ in $\mathcal{C}$ and the corresponding interval $[b_{k_j}, c_{k_j}]$ in E. In the direct product $\mathcal{C} \times E$, replace

$$[b^i_{k_j}, c^i_{k_j}] \times [b_{k_j}, c_{k_j}]$$

with $\mathsf{M}_3[b_{k_j}, c_{k_j}]$ so that

$$(x, y) \in [b^i_{k_j}, c^i_{k_j}] \times [b_{k_j}, c_{k_j}]$$

is replaced by $(x, x \wedge y, y)$ and we add the element w in the top prime square so that it forms an M_3. Of course, with this definition it is not clear whether Left is a lattice, whether it is modular, and what are the congruences of Left. It is clear, however, that this definition is the same as the more complicated one given above, so we get all these properties of the construction from the results in Chapter 6 and from Lemma 12.3.

Take a congruence $\boldsymbol{\gamma}$ of Left; then the restriction of $\boldsymbol{\gamma}$ to the base of Left is of the form $\gamma \times \alpha$, where γ is a congruence of $\mathcal{C}$ and α is a congruence of E. It is clear from Lemma 12.3 that γ and α uniquely determine $\boldsymbol{\gamma}$. Moreover, α "almost" determines $\boldsymbol{\gamma}$. In fact, $a \equiv b \pmod{\boldsymbol{\gamma}}$ is fully determined by α for $a, b \in C$. The same is true for $a, b \in \{n_0, n_1\}$. Finally, let $a \in C$ and $b \in \{n_0, n_1\}$. Then $a \equiv b \pmod{\boldsymbol{\gamma}}$ implies that all D_i are collapsed by $\boldsymbol{\gamma}$ from some i on, and so all of C is collapsed by $\boldsymbol{\gamma}$. We conclude that Left has exactly one join-irreducible congruence that is not a minimal extension of a congruence from E, namely, $\mathrm{con}((0_\mathcal{C}, 0_E), (n_0, 0_E))$, and this congruence majorizes all the join-irreducible congruences that are minimal extensions of the congruences $\mathrm{con}(b_1, c_1), \mathrm{con}(b_2, c_2), \ldots, \mathrm{con}(b_{n-1}, c_{n-1})$ of E to Left.

By the inductive assumption (I_3), every congruence of R_Q is determined by its action on E, and so we obtain that every congruence of R_Q can be extended to Left^+. Therefore, the join-irreducible congruences of Left^+ can be described as follows: they are the minimal extensions of the join-irreducible congruences of R_Q to Left^+ and the congruence $\mathrm{con}((0_\mathcal{C}, 0_E), (n_0, 0_E))$. Hence they form an ordered set isomorphic to P, and so $\mathrm{Con_J} \, \mathrm{Left}^+ \cong P$, by (I_2).

The lattice Left^+ does not satisfy all the inductive assumptions, so we will define R_P as an appropriate extension of Left^+.

Define the lattice Top as $\mathsf{M}_3[\mathcal{C}]$ and define the ideal

$$\mathcal{C}_1 = \{ (0_\mathcal{C}, 0_\mathcal{C}, x) \mid x \in \mathcal{C} \}$$

of $\mathsf{M}_3[\mathcal{C}]$. Define the lattice

$$\mathrm{Right} = [a'_Q, 1_Q] \times \mathcal{C}_1$$

and let C ($\cong \mathcal{C}$) be the filter generated by $(1_Q, 0_{\mathcal{C}_1})$ in Right. Let Right^+ be the gluing of Top and Right over $\mathcal{C}_1$ and C. Obviously, the ideal of Right^+ generated by $(1_\mathcal{C}, 0_\mathcal{C}, 0_\mathcal{C}) \in \mathrm{Top}$ is isomorphic with the filter $\mathrm{fil}(a'_Q)$ of Left^+. So we can glue Left^+ and Right^+ together to obtain R_P.

Define $a_P = (n_1, 0_E)$ ($\in$ Left) and let a'_P be its unique complement, $a'_P = (a'_Q, 0_{\mathcal{C}_1})$ ($\in$ Right). The dual atom (n_1, n_0, n_0) of Top can serve as the dual atom of R_P. Now we verify that R_P satisfies the conditions of Theorem 12.2 as well as the inductive assumptions (I_1)–(I_4).

id(a_P) is a countable well-ordered chain, namely, $\alpha_Q + \mathcal{C}$, which is a successor ordinal; the same is true of id(a'_P).

Let α be an automorphism of R_P. Since 1_Q is the smallest element $x \in R_P$ such that there is a sublattice isomorphic to M_3 from x to 1_P, it follows that $\alpha 1_Q = 1_Q$. Since R_Q is rigid, it follows that $\alpha x = x$ for all $x \in R_Q \subseteq R_P$.

Left is built on $\mathcal{C} \times E$, and E is kept fixed by α. Therefore, α maps $\mathcal{C}$ into itself. But $\mathcal{C}$ is well-ordered, so α is the identity map on $\mathcal{C}$, and so on all of Left.

Top $= [1_Q, 1_P]$ and both these elements are fixed by α. Next, consider the three atoms in the sublattice isomorphic to M_3, stretching from 1_Q to 1_P: the element $(1_{\mathcal{C}}, 0_{\mathcal{C}}, 0_{\mathcal{C}})$ is fixed because it is in R'_Q; the elements $(0_{\mathcal{C}}, 1_{\mathcal{C}}, 0_{\mathcal{C}})$ and $(0_{\mathcal{C}}, 0_{\mathcal{C}}, 1_{\mathcal{C}})$ cannot be interchanged because 1_Q has a complement in the interval $[a'_Q, (0_{\mathcal{C}}, 0_{\mathcal{C}}, 1_{\mathcal{C}})]$ but not in $[a'_Q, (0_{\mathcal{C}}, 1_{\mathcal{C}}, 0_{\mathcal{C}})]$. So α fixes the elements $(1_{\mathcal{C}}, 0_{\mathcal{C}}, 0_{\mathcal{C}}), (0_{\mathcal{C}}, 1_{\mathcal{C}}, 0_{\mathcal{C}}), (0_{\mathcal{C}}, 0_{\mathcal{C}}, 1_{\mathcal{C}})$ and also the chain $\mathcal{C}$, hence α fixes all of the lattice Top. Since we already know that in Right, α fixes id(1_Q) and fil(1_Q), it follows that α fixes every element of Right. We conclude that R_P is rigid.

It remains to verify (I_4). In R_Q, we were given the chain

$$a_Q \le b_1 < c_1 \le b_2 < c_2 \le \cdots \le b_n < c_n \le 1_Q$$

such that the join-irreducible congruences of R_Q are the principal congruences con(b_1, c_1), con$(b_2, c_2), \ldots,$ con(b_{n-1}, c_{n-1}). In R_P, we define

$$b_i^P = a_Q \vee b_i,$$
$$c_i^P = a_Q \vee c_i,$$

for $i = 1, \ldots, n-1$, and

$$b_n^P = (1_{\mathcal{C}}, 0_{\mathcal{C}}, 0_{\mathcal{C}},) \in \text{Top},$$
$$c_n^P = 1_P \in \text{Top}.$$

Then condition (I'_4) is obvious for the chain

$$a_P \le b_1^P < c_1^P \le b_2^P < c_2^P \le \cdots \le b_n^P < c_n^P = 1_P.$$

This completes the proof of Theorem 12.2.

12.4. Discussion

The Independence Theorem for Modular Lattices

In [56] (Problem II. 18), I raised the problem whether the congruence lattice and the automorphism group of a finite lattice are independent. This problem was solved affirmatively by V. A. Baranskiĭ [4], [5] and A. Urquhart [188]. This topic is discussed in Chapter 17 of this book.

It is natural to ask whether one could prove the Independence Theorem for modular lattices. This was done in G. Grätzer and E. T. Schmidt [134].

Theorem 12.4 (Independence Theorem for Modular Lattices). *Let D be a nontrivial finite distributive lattice and let G be a finite group. Then there exists a modular lattice M such that the congruence lattice of M is isomorphic to D and the automorphism group of M is isomorphic to G.*

The proof uses the following statement:

Theorem 12.5. *Let G be a finite group. Then there exists a* simple *modular lattice S with an atom p such that the automorphism group of S is isomorphic to G and every automorphism keeps p fixed.*

A weaker form of this theorem (just constructing a modular lattice) is due to E. Mendelsohn [166]. The present form is in G. Grätzer and E. T. Schmidt [130]; see also C. Herrmann [155] and E. T. Schmidt [181].

Now it is easy to prove the Independence Theorem for Modular Lattices. Let G and D be given as in the Independence Theorem. Let S be constructed as in Theorem 12.5 for G and let R be the lattice in Theorem 12.2 constructed for D. We take the ideal $\mathrm{id}(p)$ in S and the filter $\mathrm{fil}(q)$ in R; both are two-element lattices, so we can glue S and R together over $\mathrm{id}(p)$ and $\mathrm{fil}(q)$. Let M be the resulting lattice.

M is a modular lattice by Lemma 2.24. By Lemma 2.9, $\mathrm{Con}\,M \cong \mathrm{Con}\,R$, so $\mathrm{Con}\,M$ is isomorphic to D.

Now let α be an automorphism of S, and define a map $\alpha^{\dagger}$ of M into itself as follows:

$$\alpha^{\dagger}x = \begin{cases} \alpha x, & \text{if } x \in S; \\ x, & \text{if } x \in R. \end{cases}$$

Since p and 0_S are kept fixed by α, it is evident that $\alpha^{\dagger}$ is an automorphism of M. Moreover, q and p is the only pair of elements in M satisfying $q \prec p$ and $M = \mathrm{id}(p) \cup \mathrm{fil}(q)$; therefore, every automorphism of M acts as an automorphism on S and R. It is now clear that every automorphism of M is of the form $\alpha^{\dagger}$ for some automorphism α of S, hence, $\alpha \mapsto \alpha^{\dagger}$ is an isomorphism between the automorphism group of S and the automorphism group of M. It follows that the automorphism group of M is isomorphic to G.

This completes the proof of the Independence Theorem for Modular Lattices.

Two stronger results

Two results state Theorem 12.1 in stronger form.

Theorem 12.6 (R. Freese [47]). *Every finite distributive lattice D can be represented as the congruence lattice of a finitely generated modular lattice L of breadth 2.*

Since a modular lattice of breadth 2 is 2-distributive (see Section 19.5), the lattices of Freese are also 2-distributive.

Theorem 12.7 (E. T. Schmidt [178]). *Every finite distributive lattice D can be represented as the congruence lattice of a complemented modular lattice L.*

Arguesian lattices

Actually, in G. Grätzer and E. T. Schmidt [134], we prove a much stronger form of the Independence Theorem. We construct an arguesian lattice A with a given finite congruence lattice and finite automorphism group.

Recall that a lattice is *arguesian* if it satisfies the *arguesian identity*:

$$(x_1 \vee y_1) \wedge (x_2 \vee y_2) \wedge (x_3 \vee y_3) \leq ((z \vee x_2) \wedge x_1) \vee ((z \vee y_2) \wedge y_1).$$

This is a much stronger form of modularity; it is a lattice theoretic form of Desargues' Theorem. I would recommend that the reader consult Section V.5 of LTF for background.

To prove the Independence Theorem for Arguesian Lattices, we need arguesian versions of Theorems 12.2 and 12.5. In fact, in G. Grätzer and E. T. Schmidt [130], we accomplished that for Theorem 12.5. In G. Grätzer and E. T. Schmidt [134], we prove that the lattice constructed in Section 12.3 is arguesian, and that the two arguesian theorems can be combined to get the Arguesian Independence Theorem.

Problems

Part IV of this book deals with congruence-preserving extensions. Modular lattices are conspicuous with their absence in that part. Here are some obvious problems to raise:

Problem 12.1. Does every modular lattice have a proper, modular, congruence-preserving extension?

Problem 12.2. Let L be a modular lattice with a finite congruence lattice and let G be a group. Does L have a modular congruence-preserving extension whose automorphism group is isomorphic to G?

In light of Theorem 12.7, we can ask:

Problem 12.3. Does the Independence Theorem for Complemented Modular Lattices hold?

Problem 12.4. Are there any results (paralleling the results of Chapters 13 and 16) about regular, uniform, or isoform modular lattices?

Nothing is really known about congruence lattices of infinite modular lattices. It is astonishing that the following problem seems open:

Problem 12.5. Is it true that for every lattice L, there is a modular lattice M with $\operatorname{Con} L \cong \operatorname{Con} M$?

Can R. Freese's Theorem 12.6 and E. T. Schmidt's Theorem 12.7 be combined?

Problem 12.6. Can every finite distributive lattice D be represented as the congruence lattice of a finitely generated, complemented, modular lattice L?

Chapter

13

Uniform Lattices

13.1. The representation theorem

Why couldn't lattices be more like groups and rings?

We want representation theorems by lattices whose congruences behave "as in groups and rings." We will look at lattices that are regular and uniform.

Let L be a lattice. We call a congruence relation α of L *regular* if any congruence class of α determines the congruence. Let us call the lattice L *regular* if all congruences of L are regular.

Sectionally complemented lattices are regular, so Theorem 8.5 is a representation theorem for finite regular lattices.

In this chapter, we consider a concept stronger than regularity. We call a congruence relation α of a lattice L *uniform* if any two congruence classes of α are of the same size (cardinality). Let us call the lattice L *uniform* if all congruences of L are uniform. The following result was proved in G. Grätzer, E. T. Schmidt, and K. Thomsen [141]:

Theorem 13.1. *Every finite distributive lattice D can be represented as the congruence lattice of a finite uniform lattice L.*

A uniform lattice is always regular, so the lattices of Theorem 13.1 are also regular. Figure 13.1 shows the result of the construction for $D = \mathsf{C}_4$.

© Springer International Publishing Switzerland 2016
G. Grätzer, *The Congruences of a Finite Lattice*,
DOI 10.1007/978-3-319-38798-7_13

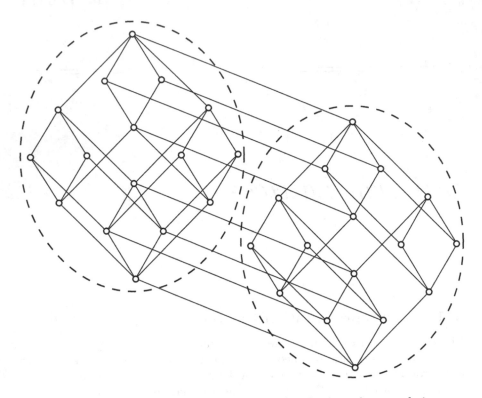

Figure 13.1: The uniform construction for the four-element chain.

13.2. *Proof-by-Picture*

Let A be a finite lattice. We introduce a new lattice, A^+.

Form $A \times \mathsf{C}_2$. Then for every $a \in A$, we get a prime interval $[(a,0),(a,1)]$. We replace this by a copy of B_2, for all $a \in A$, to obtain A^+. We identify $a \in A$ with $(a,0)$ and $b \in \mathsf{B}_2$ with the copy of b in the interval $[(0,0),(0,1)]$; so we can regard A and B_2 as sublattices of A^+.

Figure 13.2 shows the diagrams of B_1^+ and B_2^+.

It is easy to compute that A^+ is a subdirectly irreducible lattice with base congruence $\mathrm{con}((0,0),(0,1))$, and $A^+/\mathrm{con}((0,0),(0,1))$ is isomorphic to A; the isomorphism provided by

$$a \mapsto (a,0)/\mathrm{con}((0,0),(0,1)).$$

So $\mathrm{Con}\,A^+$ is isomorphic to $\mathrm{Con}\,A$ with a new zero added.

We present the *Proof-by-Picture* of Theorem 13.1 in the following form.

Theorem 13.2. *For any finite ordered set P, there exists a finite uniform lattice L such that $\mathrm{Con}_{\mathrm{J}}\,L$ is isomorphic to P, and L satisfies the following property:*

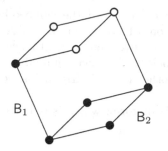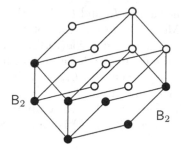

Figure 13.2: The lattices B_1^+ and B_2^+.

(A) *If* $\alpha_1, \alpha_2, \ldots, \alpha_n \in \mathrm{Con}_J L$ *are pairwise incomparable, then* L *contains atoms* $p_1, p_2, \ldots, p_n$ *that generate an ideal isomorphic to* B_n *and satisfy* $\alpha_i = \mathrm{con}(0, p_i)$ *for all* $i \leq n$.

Let P be a finite ordered set with n elements.

If $n = 1$, then $D \cong \mathsf{B}_1^+$; see Figure 13.2.

Let us assume that, for all finite distributive lattices with fewer than n join-irreducible elements, there exists a lattice L satisfying Theorem 13.2.

Let q be a minimal element of P and let $q_1, \ldots, q_k$ $(k \geq 0)$ list all upper covers of q in P. Let $P_1 = P - \{q\}$. By the inductive assumption, there exists a lattice L_1 satisfying $\mathrm{Con}_J L_1 \cong P_1$ and condition (A).

If $k = 0$, then we define $L = \mathsf{B}_1^+ \times L_1$. So we assume that $1 \leq k$.

The congruences of L_1 corresponding to the q_i's are pairwise incomparable and therefore can be written in the form $\mathrm{con}(0, p_i)$ so that the p_i's generate an ideal I isomorphic to B_k. Figure 13.3 shows the inductive step.

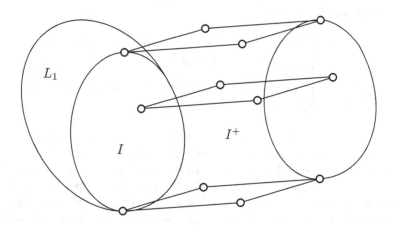

Figure 13.3: The inductive step: the chopped lattice K.

The lattices L_1 and I^+ share the ideal I, so we can form the chopped lattice $K = \mathrm{Merge}(L_1, I^+)$, in the notation of Section 5.1. We define $L = \mathrm{Id}\, K$.

The idea is that $\mathrm{con}((0,0),(0,1))$ is a join-irreducible congruence of L, below the $\mathrm{con}(0,p_i)$, for all $i \leq n$, but not below any other minimal join-irreducible congruence of L_1. And there is no other new join-irreducible congruence in L, so $\mathrm{Con_J}\, L \cong P$.

Of course, one has to compute that L is uniform, which is tedious. The induction condition (A) is easy to verify for L.

13.3.   The lattice $N(A, B)$

We introduce the A^+ construction of Section 13.2 in a more general form; see Section 13.5 for an even more general lattice construction.

The construction

Let A and B be lattices. Let us assume that A is bounded, with bounds 0 and 1, with $0 \neq 1$. Let $A^- = A - \{0, 1\}$. We introduce a new lattice construction $N(A, B)$—the A^+ of Section 13.2 is the special case $N(A, \mathsf{B}_2)$.

For $u \in A \times B$, we use the notation $u = (u_A, u_B)$; the binary relation $\leq_\times$ will denote the ordering on $A \times B$, and $\wedge_\times$, $\vee_\times$ the meet and join in $A \times B$, respectively. On the set $A \times B$, we define a binary relation $\leq_{N(A,B)}$ (denoted by $\leq_N$ if A and B are understood) as follows:

$$\leq_N \; = \; \leq_\times \; - \{ (u, v) \mid u,\, v \in A^- \times B,\; u_B \neq v_B \}.$$

We define $N(A, B) = (A \times B, \leq_N)$.

Lemma 13.3. *The set $N(A, B)$ is ordered by $\leq_N$, in fact, $N(A, B)$ is a lattice. The join and meet in $N(A, B)$ of $\leq_N$-incomparable elements can be computed using the formulas:*

$$u \vee_N v = \begin{cases} (1, u_B \vee v_B) & \text{if } u \vee_\times v \in A^- \times B \text{ and } u_B \neq v_B; \\ u \vee_\times v, & \text{otherwise.} \end{cases}$$

$$u \wedge_N v = \begin{cases} (0, u_B \wedge v_B) & \text{if } u \wedge_\times v \in A^- \times B \text{ and } u_B \neq v_B; \\ u \wedge_\times v, & \text{otherwise.} \end{cases}$$

Proof. Since $\leq_\times$ is reflexive, it follows that $\leq_N$ is reflexive, since $u_B \neq u_B$ fails for all $u \in A \times B$. Since $\leq_\times$ is antisymmetric, so is $\leq_N$.

Let $u, v, w \in A \times B$; let us assume that $u \leq_N v$ and $v \leq_N w$. Since $\leq_\times$ is transitive, we conclude that $u \leq_\times w$. So if $u \leq_N w$ fails, then $u, w \in A^- \times B$ and $u_B \neq w_B$. It follows that $v \in A^- \times B$ and either $u_B \neq v_B$ or $v_B \neq w_B$, contradicting that $u \leq_N v$ or $v \leq_N w$. So $\leq_N$ is transitive, proving that $N(A, B)$ is an order.

We prove that $N(A, B)$ is a lattice by verifying the join and meet formulas. By duality, we only have to verify one of them; we will prove the meet formula.

Let $u, v \in A \times B$ be $\leq_N$-incomparable elements, and let t be a lower bound of u and v in $N(A, B)$.

Case 1. $u \wedge_\times v$ is not a lower bound of both u and v in $N(A, B)$.

If $u \wedge_\times v$ is not a lower bound of both u and v in $N(A, B)$, say, $u \wedge_\times v \not\leq_N u$, then $u, u \wedge_\times v \in A^- \times B$ and $u_B \neq (u \wedge_\times v)_B$ (which is the same as $u_B \neq v_B$). Since $t \leq_\times u \wedge_\times v$, it follows that $t_B \leq (u \wedge_\times v)_B < u_B$, so $t \notin A^- \times B$ (otherwise, we would have $t \not\leq_N u$). We conclude that $t = (0, t_B)$, so $t_B \leq u_B \wedge v_B$, which yields that $t \leq (0, u_B \wedge v_B)$.

So in Case 1, $u \wedge_N v = (0, u_B \wedge v_B)$.

Case 2. $u \wedge_\times v$ is a lower bound of both u and v in $N(A, B)$.

If $t \not\leq_N u \wedge_\times v$, then $t, u \wedge_\times v \in A^- \times B$ and $t_B < (u \wedge_\times v)_B$, so $u \in A^- \times B$ or $v \in A^- \times B$, say, $u \in A^- \times B$. Therefore, the assumption of Case 2, namely, $u \wedge_\times v \leq_N u$, implies that $(u \wedge_\times v)_B = u_B$. So $t, u \in A^- \times B$ and $t_B \neq u_B$, contradicting that $t \leq_N u$. Thus $t \not\leq_N u \wedge_\times v$ leads to a contradiction. We conclude that $t \leq_N u \wedge_\times v$.

So in Case 2, $u \wedge_N v = u \wedge_\times v$.

Since the two cases correspond to the two clauses of the meet formula, this verifies the meet formula.   □

We will use the notation: $B_* = \{0\} \times B$, $B^* = \{1\} \times B$, and for $b \in B$, $A_b = A \times \{b\}$. Note that B_* is an ideal and B^* is a filter of $N(A, B)$.

Congruences

Let K and L be lattices, and let α be an embedding of K into L. Given a congruence α of L, we can define the congruence α_1 on K *via* α, that is for $a, b \in K$,

$$a \equiv b \pmod{\alpha_1} \quad \text{iff} \quad \alpha a \equiv \alpha b \pmod{\alpha}.$$

We will call α_1 the *restriction of α transferred* via *the isomorphism α to K*.

Let γ be a congruence relation of $N = N(A, B)$. Using the natural isomorphisms of B into $N(A, B)$ with images B_* and B^*, we define γ_* and γ^* as the restriction of γ to B_* and B^*, respectively, transferred *via* the natural isomorphisms to B. Let α_b be the restriction of γ to A_b, for $b \in B$, transferred *via* the natural isomorphisms to A.

Lemma 13.4. $\gamma_* = \gamma^*$.

Proof. Indeed, if $b_1 \equiv b_2 \pmod{\gamma_*}$, then $(0, b_1) \equiv (0, b_2) \pmod{\gamma}$. Joining both sides with $(1, b_1 \wedge b_2)$, we obtain that $(1, b_1) \equiv (1, b_2) \pmod{\gamma}$, that is, $b_1 \equiv b_2 \pmod{\gamma^*}$. This proves that $\gamma_* \leq \gamma^*$. We prove the reverse inclusion symmetrically.   □

It is easy to see that $\gamma = \gamma_* = \gamma^* \in \operatorname{Con} B$, and $\{\alpha_b \mid b \in B\} \subseteq \operatorname{Con} A$ describe γ, but it is difficult to obtain a description of the congruences of $N(A, B)$, in general. So we impose a very stringent condition on A.

Lemma 13.5. *Let A and B be lattices with $|A| > 2$ and $|B| > 1$; let A be bounded, with bounds 0 and 1. Let us further assume that A is non-separating. Then the map sending $\gamma \neq \mathbf{0}_N$ to its restriction to B_* transferred to B by the natural isomorphism is a bijection between the non-$\mathbf{0}_N$ congruences of $N(A, B)$ and the congruences of B. Therefore, $\mathrm{Con}\, N(A, B)$ is isomorphic to $\mathrm{Con}\, B$ with a new zero added.*

Proof. Let $\gamma \neq \mathbf{0}_N$ be a congruence relation of $N(A, B)$. We start with the following statement:

Claim 13.6. *There are elements $a_1 < a_2$ in A and an element $b_1 \in B$ such that*

$$(a_1, b_1) \equiv (a_2, b_1) \pmod{\gamma}.$$

Proof. Assume that $(u_1, v_1) \equiv (u_2, v_2) \pmod{\gamma}$ with $(u_1, v_1) <_N (u_2, v_2)$. We distinguish two cases.

First case. $u_1 = u_2$.

Then $v_1 < v_2$ and either $u_1 = u_2 = 0$ or $u_1 = u_2 = 1$. So either $(0, v_1) \equiv (0, v_2) \pmod{\gamma}$ or $(1, v_1) \equiv (1, v_2) \pmod{\gamma}$. By Lemma 13.4, either one of these congruences implies the other, so both of these congruences hold. Since $|A| > 2$, we can pick an $a \in A^-$. Then

$$(a, v_1) = (a, v_1) \vee (0, v_1) \equiv (a, v_1) \vee (0, v_2) = (1, v_2) \pmod{\gamma},$$

from which we conclude that $(a, v_1) \equiv (1, v_1) \pmod{\gamma}$, so the claim is verified with $a_1 = a$, $a_2 = 1$, and $b_1 = v_1$.

Second case. $u_1 < u_2$.

Since we have assumed that $(u_1, v_1) <_N (u_2, v_2)$, it follows from the definition of $\leq_N$ that either $v_1 = v_2$, or $u_1 = 0$, or $u_2 = 1$.

If $v_1 = v_2$, then $(u_1, v_1) \equiv (u_2, v_1) \pmod{\gamma}$, so the claim is verified with $a_1 = u_1$, $a_2 = u_2$, and $b_1 = v_1$.

If $u_1 = 0$, then $(0, v_2) \equiv (u_2, v_2) \pmod{\gamma}$, so the claim is verified with $a_1 = 0$, $a_2 = u_2$, and $b_1 = v_2$.

If $u_2 = 1$, then $(u_1, v_1) \equiv (1, v_1) \pmod{\gamma}$, so the claim is verified with $a_1 = u_1$, $a_2 = 1$, and $b_1 = v_1$. $\qquad\qquad\square$

Claim 13.7. *There is an element $b_2 \in B$ such that A_{b_2} is in a single congruence class of γ.*

Proof. By Claim 13.6, there are $a_1 < a_2 \in A$ and $b_1 \in B$ such that

$$(a_1, b_1) \equiv (a_2, b_1) \pmod{\gamma}.$$

Since A is non-separating, there exists $a_3 \in A$ with $0 < a_3$ and $0 \equiv a_3$ (mod $\mathrm{con}(a_1, a_2)$). Moreover, A_{b_1} is a sublattice of $N(A, B)$, so it follows that $(0, b_1) \equiv (a_3, b_1)$ (mod $\mathrm{con}((a_1, b_1), (a_2, b_1))$), and thus

$$(0, b_1) \equiv (a_3, b_1) \pmod{\gamma}.$$

Therefore, for any $b_2 \in B$ with $b_1 < b_2$, joining both sides with $(0, b_2)$, we obtain that $(0, b_2) \equiv (1, b_2)$ (mod γ), that is, A_{b_2} is in a single γ class.

This completes the proof of the claim, unless b_1 is the unit element, 1_B, of B. In this case,

$$(0, 1_B) \equiv (a_3, 1_B) \pmod{\gamma}.$$

Since A is non-separating, there exists $a_4 \in A$ with $a_4 < 1$ and $a_4 \equiv 1$ (mod $\mathrm{con}(0, a_3)$). Moreover, A_{1_B} is a sublattice of $N(A, B)$, so it follows that $(a_4, 1_B) \equiv (1, 1_B)$ (mod $\mathrm{con}((0, 1_B), (a_3, 1_B))$), and thus

$$(a_4, 1_B) \equiv (1, 1_B) \pmod{\gamma}.$$

Now choose any $b_2 < 1_B$ (recall that we have assumed that $|B| > 1$). Meeting both sides with $(1, b_2)$, we obtain that

$$(1, b_2) \equiv (0, b_2) \pmod{\gamma},$$

that is, A_{b_2} is in a single congruence class of γ.   □

Claim 13.8. *A_b is in a single congruence class of γ for each $b \in B$.*

Proof. Let $b \in B$. By Claim 13.7, there is an element $b_2 \in B$ such that A_{b_2} is in a single congruence class of γ, that is,

$$(1, b_2) \equiv (0, b_2) \pmod{\gamma}.$$

Therefore,

$$(1, b) = ((1, b_2) \vee (0, b \vee b_2)) \wedge (1, b) \equiv ((0, b_2) \vee (0, b \vee b_2)) \wedge (1, b) = (0, b) \pmod{\gamma},$$

that is, A_b is in a single congruence class of γ.   □

Now the statement of the lemma is easy to verify. It is clear that the map $\gamma \mapsto \overline{\gamma}$ is one-to-one for $\gamma \in \mathrm{Con}\, N(A, B) - \{\mathbf{0}_N\}$. It is also onto: given a congruence γ of B, define $\overline{\gamma}$ on $N(A, B)$ by

$$(u_1, v_1) \equiv (u_2, v_2) \pmod{\overline{\gamma}} \quad \text{iff} \quad v_1 \equiv v_2 \pmod{\gamma}.$$

Then $\overline{\gamma}$ is in $\mathrm{Con}\, N(A, B) - \{\mathbf{0}_N\}$ and it maps to γ.   □

Congruence classes

Now let U be a finite lattice with an ideal V isomorphic to B_n. We identify V with the ideal $(B_n)_* = \mathrm{id}((0,1))$ of $N(B_2, B_n)$ to obtain the merged chopped lattice $K = \mathrm{Merge}(U, N(B_2, B_n))$ (using the notation of Section 5.1). Let m denote the generator of $V = (B_n)_*$. We view $\mathrm{Id}\, K$ as the set of compatible pairs (u, y) with $u \in U$ and $y \in N(B_2, B_n)$ (see Section 5.2).

Lemma 13.9. *Let $u \in U$. Then*

$$\{\, y \in N(B_2, B_n) \mid (u, y) \in \mathrm{Id}\, K \,\}$$

is isomorphic to B_2.

Proof. This is clear since there are exactly four elements y of $N(B_2, B_n)$ satisfying that $u \wedge m = y \wedge m$, namely, the elements of $(B_2)_{u \wedge m}$, which form a sublattice isomorphic to B_2. Therefore,

$$\{\, y \in N(B_2, B_n) \mid (u, y) \in \mathrm{Id}\, K \,\}$$

is a four-element set, closed under coordinate-wise meets and joins; the statement follows.   □

Now let us further assume that U is uniform.

Lemma 13.10. $\mathrm{Id}\, K$ *is uniform.*

Proof. A congruence Λ of $\mathrm{Id}\, K$ can be described (see Section 5.3) as a congruence vector $(\boldsymbol{\alpha}, \boldsymbol{\gamma})$, where $\boldsymbol{\alpha}$ is a congruence of U, $\boldsymbol{\gamma}$ is a congruence of $N(B_2, B_n)$, and $\boldsymbol{\alpha}$ and $\boldsymbol{\gamma}$ restrict to the same congruence of $V = (B_n)_*$.

The trivial congruences, $\mathbf{0}_{\mathrm{Id}\, K}$ (represented by $(\mathbf{0}_U, \mathbf{0}_{N(B_2, B_n)})$) and $\mathbf{1}_{\mathrm{Id}\, K}$ (represented by $(\mathbf{1}_U, \mathbf{1}_{N(B_2, B_n)})$), are obviously uniform. Therefore, we only need to look at two cases.

First case. Λ is represented by $(\boldsymbol{\alpha}, \mathbf{0})$.
So $\boldsymbol{\alpha}]_V = \mathbf{0}_V$. Let (x, y) be an element of $\mathrm{Id}\, K$ and note that

$$(x, y)/(\boldsymbol{\alpha}, \mathbf{0}) = \{\, (t, y) \in \mathrm{Id}\, K \mid t \equiv x\,(\boldsymbol{\alpha}) \,\}.$$

If $t \equiv x \pmod{\boldsymbol{\alpha}}$, then $t \wedge m \equiv x \wedge m \pmod{\boldsymbol{\alpha}}$, but $\boldsymbol{\alpha}]_V = \mathbf{0}_V$, so $t \wedge m = x \wedge m$. We conclude that

$$(x, y)/(\boldsymbol{\alpha}, \mathbf{0}) = \{\, (t, y) \mid t \equiv x\,(\boldsymbol{\alpha}) \,\},$$
$$|(x, y)/(\boldsymbol{\alpha}, \mathbf{0})| = |x/\boldsymbol{\alpha}|.$$

So Λ is uniform; each congruence class of Λ is of the same size as a congruence class of $\boldsymbol{\alpha}$.

Second case. Λ is represented by $(\boldsymbol{\alpha}, \boldsymbol{\gamma})$, where $\boldsymbol{\gamma} \neq \mathbf{0}$.

Let (x, y) be an element of $\operatorname{Id} K$. Then

$$(x, y)/(\alpha, \gamma) = \{\, (w, z) \in \operatorname{Id} K \mid x \equiv w\,(\alpha) \text{ and } y \equiv z\,(\gamma)\,\}.$$

For a given w if (w, t_1) and $(w, t_2) \in \operatorname{Id} K$, then $t_1 \equiv t_2 \pmod{\gamma}$ because $(B_2)_w$ is in a single congruence class of γ by Lemma 13.5 (in particular, by Claim 13.8) and

$$\{\, t \in N(B_2, B_n) \mid (w, t) \in \operatorname{Id} K \,\} = (B_2)_{w \wedge m};$$

thus

$$|\{\, t \in N(B_2, B_n) \mid (w, t) \in \operatorname{Id} K \,\}| = |(B_2)_{w \wedge m}| = 4.$$

We conclude that

$$(x, y)/(\alpha, \gamma) = \{\, (w, z) \in \operatorname{Id} K \mid x \equiv w\,(\alpha),\ z \in (B_2)_{w \wedge m}\,\},$$

and

$$|(x, y)/(\alpha, \gamma)| = 4|x/\alpha|.$$

So Λ is uniform; each congruence class of Λ is four-times the size of a congruence class of α. $\qquad\square$

13.4. Formal proof

Based on the results of Section 13.3, we will prove Theorem 13.2 by induction. Let P be a finite ordered set with n elements.

If $n = 1$, then $\operatorname{Down} P \cong B_1$, so there is a lattice $L = B_1$ that satisfies Theorem 13.2.

Let us assume that for all finite ordered sets with fewer than n elements, there exists a lattice L satisfying Theorem 13.2.

Let q be a minimal element of P and let $q_1, \ldots, q_k$ ($k \geq 0$) list all upper covers of q in P. Let $P_1 = P - \{q\}$. By the inductive hypothesis, there exists a lattice L_1 satisfying $\operatorname{Con_J} L_1 \cong P_1$ and also satisfying condition (A).

If $k = 0$, then $\operatorname{Down} P \cong B_1 \times \operatorname{Down} P_1$, and so $L = B_1 \times L_1$ obviously satisfies all the requirements of the theorem. So we assume that $1 \leq k$.

The congruences of L_1 corresponding to the q_i's are pairwise incomparable; therefore, these congruences can be written in the form $\operatorname{con}(0, p_i)$ and the p_i's generate an ideal I_1 isomorphic to B_k. The lattice $N(B_2, B_k)$ also contains an ideal $(B_k)_*$ isomorphic to B_k. Merging these two lattices by identifying I_1 and $(B_k)_*$, we get the chopped lattice K and the lattice $L = \operatorname{Id} K$. The chopped lattice K is sketched in Figure 13.4. L is uniform by Lemma 13.10.

Let α be a join-irreducible congruence of L. Then we can write α as $\operatorname{con}(a, b)$, where a is covered by b. We can assume that either $a, b \in L_1$ or a, $b \in N(B_2, B_k)$. In either case, we find an atom q in L_1 or in $N(B_2, B_k)$, so that $\operatorname{con}(a, b) = \operatorname{con}(0, q)$ in L_1 or in $N(B_2, B_k)$. Obviously, q is an atom of L and $\operatorname{con}(a, b) = \operatorname{con}(0, q)$ in L.

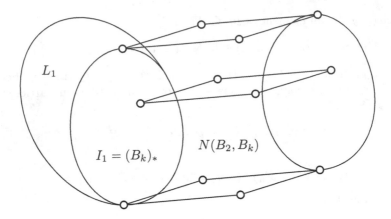

Figure 13.4: The chopped lattice K for the formal proof.

Let $\alpha_1, \alpha_2, \ldots, \alpha_t$ be pairwise incomparable join-irreducible congruences of L. To verify condition (A), we have to find atoms $p_1, p_2, \ldots, p_t$ of L satisfying $\alpha_i = \mathrm{con}(0, p_i)$, for all $i \leq t$, and such that $p_1, p_2, \ldots, p_t$ generate an ideal of L isomorphic to B_t.

Let p denote an atom in $N(B_2, B_k) - I_1$; there are two, but they generate the same congruence $\mathrm{con}(0, p)$. If $\mathrm{con}(0, p)$ is not one of $\alpha_1, \alpha_2, \ldots, \alpha_t$, then clearly we can find $p_1, p_2, \ldots, p_t$ in L_1 as required, and $p_1, p_2, \ldots, p_t$ also serves in L.

If $\mathrm{con}(0, p)$ is one of $\alpha_1, \alpha_2, \ldots, \alpha_t$, for instance, $\mathrm{con}(0, p) = \alpha_t$, then let $\{p_1, p_2, \ldots, p_{t-1}\}$ be the set of atoms establishing condition (A) for the congruences $\alpha_1, \alpha, \ldots, \alpha_{t-1}$ in L_1 and therefore, in L. Then the elements $p_1, p_2, \ldots, p_{t-1}, p$ represent the congruences $\alpha_1, \alpha_2, \ldots, \alpha_t$ and generate an ideal isomorphic to B_t. Therefore, L satisfies condition (A).

Finally, it is clear from this discussion that $\mathrm{Con_J}\, K$ has exactly one more element than $\mathrm{Con_J}\, L_1$, namely, $\mathrm{con}(0, p)$, and this congruence relates to the congruences in $\mathrm{Con_J}\, L_1$ exactly as q relates to the elements of P. Therefore, $P \cong \mathrm{Con_J}\, L$.

13.5. Discussion

Isoform lattices

The next logical step is to require that that any two congruence classes be not only of the same size but isomorphic as lattices. We called such lattices *isoform*. The following result was proved in G. Grätzer and E. T. Schmidt [136]:

Theorem 13.11. *Every finite distributive lattice D can be represented as the congruence lattice of a finite isoform lattice L.*

Since the congruence classes of isomorphic lattices are of the same size, Theorem 13.11 is a much stronger version of Theorem 13.1. Figure 13.1 shows that the lattice we obtained in this chapter for $D = \mathsf{C}_4$ is not isoform. Indeed, the congruence marked by the two dotted ovals has two congruence classes of 16 elements each, but they are not isomorphic; in the top lattice, two dual atoms are join-irreducible, and in the bottom lattice, none are join-irreducible.

Note that isoform lattices are not "like groups and rings." The isoform concept is special to idempotent algebras.

The $N(A, B, \alpha)$ construction

We could prove Theorem 13.11 based on the following generalization of the $N(A, B)$ construction.

Let $P = (P, \leq_P)$ be a finite ordered set. Then the ordering $\leq_P$ on P is the reflexive-transitive closure of $\prec_P$, the covering relation in $(P, \leq_P)$, in formula: $\mathrm{rt}(\prec_P) = \leq_P$. Now take a subset H of $\prec_P$, and take the reflexive-transitive extension $\mathrm{rt}(H)$ of H. Then $(P, \mathrm{rt}(H))$ is also an ordered set; we call it a *pruning* of P. If you think of P in terms of its diagram, then the terminology is easy to picture: We obtain the diagram of $(P, \mathrm{rt}(H))$ from the diagram of P by cutting out (pruning) some edges (each representing a covering) but not deleting any elements. For instance, the lattice of Figure 13.1 is a pruning of the boolean lattice B_2^5.

Let L be a finite lattice. We call L *discrete-transitive* if for any congruence γ of L and for $a < b < c$ in L, whenever γ is discrete on $[a, b]$ and $[b, c]$, then γ is discrete on $[a, c]$.

Let A be a nontrivial finite lattice with bounds 0 and 1; let $|A| > 2$. Set $A^- = A - \{0, 1\}$. Let B be a nontrivial finite lattice with a discrete-transitive congruence α.

To prune $A \times B$, we define the set:

$$\mathrm{Prune}(A, B, \alpha) = \{\, ((a, b_1), (a, b_2)) \mid a \in A^-,\ b_1 \prec b_2 \text{ in } B,\ b_1 \equiv b_2\,(\alpha)\,\}.$$

$\mathrm{Prune}(A, B, \alpha)$ is a subset of $\prec_\times$, so we can define $H = \prec_\times - \mathrm{Prune}(A, B, \alpha)$. Now we take the reflexive-transitive extension $\mathrm{rt}(H)$ of H. The set $A \times B$ with the ordering $\mathrm{rt}(H)$ is $N(A, B, \alpha)$. It is clear that $N(A, B, \mathbf{0})$ is the direct product $A \times B$ and $N(A, B, \mathbf{1}) = N(A, B)$.

We describe in G. Grätzer and E. T. Schmidt [136] the order $N(A, B, \alpha)$: $u \leq v$ in $N(A, B, \alpha)$ iff

$$u_A,\ v_A \in A^- \text{ and } [u_B, v_B] \text{ is } \alpha\text{-discrete},$$

or

$$u_A \text{ or } v_A \notin A^-,$$

and we prove that $N(A, B, \alpha)$ is a lattice under this ordering.

Pruning seldom produces a lattice. Note the very strong condition imposed on α to make the pruned order a lattice.

We use the $N(A, B, \alpha)$ construction to obtain the following version of Theorem 13.11 (G. Grätzer and E. T. Schmidt [136]):

Theorem 13.12. *Every finite distributive lattice D can be represented as the congruence lattice of a finite lattice L with the following properties:*

(i) *L is isoform.*

(ii) *For every congruence α of L, the congruence classes of α are projective intervals.*

(iii) *L is a finite pruned boolean lattice.*

(iv) *L is discrete-transitive.*

By Properties (i) and (ii), for every congruence relation α of L and for any two congruence classes U and V of α, the congruence classes U and V are required to be isomorphic and projective intervals, but we do not require that there be a projectivity that is also an isomorphism.

We refer the reader to G. Grätzer and E. T. Schmidt [136] for a proof of Theorem 13.12. A stronger form of Theorem 13.12 will be proved in Chapter 16.

Problems

Let us start with some general problems on uniformity.

Problem 13.1. Develop a theory of uniform lattices.

Problem 13.2. Prove that uniformity is not a first-order property.

The answer to the following question is likely to be in the negative.

Problem 13.3. Are infinite relatively complemented lattices uniform?

The congruence classes of a congruence of the lattice L in Theorem 13.12 are both isomorphic and projective.

Problem 13.4. Is it possible to sharpen Theorem 13.12 so that the isomorphisms of the congruence classes be projectivities?

Can we combine the main result of this chapter with the results of the previous chapters? In other words,

Problem 13.5. Can every finite distributive lattice D be represented as the congruence lattice of a finite uniform (resp., isoform) lattice L with some additional property: L be semimodular, sectionally complemented, or 2-distributive, and so on?

Problem 13.6. Can every finite distributive lattice D be represented as the congruence lattice of a uniform (resp., isoform) modular lattice L?

What happens in the infinite case?

Problem 13.7. Is there an analogue of Theorem 13.1 (resp., Theorem 13.11) for infinite lattices?

Let **Uniform** and **Isoform** denote the class of uniform and isoform lattices, respectively. (Recall that the function **mcr** was introduced in Section 8.5.)

Problem 13.8. What is $\mathbf{mcr}(n, \mathbf{Uniform})$? What is $\mathbf{mcr}(n, \mathbf{Isoform})$? Can one get a good result for $O(\mathbf{mcr}(n, \mathbf{Uniform}))$ and $O(\mathbf{mcr}(n, \mathbf{Isoform}))$?

Let L be a finite uniform lattice and let γ be an isomorphism between a finite distributive lattice D and $\operatorname{Con} L$. Then we can introduce a function $s = s(L, \gamma)$ from D to the natural numbers, as follows: Let $d \in D$; then $\gamma(d)$ is a congruence of L. Since L is uniform, all congruence classes of $\gamma(d)$ are of the same size. Let $s(d)$ be the size of the congruence classes.

The function s has the following obvious properties:

(s$_1$) $s(0) = 1$.

(s$_2$) If $a < b$ in D, then $s(a) < s(b)$.

Problem 13.9. Characterize the function s. In other words, let f be a function from a finite distributive lattice D to the natural numbers that satisfies conditions (s$_1$) and (s$_2$) above. What additional conditions on the function f are required for $f = s(L, \gamma)$, for some finite uniform lattice L and isomorphism $\gamma \colon D \to \operatorname{Con} L$?

This problem may be too difficult to solve in its full generality. The following lists some special cases that may be easier to attack.

Problem 13.10. Characterize the function s for some special classes of finite distributive lattices:

(i) finite boolean lattices;

(ii) finite chains;

(iii) "small" distributive lattices.

Part IV

Congruence Lattices and Lattice Extensions

Sectionally Complemented Lattices

14.1. The extension theorem

As outlined in the Introduction, Part IV deals with congruence-preserving extension theorems. We start with sectionally complemented lattices, as we began Part III. The reader should compare the relatively small constructs and shorter proofs of Part III with the much more complex ones in this part.

For instance, to represent the distributive lattice $C_1 + B_2$ as the congruence lattice of a finite sectionally complemented lattice, by Figure 8.3, we take the ideal lattice L of the chopped lattice M_V, and we are done. The lattice L has 10 elements.

Now start with the lattice K of Figure 14.1, a small seven-element lattice, whose congruence lattice is $C_1 + B_2$. Even though K was chosen to minimize the size of the sectionally complemented congruence-preserving extension L, we find this L much larger than the L of the representation theorem. In general, the extension L we construct is $O(|K|^{|K|})$ in size.

In [129], G. Grätzer and E. T. Schmidt proved the congruence-preserving extension theorem for sectionally complemented lattices.

Theorem 14.1. *Every finite lattice K has a finite, sectionally complemented, congruence-preserving extension L.*

© Springer International Publishing Switzerland 2016
G. Grätzer, *The Congruences of a Finite Lattice*,
DOI 10.1007/978-3-319-38798-7_14

The proof—as most proofs in Part IV—has two major ingredients: the representation theorem from Part III and the cubic extension from Part II, with a little bit of chopped lattices thrown in.

Since every finite sectionally complemented lattice is atomistic, this result contains the first published result on congruence-preserving extension, namely, the theorem of M. Tischendorf [185].

Theorem 14.2. *Every finite lattice has a finite atomistic congruence-preserving extension.*

14.2. *Proof-by-Picture*

Let us find a finite, sectionally complemented, congruence-preserving extension for the lattice K of Figure 14.1. The figure also shows $\operatorname{Con} K$, $\operatorname{Con}_J K$, and $\operatorname{Con}_M K$; the congruence γ_1 splits K into two classes, marked by the dotted line (γ_2 is symmetric). Note that $\operatorname{Con}_J K \cong \operatorname{Con}_M K$ (Theorem 2.19).

So K has three subdirectly irreducible quotients, $K \cong K/0$, $K/\gamma_1 \cong \mathsf{C}_2$, and $K/\gamma_2 \cong \mathsf{C}_2$.

To form a cubic extension (see Section 7), we embed the subdirectly irreducible quotients into simple lattices; C_2 being simple, we only have to embed K. Figure 14.2 shows the $\operatorname{Simp} K$ we choose (compare it with the construct of Lemma 14.3), adding a single element a to K. So the cubic extension of K is

$$\operatorname{Cube} K = \operatorname{Simp} K \times \mathsf{C}_2 \times \mathsf{C}_2.$$

This lattice is not easy to draw, but we easily draw the chopped lattice M of Figure 14.3, whose ideal lattice is $\operatorname{Cube} K$.

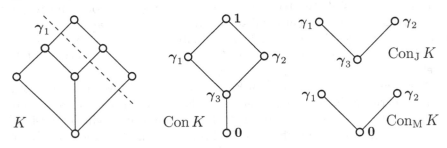

Figure 14.1: The example lattice K.

In this example, the boolean ideal B—of Theorem 7.2—is generated by the atoms a, s_{γ_1}, and s_{γ_2}.

Now recall Theorem 8.9: For $D = \operatorname{Con} K$, there is a a finite sectionally complemented lattice L_0 with $\operatorname{Con} L_0 \cong D$ and with a boolean ideal B_0, the eight-element boolean lattice with atoms p_{γ_1}, p_{γ_2}, and p_{γ_3} so that $\operatorname{con}(0, p_{\gamma_i})$ represents γ_i for $i = 1, 2, 3$; see Figure 14.4.

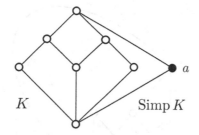

Figure 14.2: A simple extension of K.

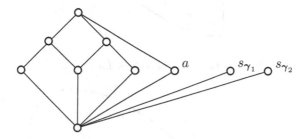

Figure 14.3: The chopped lattice M.

We merge Simp K and L_0 by identifying the zeros and the atom a of Simp K with the atom of p_{γ_3} of L_0 (and of B_0), see Figure 14.5. We view M as being a part of the merged chopped lattice by identifying s_{γ_1} with $p_{\gamma_1^\dagger} = p_{\gamma_2}$ and s_{γ_2} with $p_{\gamma_2^\dagger} = p_{\gamma_1}$; Figure 14.5 shows the merged chopped lattice, the black-filled elements form M. Basically, we identify the boolean ideal B of Cube K with the boolean ideal B_0 of L_0.

We define L as the ideal lattice of this chopped lattice. By the Atom Lemma—Lemma 5.8—L is a finite sectionally complemented lattice. The diagonal embedding Diag of K into Id M of Section 7.1 also embeds K into L, so (after identifying $a \in K$ with Diag(a)) the lattice L can be regarded as an extension of K.

What about the congruences of L? Clearly, merging L_0 with a simple lattice, we obtain a chopped lattice that is a congruence-preserving extension of L_0, so L is a congruence-preserving extension of L_0 and Con $L \cong$ Con $L_0 \cong$ Con $K \cong D$.

A congruence of L can be described as a compatible congruence vector (α, γ) of the chopped lattice, where α is a congruence of Cube K and γ is a congruence of L_0 agreeing on $\{o, a\} = \{0, p_{\gamma_3}\}$. The congruence α is determined on B and the congruence γ is determined on B_0. In L, we identified B and B_0, so it follows that $\alpha = \gamma$. Moreover, (α, α) is a compatible congruence vector, because α acts on B by collapsing p_γ and 0 iff $\gamma \leq \alpha$, while α acts on B_0 by collapsing $p_{\gamma^\dagger}$ and 0 iff $\gamma^\dagger \not\leq \alpha$; also, $\gamma \leq \alpha$ is equivalent to $\gamma^\dagger \not\leq \alpha$.

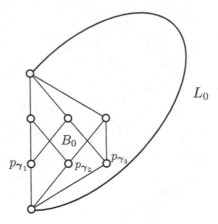

Figure 14.4: The lattices L_0 and B_0.

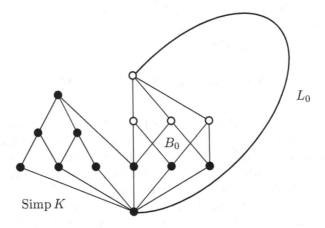

Figure 14.5: Merging L_0 and M.

Since K has the congruence extension property in Cube K, we conclude that $\gamma\colon \alpha \mapsto \mathrm{con}(\alpha)$ (in L), for $\alpha \in \mathrm{Con}\,K$, is a bijection. On the other hand, $\mathrm{Con}\,L \cong \mathrm{Con}\,L_0 \cong \mathrm{Con}\,K$, so γ is an isomorphism between $\mathrm{Con}\,K$ and $\mathrm{Con}\,L$, that is, L is a congruence-preserving extension of K.

14.3. Simple extensions

To start out in the proof, we construct sectionally complemented cubic extensions. So we need the following statement.

Lemma 14.3. *For every finite lattice K, there is a finite, simple, sectionally complemented $\{0, 1\}$-extension L.*

Proof. We proceed by induction. Let 0 and 1 be the zero and unit element of L, respectively. If $|K| \leq 2$, then we can take $L = K$. So we can assume that $|K| > 2$. We distinguish two cases.

Case 1: 1 *is join-irreducible.* Let a be the unique dual atom of K. By the induction hypothesis, $\mathrm{id}(a)$ has a finite, simple, sectionally complemented $\{0, 1\}$-extension L'. We define $L = L' \cup \{1, u, v\}$, where u and v are common complements to all elements of $L' - \{0\}$ and are complementary, that is, $u \wedge x = 0$ and $u \vee x = 1$ for all $x \in (L' \cup \{v\}) - \{0\}$; and symmetrically for v. It is easy to see that L satisfies all the requirements.

Case 2: 1 *is join-reducible.* So there are elements $x_1, x_2 \in K - \{1\}$ satisfying $x_1 \vee x_2 = 1$. Let

$$N(K) = K - \{\, x \mid x = 0 \text{ or } x \text{ is an atom} \,\}.$$

For every $a \in N(K)$, we adjoin an atom $p_a < a$ so that if $a \neq b$, then $p_a \neq p_b$. We define on

$$\mathrm{Simp}\, K = K \cup \{\, p_a \mid a \in N(K) \,\}$$

(a disjoint union) an ordering by the following rules:

(α) K is a suborder of $\mathrm{Simp}\, K$;

(β)   if $a \in N(K)$ and $x \in K$, then

$$\begin{aligned} x < p_a &\quad\text{iff}\quad x = 0, \\ p_a < x &\quad\text{iff}\quad a \leq x; \end{aligned}$$

(γ)   if a and $b \in N(K)$, then

$$p_a \leq p_b \quad\text{iff}\quad a = b.$$

It is easy to check that $\mathrm{Simp}\, K$ is a lattice.

Obviously, $\mathrm{Simp}\, K$ is a finite lattice; 0 and 1 are the zero and unit elements of $\mathrm{Simp}\, K$, respectively. $\mathrm{Simp}\, K$ is an extension of K. To show that $\mathrm{Simp}\, K$ is sectionally complemented, take $0 < u < v$ in $\mathrm{Simp}\, K$.

If $u \in K$, then $v \in N(K)$ and p_v is a sectional complement of u in $[0, v]$.

If $u \notin K$, then $u = p_a$, for some $a \in N(K)$, and $v \in K$ satisfies $a \leq v$. If $a = v$, then any x satisfying $0 < x < a$ is a sectional complement of u in $[0, v]$; and there is such an x because $a \in N(K)$. If $a < v$, then p_v is a sectional complement of u in $[0, v]$.

Finally, L is simple. Indeed, let $\boldsymbol{\alpha} > \mathbf{0}$ be a congruence of K. We verify that there is an $a > 0$ in L, such that $a \equiv 0 \pmod{\boldsymbol{\alpha}}$. Indeed, since $\boldsymbol{\alpha} > \mathbf{0}$, there are $u, v \in K$ such that $u < v$ and $u \equiv v \pmod{\boldsymbol{\alpha}}$. If $u = 0$, then $a = v$ satisfies the requirements. If $0 < u$, then $v \in N(K)$, so there is an element $p_v \in L$.

If $u \in K$, then $u \equiv v \pmod{\alpha}$ implies that $p_v \equiv 0 \pmod{\alpha}$, so we can take $a = p_v$.

If $u \in \operatorname{Simp} K - K$, then $u = p_x$ for some $x \in N(K)$. Obviously, $v \in K$ and $x \leq v$. Since $x \in N(K)$, there is an $a \in K$ satisfying $0 < a < x$. Now $u \equiv v \pmod{\alpha}$ implies that $p_x \equiv x \pmod{\alpha}$ and so $a \equiv 0 \pmod{\alpha}$.

Using the congruence $a \equiv 0 \pmod{\alpha}$, we conclude that $p_1 \equiv 1 \pmod{\alpha}$. Meeting with x_1, we obtain that $x_1 \equiv 0 \pmod{\alpha}$, and similarly $x_2 \equiv 0 \pmod{\alpha}$. Joining these two congruences, we obtain that $0 \equiv 1 \pmod{\alpha}$, that is, $\alpha = 1$. □

We should point out that since a finite partition lattice is simple and sectionally complemented by O. Ore [167], Lemma 14.3 follows from the following very deep result of P. Pudlák and J. Tůma [169]: *Every finite lattice can be embedded into a finite partition lattice.* It can also be derived from a much earlier result of R. P. Dilworth (first published in P. Crawley and R. P. Dilworth [13]): *Every finite lattice can be embedded into a finite geometric lattice.*

14.4. Formal proof

Let K be a finite lattice. In this section we construct a finite, sectionally complemented, congruence-preserving extension L of K, as required by Theorem 14.1.

Using Lemma 14.3, for every $\gamma \in \operatorname{Con_M} K$, we select a finite, sectionally complemented, simple extension $K_\gamma = \operatorname{Simp} K/\gamma$ of K/γ, with an atom, s_γ, and zero, 0_γ.

Let $D = \operatorname{Con} K$, and let L_0 be a lattice whose existence was stated in Theorem 8.9 for the finite distributive lattice D.

For a congruence α of K, let α_0 denote the congruence of L_0 determined by the set $\{\, q_\gamma \mid \gamma \leq \alpha \,\}$, that is,

$$\alpha_0 = \bigvee (\operatorname{con}(q_\gamma, 0_{L_0}) \mid \gamma \leq \alpha);$$

obviously, $\alpha \mapsto \alpha_0$ sets up an isomorphism $\operatorname{Con} K \to \operatorname{Con} L_0$.

Now we inductively construct the lattice L of Theorem 14.1. Let $\gamma_1, \dots, \gamma_n$ list the meet-irreducible congruences of K. We apply the Atom Lemma, see Lemma 5.8, with $A = L_0$, $B = \operatorname{Simp} K/\gamma_1$, and the atoms $p_{\gamma_1^\dagger}$ of L_0 and s_{γ_1} of $\operatorname{Simp} K/\gamma_1$, to obtain the chopped lattice M_1 and $L_1 = \operatorname{Id} M_1$. From Corollary 5.7, it follows that L_1 is a congruence-preserving extension of L_0 with the same boolean ideal B_0. Moreover, L_1 is finite and sectionally complemented by the Atom Lemma. So we can apply again the Atom Lemma with $A = L_1$, $B = \operatorname{Simp} K/\gamma_2$, and the atoms $p_{\gamma_2^\dagger}$ of L_1 and s_{γ_2} of $\operatorname{Simp} K/\gamma_2$, obtaining the chopped lattice M_2 and $L_2 = \operatorname{Id} M_2$. In n steps, we construct M_n and $L = L_n = \operatorname{Id} M_n$.

It is clear that L is a finite sectionally complemented lattice. It is also evident that L is a congruence-preserving extension of L_0, hence, $\mathrm{Con}\, L \cong D$. So to complete the proof of Theorem 14.1, we have to show that K is congruence reflecting in L.

In the next step, we need the following statement:

Lemma 14.4. *Let A, B, C be pairwise disjoint finite lattices with zeros $0_A, 0_B$, 0_C, and atoms $s_A \in A$, $s_B, s'_B \in B$, $s_B \neq s'_B$, $s_C \in C$. We construct some chopped lattices by merging.*

(i) *Let N_1 be the chopped lattice obtained by forming the disjoint union of A and B and identifying 0_A with 0_B and s_A with s_B.*

(ii) *Let N be the chopped lattice obtained by forming the disjoint union of A, B, C and identifying 0_A with 0_B and 0_C, s_A with s_B, and s'_B with s_C.*

(iii) *Form the ideal lattice $\mathrm{Id}\, N_1$, which we consider an extension of N_1, so it has atoms $s_A = s_B$ and s'_B and let N_2 be the chopped lattice that is the disjoint union of $\mathrm{Id}\, N_1$ and C with the zeros identified and also the atom $s'_B \in \mathrm{Id}\, N_1$ is identified with the atom $s_C \in C$.*

Then $\mathrm{Id}\, N_2$ is isomorphic to $\mathrm{Id}\, N$.

Proof. The elements of $\mathrm{Id}\, N_1$ are compatible pairs $(a, b) \in A \times B$. The elements of $\mathrm{Id}\, N_2$ are compatible pairs $((a, b), c)$; note that $(a, b) \in \mathrm{Id}\, N_1$ iff $a \wedge s_A = b \wedge s_B$ and note also that $(a, b) \wedge s'_B = b \wedge s'_B$. On the other hand, the elements of $\mathrm{Id}\, N$ are compatible triples $(a, b, c) \in A \times B \times C$. It should now be obvious that $((a, b), c) \mapsto (a, b, c)$ is the required isomorphism $\mathrm{Id}\, N_2 \to \mathrm{Id}\, N$. $\qquad\square$

Another view of L is the following. We form the chopped lattice M_1; instead of proceeding to $L_1 = \mathrm{Id}\, M_1$, let us now form by merging the chopped lattice M'_2 from $M_1 = M'_1$ and $\mathrm{Simp}\, K/\gamma_2$ by identifying the zeros of M'_1 and $\mathrm{Simp}\, K/\gamma_2$ and the atom $p_{\gamma_2^\dagger}$ of M'_1 with the atom s_{γ_2} of $\mathrm{Simp}\, K/\gamma_2$. Observe that M'_2 is the union of L_0, $\mathrm{Simp}\, K/\gamma_1$ and $\mathrm{Simp}\, K/\gamma_2$; also, $\mathrm{Simp}\, K/\gamma_1 \cap \mathrm{Simp}\, K/\gamma_2 = \{0\}$. By Lemma 14.4, $\mathrm{Id}\, M'_2 \cong L_2$. Proceeding thus, we obtain the chopped lattice

$$M'_n = L_0 \cup \mathrm{Simp}\, K/\gamma_1 \cup \mathrm{Simp}\, K/\gamma_2 \cup \cdots \cup \mathrm{Simp}\, K/\gamma_n,$$

where, for all $1 \leq i < j \leq n$, we have $\mathrm{Simp}\, K/\gamma_i \cap \mathrm{Simp}\, K/\gamma_j = \{0\}$, and $L \cong \mathrm{Id}\, M'_n$, again by Lemma 14.4. So the chopped sublattice

$$\mathrm{Simp}\, K/\gamma_1 \cup \mathrm{Simp}\, K/\gamma_2 \cup \cdots \cup \mathrm{Simp}\, K/\gamma_n$$

of M_n' is just the lattices $\mathrm{Simp}\,K/\gamma_1$, $\mathrm{Simp}\,K/\gamma_2$, $\ldots$, $\mathrm{Simp}\,K/\gamma_n$ merged by their zeros, and, therefore,

$$\mathrm{Id}(\mathrm{Simp}\,K/\gamma_1 \cup \mathrm{Simp}\,K/\gamma_2 \cup \cdots \cup \mathrm{Simp}\,K/\gamma_n) \cong \mathrm{Cube}\,K.$$

Thus we can view L as the ideal lattice of the chopped lattice obtained by gluing together L_0 and $\mathrm{Cube}\,K$ over the boolean lattices B_0 in L_0 and B in $\mathrm{Cube}\,K$. A congruence α of L is then determined by a congruence α_{L_0} on L_0 and a congruence $\alpha_{\mathrm{Cube}\,K}$ on $\mathrm{Cube}\,K$ that agree on the boolean lattice. Since $\mathrm{Cube}\,K$ is a congruence-preserving extension of $K \cong \mathrm{Diag}(K)$, it follows now that L is a congruence-preserving extension of K, concluding the proof of Theorem 14.1.

14.5.   Discussion

Problem 14.1. By Lemma 14.3, every finite lattice K has a finite, simple, sectionally complemented extension L. What is the minimum size of such an L?

Problem 14.2. What is the size of the lattice L we construct for Theorem 14.1? What is the minimum size of a lattice L satisfying Theorem 14.1?

Can we do better than Theorem 14.1?

Problem 14.3. Is there a natural subclass **S** of **SecComp** for which Theorem 14.1 holds (that is, every finite lattice K has a finite, sectionally complemented, congruence-preserving extension $L \in \mathbf{S}$)?

For infinite lattices, almost nothing is known, as illustrated by the following question:

Problem 14.4. Does every sectionally complemented lattice have a proper, sectionally complemented, congruence-preserving extension?

Semimodular Lattices

15.1. The extension theorem

In Part III first we considered finite sectionally complemented lattices, then finite semimodular lattices. We do the same in this part.

In G. Grätzer and E. T. Schmidt [132], we follow the same path as in Chapter 14. First, we have to find a semimodular cubic extension; luckily, P. Crawley and R. P. Dilworth [13] comes to the rescue. Then we utilize the representation theorem for semimodular lattices, Theorem 10.1, to order the congruences. These two lattices have to be connected since there is no obvious way of amalgamating the two; we do this with the "conduit" construction. Thus we obtain the result of [132]:

Theorem 15.1. *Every finite lattice K has a congruence-preserving embedding into a finite semimodular lattice L.*

15.2. *Proof-by-Picture*

Let us start with a finite lattice K of Figure 15.1 (the dual of N_6, introduced in Section 7.1; see Figure 7.1), with $\operatorname{Con} K = \{0, \gamma, 1\}$, the three-element chain. The congruence γ splits K into two classes, marked in Figure 15.1 with the dashed line. First, we define a semimodular extension $\operatorname{Cube} K$. To do this, we have to specify the function Simp. Since $\operatorname{Con}_M K = \{\gamma, 0\}$, we have to define Simp $\mathbf{K}/\gamma$ and Simp $\mathbf{K}/\mathbf{0}$. The subdirectly irreducible quotient, $\mathbf{K}/\gamma$, is isomorphic to C_2, a simple semimodular lattice, so we can define

© Springer International Publishing Switzerland 2016

G. Grätzer, *The Congruences of a Finite Lattice*,

DOI 10.1007/978-3-319-38798-7_15

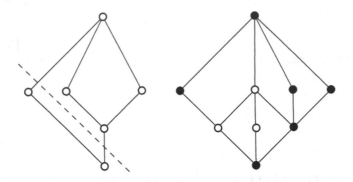

Figure 15.1: The lattice K and a simple semimodular extension $\mathrm{Simp}\,K$.

$\mathrm{Simp}\,\mathsf{C}_2 = \mathsf{C}_2$. We embed the lattice $K/\mathbf{0} \cong K$ into a simple semimodular lattice $\mathrm{Simp}\,K$ as in Figure 15.1. Therefore, the semimodular cubic extension, $\mathrm{Cube}\,K$, is defined as $\mathsf{C}_2 \times \mathrm{Simp}\,K$. We thereby obtain a semimodular congruence-reflecting extension $\mathrm{Cube}\,K$; see Figure 15.2. This figure also shows a "sketch" of the diagram, which we use in Figure 15.5 to visualize the gluings.

The black-filled elements of $\mathrm{Cube}\,K$ in Figure 15.2 form the lattice K, identified with $\mathrm{Diag}(K)$. To see how the congruences of K extend to $\mathrm{Cube}\,K$, we only have to describe the extension $\mathrm{Cube}(\gamma)$ of γ, which splits $\mathrm{Cube}\,K$ into two classes as shown by the dashed line in Figure 15.2. The shaded area in Figure 15.2 is (one option for) the filter F (of Corollary 7.3), which is a congruence-preserving sublattice of $\mathrm{Cube}\,K$.

We color the filter F with two colors, p_1 and p_2. In $\mathrm{Cube}\,K$, we have $\mathrm{con}(p_1) = \mathrm{Cube}(\gamma)$. Also, $\mathrm{Con}(\mathrm{Cube}\,K) = \{\mathbf{0}, \mathrm{con}(p_1), \mathrm{con}(p_2), \mathbf{1}\}$.

We now have to construct a congruence-determining extension of $\mathrm{Cube}\,K$ in which $\mathrm{con}(p_1) = \mathbf{1}$, so that $\mathrm{con}(p_1) > \mathrm{con}(p_2)$. We borrow from Section 10.2 the semimodular lattice S; see Figure 15.3. We utilize the special circumstances to make S a little smaller. The lattice S contains an ideal N that is a chain of length 2; the two prime intervals colored by p_1 and p_2. The ideal N is a congruence-determining sublattice of S. This figure also shows a sketch of the diagram. Note that in S, we have $\mathrm{con}(p_1) > \mathrm{con}(p_2)$.

We have to patch $\mathrm{Cube}\,K$ and S together, respecting the colors. We utilize for this the modular lattice M—which we call the 'conduit'—shown in Figure 15.4 with a sketch. The lattice M has an ideal B, which is a congruence-preserving sublattice isomorphic to F and a filter J, which is a congruence-preserving sublattice isomorphic to N. We color M as shown in Figure 15.4, so that there are color preserving isomorphisms $\gamma\colon B \to F$ and $\delta\colon J \to N$.

We glue $\mathrm{Cube}\,K$ and the conduit over the filter F and ideal B with respect to γ; the resulting lattice T has filter J. We glue T and S over the filter J and ideal N with respect to δ to obtain the lattice L. The two gluings are sketched in Figure 15.5.

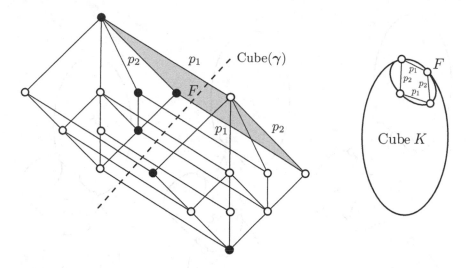

Figure 15.2: The semimodular cubic extension Cube K with sketch.

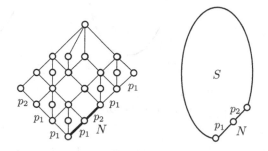

Figure 15.3: The lattice S with sketch.

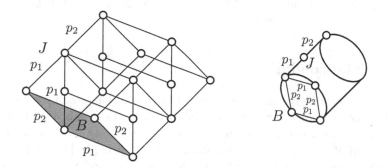

Figure 15.4: The conduit with sketch.

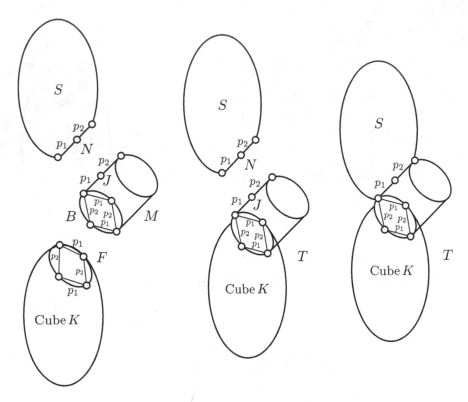

Figure 15.5: Two gluings: before and after; sketches.

The semimodularity of L follows by two applications of Lemma 2.24.

This semimodular congruence-preserving extension L of the six-element lattice K has 57 elements; compare this with the size of the lattice from Section 10.2: S has 24 elements.

15.3. The conduit

In this section we give the general construction of the conduit, a modular lattice, which we constructed in a special case in Section 15.2; see Figure 15.4.

Let $n \geq 1$ be a natural number, and for every i, with $1 \leq i \leq n$, we take a copy T_i of M_3, with atoms p_i, q_i, and r_i. We form $\mathsf{M}_3^n = \prod(\, T_i \mid 1 \leq i \leq n\,)$ with zero 0, and regard T_i as an ideal of M_3^n, so p_i, q_i, and r_i are atoms of M_3^n for $1 \leq i \leq n$.

Let B be the sublattice of M_3^n generated by $\{\, p_i \mid 1 \leq i \leq n\,\}$; let $1_B = \bigvee(\, p_i \mid 1 \leq i \leq n\,)$. Obviously, B is a 2^n-element boolean lattice, an ideal of M_3^n.

Define

$$\bar{q}_i = \bigvee(q_j \mid 1 \le j \le i),$$

for $1 \le i \le n$, and set $E = \{\bar{q}_i \mid 0 \le i \le n\}$, where $\bar{q}_0$ is the zero of M_3^n. Obviously, E is a maximal chain (of length n) in $\mathrm{id}(\bar{q}_n)$ of M_3^n and $b \wedge e = 0$ for all $b \in B$ and $e \in E$.

By Theorem 2.11, the sublattice A of M_3^n generated by B and E is isomorphic to $B \times E$ under the isomorphism $b \vee e \mapsto (b, e)$.

Let $b \in B$ and let i, with $0 \le i < n$, satisfy the condition: $p_{i+1} \not\le b$. Define the element of M_3^n:

$$r(b, i) = b \vee \bar{q}_i \vee r_{i+1},$$

and the subset M of M_3^n:

$$M = A \cup \{r(b, i) \mid b \in B, \ 0 \le i < n, \ p_{i+1} \not\le b\}.$$

M is a sublattice of M_3^n, hence M is a modular lattice. The lattice M contains B and E as ideals.

Let α be a congruence of B. Let α^E be the congruence on E satisfying $\bar{q}_i \equiv \bar{q}_{i+1} \pmod{\alpha^E}$ in E iff $p_{i+1} \equiv 0 \pmod{\alpha}$ in B for $0 \le i < n$. Then $\alpha \times \alpha^E$ is a congruence on $B \times E$. We extend this to a congruence α^M of M as follows: let $r(b, i)$ be defined (that is, $b \in B$ and $p_{i+1} \not\le b$); if $b \equiv b \vee p_{i+1} \pmod{\alpha}$ in B, then $r(b, i) \in (b \vee \bar{q}_i)/\alpha^M$, otherwise, $\{r(b, i)\}$ is a congruence class.

The map $\alpha \mapsto \alpha^M$ is an isomorphism between $\mathrm{Con}\,B$ and $\mathrm{Con}\,M$. In fact, M is a congruence-preserving extension of both B and E. Let

$$J = \{1_B \vee e \mid e \in E\}.$$

Obviously, E and J are isomorphic chains and J is a filter of M, a congruence-preserving sublattice of M.

To summarize:

Lemma 15.2. *For each $n \ge 1$, M_3^n has sublattices B, J, and M satisfying the following conditions:*

(i) *B is an ideal of M and it is isomorphic to the boolean lattice B_n.*

(ii) *J is a filter of M and length $J = n$.*

(iii) *M is a congruence-preserving extension of both B and J.*

Note that $\mathrm{Con}\,M$ is a boolean lattice and $\mathrm{Con}\,M \cong \mathrm{Con}\,B \cong \mathrm{Con}\,J$.

15.4. The construction

We are given a finite lattice K. To prove the Theorem 15.1, we have to construct a finite, semimodular, congruence-preserving extension L of K.

We glue L together from three lattices, sketched in Figure 15.5.

The first lattice is S of Section 10.3 (denoted there by L)—see Figure 10.4—with the ideal N, defined at the end of Section 10.3, which is a chain consisting of the gray-filled elements, a congruence-determining sublattice of S.

Let $D = \operatorname{Con} K$; in Section 10.3, we construct a finite semimodular lattice S with $D \cong \operatorname{Con} S$. The construction of S in Section 10.3 starts out by representing D as $\operatorname{Down} P$ for $P = \mathrm{J}(D)$. By Theorem 2.19, the ordered sets $\mathrm{J}(D)$ and $\mathrm{M}(D)$ are isomorphic, so we can start the construction of S with $P = \mathrm{M}(D)$. Accordingly, we color N with $P = \mathrm{M}(D)$; there is a one-to-one correspondence between the prime intervals of N and the elements of P, the meet-irreducible congruences of K.

The second lattice is the conduit lattice M of Section 15.3, with the ideal B (which is boolean) and the filter J (which is a chain), constructed so that J and N are isomorphic. This (unique) isomorphism allows us to color J so that the isomorphism preserves colors. The coloring of J extends to a coloring of all of M, in particular, to a coloring of B. Both B and J are congruence-preserving sublattices of M.

The third lattice is a cubic extension, so we have to define the function Simp. For a finite lattice A, let $\operatorname{Simp} A$ be a finite partition lattice extending A, a semimodular lattice by Lemma 2.25. Such an extension exists by P. Pudlák and J. Tůma [169]—alternatively, use the more accessible result of P. Crawley and R. P. Dilworth [13] (Theorem 14.1) to obtain a simple semimodular extension. In the cubic extension $\operatorname{Cube} K$, we have the filter F generated by dual atom $t_\gamma \in \operatorname{Simp} K_\gamma$, for $\gamma \in \operatorname{Con}_{\mathrm{M}} K = P$, by Corollary 7.3. We color $[t_\gamma, 1_F]$ by γ, for $\gamma \in \operatorname{Con}_{\mathrm{M}} K = P$, and thereby color all of F.

Observe that B and F are isomorphic finite boolean lattices, both are colored by P, so there is an isomorphism (unique!) α between B and F that preserves colors.

First gluing. We glue M and S together over J and N to obtain the lattice T.

The lattice T is also colored by P and T has an ideal B, a congruence-determining sublattice.

The lattice T is obtained by gluing together two semimodular lattices, hence it is semimodular by Lemma 2.24.

Second gluing. We have the color preserving isomorphism between the boolean ideal B of the lattice T of the first step and the filter F in $\operatorname{Cube} K$; we do the second gluing of T and B with $\operatorname{Cube} K$ and F with respect to this isomorphism α, to obtain the lattice L. The lattice L is obtained by gluing together two semimodular lattices, hence it is semimodular by Lemma 2.24.

15.5.   Formal proof

To prove Theorem 15.1, we have to verify that L is a finite, semimodular, congruence-preserving extension of K. We already know that L is finite and semimodular.

We view K as a sublattice of Cube K (as in Section 7.1), so K is a sublattice of L. Let κ be a congruence of K. We have to show that κ has one and only one extension to L.

Let $\gamma \in \mathrm{Con}_M K$ be the color of the prime interval $\mathfrak{p}$ of L. Let $\mathfrak{q}_\gamma$ be the unique prime interval in N of color γ.

We claim that $\mathrm{con}(\mathfrak{p}) = \mathrm{con}(\mathfrak{q}_\gamma)$. To prove this claim, let $\mathfrak{p}$ be a prime interval in L. Then $\mathfrak{p}$ is a prime interval in S, or M, or Cube K. So we have three cases to consider.

First, if $\mathfrak{p}$ is a prime interval in S, this follows from the fact that N is a congruence-determining sublattice of S; see Lemma 3.21 and the last paragraph of Section 10.3.

Second, if $\mathfrak{p}$ is a prime interval in M, this follows from the fact that J is a congruence-preserving sublattice of M, by Lemma 15.2.

Third, if $\mathfrak{p}$ is a prime interval in Cube K, then there is a prime interval $\mathfrak{p}_1$ in F with $\mathrm{con}(\mathfrak{p}) = \mathrm{con}(\mathfrak{p}_1)$, since F is a congruence-determining sublattice of Cube K. Because of the gluing, the prime interval $\mathfrak{p}_1$ can be regarded as a prime interval of M, and now the claim follows from the second case. This completes the proof of the claim.

It follows from this claim that a congruence α of L can be described by the set Σ of prime intervals of N collapsed by α, or equivalently, by the set of colors of the prime intervals: $C = \{\, \gamma \mid p_\gamma \in \Sigma \,\}$. Now by the construction of S (see Section 10.3), the set C is a down set of $P = \mathrm{Con}_M K$. And conversely, if C is a down set of P, then it defines a congruence α_C on S, as computed in Section 10.3. This congruence α_C spreads uniquely to M because J is a congruence-preserving sublattice of M. Finally, from the description of Cube $\mathbf{0}$ in Section 7.2, we see that the congruence uniquely spreads to Cube K. So there is a one-to-one correspondence between congruences of L and down sets of P.

For $\kappa \in \mathrm{Con} K$, its extension Cube$(\kappa)$ to Cube K corresponds to the down set $\{\, \gamma \mid \kappa \nleq \gamma \,\}$. So every congruence of K extends. But since there are no more congruences than those that correspond to down sets of P, the extension must be unique. This completes the proof of Theorem 15.1.

15.6.   Discussion

Using P. Crawley and R. P. Dilworth [13], we get a very large simple semimodular lattice extending the lattice K. Of course, from P. Pudlák and J. Tůma [169], we get an even larger lattice. So it is natural to ask:

Problem 15.1. Let $f(n)$ be the smallest integer with the property that every lattice K of n elements can be embedded into a simple semimodular lattice of size $f(n)$. Give an asymptotic estimate of $f(n)$.

Problem 15.2. What is the size of the lattice L we construct for Theorem 15.1 expressed in terms of $f(n)$ of Problem 15.1? What is the minimum size of a lattice L satisfying Theorem 15.1?

Can we do better than Theorem 15.1?

Problem 15.3. Is there a natural subclass **S** of the class of semimodular lattices for which Theorem 15.1 holds (that is, every finite lattice K has a finite, semimodular, congruence-preserving extension $L \in$ **S**)?

How about planar lattices?

Problem 15.4. Does every planar lattice K have a congruence-preserving embedding into a finite, planar, semimodular lattice L?

Isoform Lattices

16.1. The result

In Chapter 13, we proved the representation theorem for finite uniform lattices. In Section 13.5, we discussed the representation theorem for finite isoform lattices, but we did not prove it, for a reason I am about to explain.

In G. Grätzer, R. W. Quackenbush, and E. T. Schmidt [113], we proved the following:

Theorem 16.1. *Every finite lattice K has a congruence-preserving extension to a finite isoform lattice L.*

This problem was raised for uniform lattices as Problem 1 in G. Grätzer, E. T. Schmidt, and K. Thomsen [141] and for isoform lattices as Problem 2 in G. Grätzer and E. T. Schmidt [136]. (See also Problem 9 in G. Grätzer and E. T. Schmidt [138].)

The proof of this theorem breaks the earlier pattern: we start constructing a congruence-preserving extension by a cubic extension and then we utilize the representation theorem. For isoform lattices, we again start with a cubic extension, but we do not know how to proceed utilizing the representation theorem—this is why it was not important to have the representation theorem proved in Chapter 13. Instead, we use "pruning." Special cases of pruning are the A^+ construction of Section 13.2, the $N(A, B)$ construction of Section 13.3, and the $N(A, B, \alpha)$ construction of Section 13.5.

© Springer International Publishing Switzerland 2016
G. Grätzer, *The Congruences of a Finite Lattice*,
DOI 10.1007/978-3-319-38798-7_16

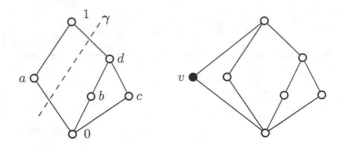

Figure 16.1: The lattices K and S_0.

16.2.   *Proof-by-Picture*

In this section we construct the lattice L of Theorem 16.1 for the lattice $K = N_6$ of Figure 16.1.

As always, when we construct a congruence-preserving extension, we start with a cubic extension. To construct the cubic extension of K, we have to define the function Simp on the subdirectly irreducible quotients. The lattice K has only one nontrivial congruence γ, splitting K into two classes, indicated by the dashed line; it has two meet-irreducible congruences, γ and $\mathbf{0}$. We use the notation: Simp $K/\mathbf{0} = S_0$ and Simp $K/\gamma = S_\gamma$. The choice for S_0 is easy; add a single element v as in Figure 16.1 (the black-filled element). The lattice K/γ is simple, so we can choose $S_\gamma = C_2$.

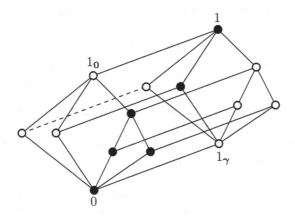

Figure 16.2: The cubic extension S of K.

Figure 16.2 shows the cubic extension $S = S_0 \times S_\gamma$ of K; the black-filled elements represent K. On the diagram, we labeled four elements: the 0 and 1 of S, and the unit elements of the two direct factors: 1_0 and 1_γ.

The congruences of S are in one-to-one correspondence with subsets of $\{1_0, 1_\gamma\}$, and Cube γ (the extension of γ to S) corresponds to $\{1_0\}$, that is, Cube $\gamma = \mathrm{con}(0, 1_0)$. So we have to "kill" the congruence $\mathrm{con}(0, 1_\gamma)$; in other words, we have to make sure that $1_\gamma \equiv 0$ implies that $1_0 \equiv 0$.

In Chapters 14 and 15, such a congruence-forcing was accomplished by a suitable extension of S utilizing the appropriate representation theorems. The inspiration for the isoform result comes from the representation theorem: the use of pruning, although the representation theorem itself is not needed.

We construct L from S by pruning (deleting) a single edge, the dashed edge of Figure 16.2; the resulting ordered set L is depicted in Figure 16.3.

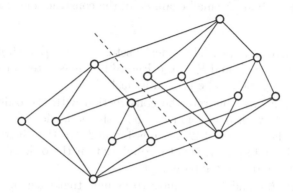

Figure 16.3: The lattice L.

It has to be checked that L is a lattice (pruning an edge may not define a lattice). Then we have to see that $1_\gamma \equiv 0$ implies that $1_0 \equiv 0$ in L, so L has only one nontrivial congruence and it splits L into two classes as indicated by the dashed line. The two classes are both isomorphic to S_0, so L is isoform.

Computing this example in detail, one comes to the conclusion that everything works out because the element v in S_0 is so special. We axiomatize the important property of v as follows. For a finite lattice A, let us call an element v a *separator* if $0 \prec v \prec 1$. A lattice A with a separator is called *separable*; we also allow C_2 as a separable lattice.

The general construction starts with a finite ordered set P (which plays the role of $\mathrm{Con}_M K$), and a family S_p, for $p \in P$, of separable lattices. We form the set $S = \prod(S_p \mid p \in P)$. In each $S_p \neq C_2$, with $p \in P$, we fix a separator v_p.

Let us describe the edges we delete from S. Let

$$\mathbf{a} = (a_p)_{p \in P} \prec \mathbf{b} = (b_p)_{p \in P}$$

in S. Then there is exactly one $q \in P$ with $a_q \prec b_q$ in S_q, and $a_p = b_p$ for all $p \neq q$. *We delete the edge* $\mathbf{a}, \mathbf{b}$ *if there is a* $p < q$ *such that* S_p *has a separating element* v_p *and* $v_p = a_p = b_p$. The pruned diagram defines the ordered set L.

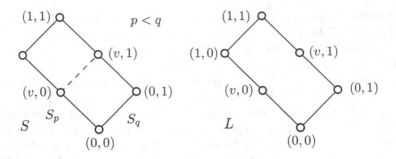

Figure 16.4: A small example of the construction of L.

Here is a small example. The ordered set is $P = \{p, q\}$ with $p < q$ and the two lattices are $S_p = C_3$ and $S_q = C_2$. Figure 16.4 illustrates the construction; the edge of $S = C_3 \times C_2$ to be pruned is dashed.

In Figure 16.5, we illustrate the construction with the ordered set $P = \{p, q, r\}$ (the hat) with $p < q$ and $r < q$, with the lattices $S_p = S_q = S_r = C_3 = \{0, v, 1\}$. Four edges of C_3^3 are missing in L; on the diagram, these are marked with dashed lines. It is not so obvious that the ordered set L defined by the pruned diagram is a lattice.

In Sections 16.4 and 16.5 we have to assume that each S_p is simple and $S_p \neq C_2$. So even the smallest appropriate example is fairly large. To con-

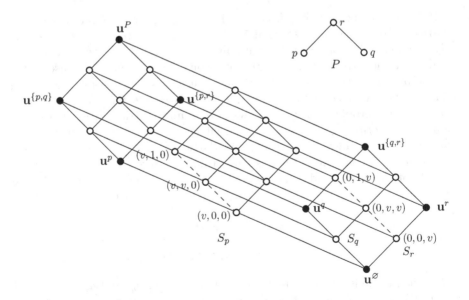

Figure 16.5: A nontrivial example of the construction.

struct the smallest example, let $P = \{p, q\}$ with $p < q$ and $S_p = S_q = \mathsf{M}_3 = \{0, a, b, v, 1\}$. Figure 16.6 illustrates what we get; note that the diagram is turned to its side. In this diagram, S_q is on the right, its elements are labeled $0, a, b, v, 1$; its unit element is $\mathbf{u}^q = (0, 1)$. The lattice S_p is on the left, its elements are all labeled with 0; its unit element is $\mathbf{u}^p = (1, 0)$. Five edges of M_3^2 are missing in L; on the diagram these are marked with dashed lines. It is easy to see that the ordered set L defined by the pruned diagram is a lattice.

This is how far we can get with pictures. We can illustrate the basic construction, but have no *Proof-by-Picture* that L is a lattice. This would have to be computed; we have no special insight why the computation works. It is even worse to verify the isoform property...

16.3.   Formal construction

In this section for a finite ordered set P, and a family S_p, for $p \in P$, of separable lattices, we form the set $S = \prod(S_p \mid p \in P)$. In S, $\vee_S, \wedge_S, \leq_S, \prec_S$ denotes the join, meet, ordering relation, and covering relation, respectively. If there is no ambiguity, subscripts will be dropped.

The elements of S will be boldface lower case letters. An element $\mathbf{s} \in S$ is of the form $(s_p)_{p \in P}$, and we write $\mathbf{s}_p$ for s_p. For $q \in P$, define the elements $\mathbf{u}^q \in S$ and $\mathbf{v}^q \in S$ as follows:

$$(\mathbf{u}^q)_p = \begin{cases} 1 & \text{if } p = q; \\ 0 & \text{if } p \neq q; \end{cases}$$

$$(\mathbf{v}^q)_p = \begin{cases} v_p & \text{if } p = q; \\ 0 & \text{if } p \neq q. \end{cases}$$

Let B be the sublattice of S generated by $\{\mathbf{u}^p \mid p \in P\}$. We call B the *skeleton* of S; it is a boolean sublattice of S with n atoms. For a subset Q of P, set $\mathbf{u}^Q = \bigvee_S \{\mathbf{u}^p \mid p \in Q\}$ with complement $(\mathbf{u}^Q)' = \mathbf{u}^{P-Q}$ in B (and in S). The elements of B are black-filled in Figure 16.5.

Now we come to the crucial definition of this chapter. We define a binary relation $\leq$ on the set S. For the elements $\mathbf{a} = (a_p)_{p \in P}$ and $\mathbf{b} = (b_p)_{p \in P}$ of S:

(P) $\mathbf{a} \leq \mathbf{b}$ iff $\mathbf{a} \leq_S \mathbf{b}$ and

$$p < p' \text{ in } P, \ S_p \neq \mathsf{C}_2, \text{ and } a_p = v_p = b_p \quad \text{imply that} \quad a_{p'} = b_{p'}.$$

In this section and in the next, when we write "$p \in P$, and $a_p = v_p = b_p$", we will mean implicitly that $S_p \neq \mathsf{C}_2$ (so that v_p be defined). In Sections 16.4 and 16.5 we assume that each S_p is simple and $S_p \neq \mathsf{C}_2$, so this implicit assumption always holds for all $p \in P$.

Let L be the relational system $(S, \leq)$. In Theorem 16.2, we show that L is an ordered set and in Theorem 16.4, we verify that L is a lattice.

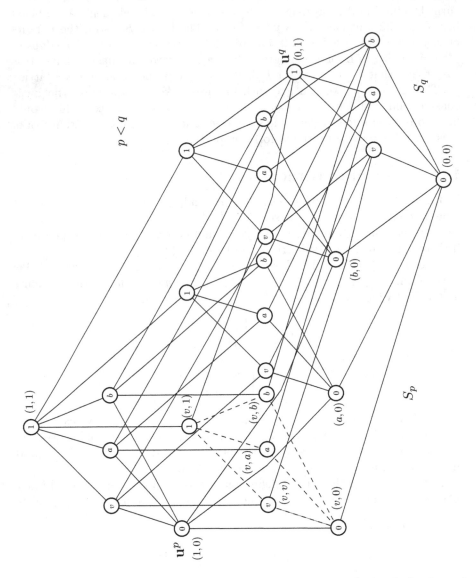

Figure 16.6: The smallest example of the construction with simple lattices of more than two elements—sideways.

Note that if P is an antichain, then $\leq$ is the same as $\leq_S$, so $L = S$ as a relational system. Figure 16.4 illustrates the smallest example where S and L differ.

Now we proceed to prove that L is a lattice and verify some properties of the covering relation in L.

Theorem 16.2. *L is an order.*

Proof. The relation $\leq$ is obviously reflexive. By the definition of $\leq$:

If $\mathbf{a} \leq \mathbf{b}$, then $\mathbf{a} \leq_S \mathbf{b}$.

So the anti-symmetry of $\leq_S$ implies the anti-symmetry of $\leq$.

To prove the transitivity of $\leq$, let $\mathbf{a}, \mathbf{b}, \mathbf{c} \in L$, and let $\mathbf{a} \leq \mathbf{b} \leq \mathbf{c}$. Then $\mathbf{a} \leq_S \mathbf{b} \leq_S \mathbf{c}$, so by the transitivity of $\leq_S$, we get that $\mathbf{a} \leq_S \mathbf{c}$. To verify (P) for $\mathbf{a}$ and $\mathbf{c}$, let $p < p'$ in P, and assume that $a_p = v = c_p$ (recall our implicit assumption: $S_p \neq C_2$). Since $\mathbf{a} \leq_S \mathbf{b} \leq_S \mathbf{c}$, it follows that $a_p \leq b_p \leq c_p$, and we conclude that $a_p = b_p = c_p = v$. Since $\mathbf{a} \leq \mathbf{b}$ and $\mathbf{b} \leq \mathbf{c}$, applying (P) twice, we obtain the equalities $a_{p'} = b_{p'}$, and $b_{p'} = c_{p'}$. It follows that $a_{p'} = c_{p'}$, verifying (P) for $\mathbf{a}$ and $\mathbf{c}$, hence proving that $\mathbf{a} \leq \mathbf{c}$. Therefore, $\leq$ is transitive. $\square$

We will need the following properties of the ordered set L.

Lemma 16.3.

(1) *If $\mathbf{a} \prec \mathbf{b}$ in L, then $\mathbf{a} \prec_S \mathbf{b}$ in S.*

(2) *If $\mathbf{a} \prec_S \mathbf{b}$ in S, then $\mathbf{a} \prec \mathbf{b}$ in L unless there are $q < q'$ in P with $a_q = v = b_q$ and $a_{q'} \prec b_{q'}$, in which case, $\mathbf{a} \parallel \mathbf{b}$ in L.*

Proof of (1). We start by observing that $\mathbf{a} \prec_S \mathbf{b}$ iff there is a unique $q \in P$ such that $a_q \prec b_q$ in S_q, and $a_r = b_r$, for all $r \in P$, otherwise.

Let us assume that $\mathbf{a} \prec \mathbf{b}$ in L. Then $\mathbf{a} <_S \mathbf{b}$, by (P). Thus there is $q \in P$ such that $a_q < b_q$.

We claim that $a_q \prec b_q$ in S_q.

Indeed, if $a_q < c < b_q$, for some element $c \in S_q$, then define $\mathbf{c} = (c_p)_{p \in P}$ as follows:

$$c_p = \begin{cases} c, & \text{if } p = q; \\ v, & \text{if } a_p = v = b_p; \\ x, & \text{with } x \in \{a_p, b_p\} - \{v\}, \text{ otherwise.} \end{cases}$$

Obviously, $\mathbf{a} <_S \mathbf{c} <_S \mathbf{b}$.

We now verify that $\mathbf{a} < \mathbf{c}$. Let $p < p'$ in P and let $a_p = c_p = v$. Notice that $a_p = c_p = b_p = v$. Since $\mathbf{a} < \mathbf{b}$, by (P), it follows that $a_{p'} = b_{p'}$. But then $a_{p'} = c_{p'}$, verifying $\mathbf{a} < \mathbf{c}$ by (P).

We can verify that $\mathbf{c} < \mathbf{b}$, similarly.

Therefore, $\mathbf{a} < \mathbf{c} < \mathbf{b}$, contradicting the assumption that $\mathbf{a} \prec \mathbf{b}$. This proves the claim. Thus we may assume that $a_q \prec b_q$.

We now claim that there is only one q with $a_q < b_q$.

Indeed, let us assume that there are $q \neq r$ in P with $a_q < b_q$ and $a_r < b_r$. We assume that q is minimal in P with the property: $a_q < b_q$. There are two cases to consider.

Case 1: $b_q = v$.

Since v is a separator, it follows that $a_q = 0$. Define $\mathbf{c} = (c_p)_{p \in P}$ by

$$c_p = \begin{cases} a_p & \text{for } p = q; \\ b_p & \text{for } p \neq q. \end{cases}$$

Obviously, $\mathbf{a} <_S \mathbf{c} <_S \mathbf{b}$. Since $a_q = 0$ and $c_p = b_p$, for $p \neq q$, therefore, $c_p = v$ iff $b_p = v$. We now verify that $\mathbf{a} < \mathbf{c}$. Let $p < p'$ in P and let $a_p = c_p = v$. We cannot have $p = q$, because $a_q = 0$. Thus $p \neq q$, and so $a_p = b_p = c_p = v$. Now (P) and $\mathbf{a} < \mathbf{b}$ imply that $a_{p'} = b_{p'}$. For any p', either $a_{p'}$ or $b_{p'}$ equals $c_{p'}$. Therefore, $a_{p'} = b_{p'} = c_{p'}$, concluding that $\mathbf{a} < \mathbf{c}$.

We next verify that $\mathbf{c} < \mathbf{b}$. Let $p < p'$ in P and let $c_p = b_p = v$ and assume to the contrary that $c_{p'} < b_{p'}$. By the definition of $\mathbf{c}$, this is possible only if $p' = q$, in which case the minimality of q forces $a_p = b_p = v$, contradicting $\mathbf{a} < \mathbf{b}$. Thus $\mathbf{c} < \mathbf{b}$.

We conclude that $\mathbf{a} < \mathbf{c} < \mathbf{b}$, contradicting that $\mathbf{a} \prec \mathbf{b}$.

Case 2: $b_q \neq v$.

Define $\mathbf{c} = (b_p)_{p \in P}$ by

$$c_p = \begin{cases} b_p & \text{for } p = q; \\ a_p & \text{for } p \neq q. \end{cases}$$

Obviously, $\mathbf{a} <_S \mathbf{c} <_S \mathbf{b}$. We now verify that $\mathbf{a} < \mathbf{c}$. Let $p < p'$ in P and let $a_p = c_p = v$. Assume, to the contrary, that $a_{p'} \neq c_{p'}$. By the definition of $\mathbf{c}$, then $p' = q$. Since q is minimal with respect to $a_q < b_q$, it follows from $p < p' = q$ that $a_p = b_p$. So $a_p = b_p = v$ and $a_{p'} < b_{p'}$, which by (P) contradicts the assumption $\mathbf{a} < \mathbf{b}$. This proves that $\mathbf{a} < \mathbf{c}$.

We next verify that $\mathbf{c} < \mathbf{b}$. Let $p < p'$ in P and let $c_p = b_p = v$. By the assumption of Case 2, $b_q \neq v$, and so $p \neq q$. We conclude that $c_p = b_p = a_p = v$. Since $\mathbf{a} < \mathbf{b}$, by (P), we obtain that $a_{p'} = b_{p'}$. Therefore, $p' \neq q$. By the definition of $\mathbf{c}$, we have that $a_{p'} = c_{p'}$, so $a_{p'} = b_{p'} = c_{p'}$, verifying $\mathbf{c} < \mathbf{b}$.

We conclude that $\mathbf{a} < \mathbf{c} < \mathbf{b}$, contradicting that $\mathbf{a} \prec \mathbf{b}$.

This completes the proof of (1). $\qquad\Box$

Proof of (2). Now assume that $\mathbf{a} \prec_S \mathbf{b}$. Then there is a unique $q \in P$ with $a_q \prec b_q$, and otherwise, $a_r = b_r$ for $r \in P$. Consequently, if $\mathbf{a} < \mathbf{b}$, then $\mathbf{a} \prec \mathbf{b}$ in L. If $\mathbf{a} < \mathbf{b}$ fails, then by (P), there are elements $p < p'$ in P with $a_p = v = b_p$ and $a_{p'} < b_{p'}$, implying that $\mathbf{a} \parallel \mathbf{b}$ in L. $\qquad\Box$

Now we prove that we have constructed a lattice and describe the lattice operations. To facilitate this, we introduce the following terminology: Let $\mathbf{a} = (a_p)_{p \in P}, \mathbf{b} = (b_p)_{p \in P} \in L$, and let $q \in P$; we will call the element q an $\{\mathbf{a}, \mathbf{b}\}$-*fork* if $a_q = b_q = v$ and $a_{q'} \neq b_{q'}$ for some $q' > q$.

Theorem 16.4. *L is a lattice. Let* $\mathbf{a} = (a_p)_{p \in P}, \mathbf{b} = (b_p)_{p \in P} \in L$. *Then*

$$(\mathbf{a} \vee \mathbf{b})_p = \begin{cases} 1 & \text{if } a_p \vee b_p = v \text{ and} \\ & (1_\vee) \ p' \text{ is an } \{\mathbf{a}, \mathbf{b}\}\text{-fork, or} \\ & (2_\vee) \ b_p \leq a_p \text{ and } b_{p'} \nleq a_{p'}, \text{ or} \\ & (3_\vee) \ a_p \leq b_p \text{ and } a_{p'} \nleq b_{p'} \text{ for some } p' \geq p; \\ a_p \vee b_p, & \text{otherwise}; \end{cases}$$

and

$$(\mathbf{a} \wedge \mathbf{b})_p = \begin{cases} 0 & \text{if } a_p \wedge b_p = v \text{ and} \\ & (1_\wedge) \ p' \text{ is an } \{\mathbf{a}, \mathbf{b}\}\text{-fork, or} \\ & (2_\wedge) \ b_p \geq a_p \text{ and } b_{p'} \ngeq a_{p'}, \text{ or} \\ & (3_\wedge) \ a_p \geq b_p \text{ and } a_{p'} \ngeq b_{p'} \text{ for some } p' \geq p; \\ a_p \wedge b_p & \text{otherwise}. \end{cases}$$

Proof. Recall that v is separable, and therefore join-irreducible, hence $a_p \vee b_p = v$ is equivalent to $(a_p, b_p) \in \{(0, v), (v, 0), (v, v)\}$; and similarly, $a_p \wedge b_p = v$ is equivalent to $(a_p, b_p) \in \{(1, v), (v, 1), (v, v)\}$. Also notice that if $a_p = b_p = v$ and Case $(1_\vee)$ occurs, then p is an $\{\mathbf{a}, \mathbf{b}\}$-fork; in Cases $(2_\vee)$ and $(3_\vee)$, we must have $p' > p$. Let $\mathbf{c}$ be the element defined in the join formula. To see that $\mathbf{a} \leq \mathbf{c}$ (using the definition (P) of $\leq$), first note that $\mathbf{a} \leq_S \mathbf{a} \vee_S \mathbf{b} \leq_S \mathbf{c}$, so that $\mathbf{a} \leq_S \mathbf{c}$. Let $a_p = c_p = v$ and $p' > p$; we must show that $a_{p'} = c_{p'}$.

Now $a_p = c_p = v$ implies that $a_p = a_p \vee b_p = c_p = v$, and so $b_p \leq a_p = v$. Hence, for all $q \geq p$, none of Cases $(1_\vee)$–$(3_\vee)$ occur. That is, for all $q \geq p$, q is not an $\{\mathbf{a}, \mathbf{b}\}$-fork (Case $(1_\vee)$) and $b_q \leq a_q$ (Case $(2_\vee)$). In Case $(3_\vee)$, using that $b_p \leq a_p = v$ if $a_p \leq b_p$, then $a_p = b_p = v$, and since p is not an $\{\mathbf{a}, \mathbf{b}\}$-fork, $a_q = b_q$ for $q \geq p$.

Consequently, $b_{p'} \leq a_{p'}$. Hence, $a_{p'} = a_{p'} \vee b_{p'} \leq c_{p'}$, with equality holding unless $v = a_{p'} = a_{p'} \vee b_{p'}$ and one of Cases $(1_\vee)$–$(3_\vee)$ holds for some $p'' \geq p'$. But Case $(1_\vee)$ cannot occur because $p'' = q \geq p$ is not an $\{\mathbf{a}, \mathbf{b}\}$-fork, and Case $(2_\vee)$ cannot occur since $p'' \geq p$ forces $b_{p''} \leq a_{p''}$. For Case $(3_\vee)$, note that $b_{p'} \leq a_{p'}$, since $p' > p$. Hence, $a_{p'} \leq b_{p'} = v$ forces $a_{p'} = b_{p'} = v$. But $p' > p$ is not an $\{\mathbf{a}, \mathbf{b}\}$-fork, so $a_{p''} = b_{p''}$, and so Case $(3_\vee)$ cannot occur. Thus $a_{p'} = c_{p'}$ and $\mathbf{a} \leq \mathbf{c}$. Similarly, $\mathbf{b} \leq \mathbf{c}$.

It remains to show that $\mathbf{c}$ is the least upper bound of $\mathbf{a}, \mathbf{b}$ in L. So let $\mathbf{a}, \mathbf{b} \leq \mathbf{d}$. Then $\mathbf{a}, \mathbf{b} \leq_S \mathbf{d}$, so $\mathbf{a} \vee_S \mathbf{b} \leq_S \mathbf{d}$. We first show that $\mathbf{c} \leq_S \mathbf{d}$. For $p \in P$ if $a_p \vee b_p = c_p$, then $c_p \leq d_p$. Let $a_p \vee b_p < c_p$; as in the previous paragraph, it follows that $a_p \vee b_p = v < 1 = c_p$. Therefore, $v \leq d_p \leq 1$. If $d_p = 1$, then $c_p \leq d_p$, as required. Now assume that $d_p = v$. Since $a_p \vee b_p = v < 1 = c_p$, one of Cases $(1_\vee)$–$(3_\vee)$ must occur; in each case, we will arrive at a contradiction. If Case $(1_\vee)$ occurs, then $a_p = b_p = d_p = v$ and, for some $p' > p$, we have $a_{p'} \neq b_{p'}$. We conclude that $d_{p'}$ cannot equal both $a_{p'}$ and $b_{p'}$, contradicting that $\mathbf{d}$ is a common upper bound of $\mathbf{a}$ and $\mathbf{b}$. If Case $(2_\vee)$ occurs, then $b_p \leq a_p = v = d_p$ and $b_{p'} \not\leq a_{p'}$ for some $p' > p$. Then $d_{p'} \geq a_{p'} \vee b_{p'} > a_{p'}$, contradicting that $\mathbf{a} \leq \mathbf{d}$. Symmetrically, Case $(3_\vee)$ leads to a contradiction. Thus $\mathbf{c} \leq_S \mathbf{d}$.

To prove that $\mathbf{c} \leq \mathbf{d}$, let $c_p = d_p = v$ and choose $p < p'$; we must show that $c_{p'} = d_{p'}$. Since $c_p = d_p = v$, the join formula tells us that $a_p \vee b_p = c_p = v$; without loss of generality, we may assume that $a_p = v$ so that $a_p = v = d_p$. As $\mathbf{c} \leq_S \mathbf{d}$, either $c_{p'} = d_{p'}$ or $c_{p'} < d_{p'}$. In the latter case, we have $a_{p'} \leq c_{p'} < d_{p'}$, contradicting that $\mathbf{a} \leq \mathbf{d}$. Hence, $c_{p'} = d_{p'}$. This proves that $\mathbf{c} \leq \mathbf{d}$ and therefore $\mathbf{c}$ is the least upper bound of $\mathbf{a}, \mathbf{b}$ in L.

The proof of the meet formula is similar, *mutatis mutandis*. $\square$

From now on, for $\mathbf{a}, \mathbf{b} \in L$, $\mathbf{a} \leq \mathbf{b}$, $\mathbf{a} \vee \mathbf{b}$, $\mathbf{a} \wedge \mathbf{b}$ refer to the ordering and operations in L.

We will need two lemmas that immediately follow from the join and meet formulas.

For $p \in P$, the element v is doubly irreducible in S_p. Therefore, each $S'_p = S_p - \{v\}$ is a sublattice of S_p. Let S' be the direct product of the S'_p, $p \in P$. Obviously, S' is a sublattice of S.

Lemma 16.5.

(1) S' is a sublattice of L.

(2) For $p \in P$, the interval $[0, \mathbf{u}^p]$ is a sublattice of both L and S; in fact, the same sublattice, that is, $(\leq)\rceil_{[0, \mathbf{u}^p]} = (\leq_S)\rceil_{[0, \mathbf{u}^p]}$.

Proof. The proof follows from the join and meet formulas in Theorem 16.4. $\square$

Lemma 16.6. *Let $\mathbf{a} \in L$ and let D be a down set of P. Then*

$$\mathbf{a} \vee \mathbf{u}^D = \mathbf{a} \vee_S \mathbf{u}^D,$$

and

$$\mathbf{a} \wedge (\mathbf{u}^D)' = \mathbf{a} \wedge_S (\mathbf{u}^D)'.$$

Proof. By the join formula, $(\mathbf{a} \vee \mathbf{u}^D)_p = (\mathbf{a} \vee_S \mathbf{u}^D)_p$ unless $a_p = v$, $p \notin D$, and $a_{p'} < 1$ for some $p < p'$ with $p' \in D$. But this cannot happen since D is a down set. By the meet formula, $(\mathbf{a} \wedge (\mathbf{u}^D)')_p = (\mathbf{a} \wedge_S (\mathbf{u}^D)')_p$ unless $a_p = v$, $p \notin D$, and $a_{p'} > 0$ for some $p < p'$ with $p' \in D$. But this cannot happen either since D is a down set. $\square$

16.4. The congruences

Now let each S_p be simple and let each $S_p \neq C_2$; then S has a congruence lattice isomorphic to B such that $\mathbf{u}^Q \in B$ *is associated with the smallest congruence on S collapsing 0 and* $\mathbf{u}^Q$, *denoted* α_Q. *Clearly, S is isoform. We are going to prove that the congruences of L are just the congruences* α_D *of S for D a down set of P. As an illustration, in Figure 16.6, the lattice has exactly one nontrivial congruence, "projecting" L onto M_q. The five congruence classes have five elements each, labeled x, with $x \in \{0, v, a, b, 1\}$.*

For $p \in P$, the lattice S_p is assumed to be simple, so the set $S_p - \{0, 1, v\} \neq \varnothing$. Therefore, we can select $w_p \in S_p - \{0, 1, v\}$; we will write w for w_p if the index is understood. We define $\mathbf{w}^q$ by $(\mathbf{w}^q)_q = w$ and otherwise, $(\mathbf{w}^q)_p = 0$.

Lemma 16.7. *Let α be a congruence of L, and let $0 \equiv \mathbf{u}^p \pmod{\alpha}$ for some $p \in P$. If $q < p$, then $0 \equiv \mathbf{u}^q \pmod{\alpha}$.*

Proof. From $0 \equiv \mathbf{u}^p \pmod{\alpha}$ and $q < p$, the join formula yields the congruence $\mathbf{v}^q = 0 \vee \mathbf{v}^q \equiv \mathbf{u}^p \vee \mathbf{v}^q = \mathbf{u}^{\{p,q\}} \pmod{\alpha}$. Similarly, we get the congruence $0 = \mathbf{w}^q \wedge \mathbf{v}^q \equiv \mathbf{w}^q \wedge \mathbf{u}^{\{p,q\}} = \mathbf{w}^q \pmod{\alpha}$. Lemma 16.5 implies that $[0, \mathbf{u}^q]$ is a simple sublattice of L, so we obtain that $0 \equiv \mathbf{u}^q \pmod{\alpha}$. □

Lemma 16.8. *Let α be a congruence of L and D a down set of P. Let $\mathbf{a}, \mathbf{b} \in L$ with $a_p = 0$ and $b_p = 1$, for $p \in D$, and $a_p = b_p$ for $p \notin D$. Then $\mathbf{a} \leq \mathbf{b}$ and $0 \equiv \mathbf{u}^D \pmod{\alpha}$ iff $\mathbf{a} \equiv \mathbf{b} \pmod{\alpha}$.*

Proof. Clearly, $\mathbf{b} = \mathbf{a} \vee_S \mathbf{u}^D = \mathbf{a} \vee \mathbf{u}^D$, by Lemma 16.5. Thus $\mathbf{a} \leq \mathbf{b}$. Let $0 \equiv \mathbf{u}^D \pmod{\alpha}$. Then $\mathbf{a} = 0 \vee \mathbf{a} \equiv \mathbf{u}^D \vee \mathbf{a} = \mathbf{b} \pmod{\alpha}$. Conversely, if $\mathbf{a} \equiv \mathbf{b} \pmod{\alpha}$, then $0 = \mathbf{a} \wedge \mathbf{u}^D \equiv \mathbf{b} \wedge \mathbf{u}^D = \mathbf{u}^D \pmod{\alpha}$. □

Lemma 16.9. *Let α be a congruence of L and let $\mathbf{a} \prec \mathbf{b}$ in L with $\mathbf{a} \equiv \mathbf{b} \pmod{\alpha}$. Then there is a unique $p \in P$ with $a_p \prec b_p$ and $a_q = b_q$ for $p \neq q$; moreover, $0 \equiv \mathbf{u}^p \pmod{\alpha}$.*

Proof. If $\mathbf{a} \prec \mathbf{b}$, then from Lemma 16.3, there is a unique $p \in P$ with $a_p \prec b_p$ and $a_q = b_q$ otherwise. Moreover, if $a_q = b_q = v$, then $q \nleq p$. If $b_p = v$ so that $a_p = 0$, then we get $(\mathbf{a} \vee \mathbf{w}^p)_p = w < 1 = (\mathbf{b} \vee \mathbf{w}^p)_p$, and otherwise $(\mathbf{a} \vee \mathbf{w}^p)_q = a_q = b_q = (\mathbf{b} \vee \mathbf{w}^p)_q$ since $a_q = b_q = v$ implies that $q \nleq p$, so that $(\mathbf{w}^p)_r = 0$, for all $q < r$. This yields the covering $\mathbf{a} \vee \mathbf{w}^p \prec \mathbf{b} \vee \mathbf{w}^p$ and the congruence $\mathbf{a} \vee \mathbf{w}^p \equiv \mathbf{b} \vee \mathbf{w}^p \pmod{\alpha}$. Thus we may assume that $b_p \neq v$. Then $0 \leq \mathbf{a} \wedge \mathbf{u}^p < \mathbf{b} \wedge \mathbf{u}^p \leq \mathbf{u}^p$ and $\mathbf{a} \wedge \mathbf{u}^p \equiv \mathbf{b} \wedge \mathbf{u}^p \pmod{\alpha}$. By Lemma 16.5, the interval $[0, \mathbf{u}^p]$ is a simple sublattice of L. Hence, $0 \equiv \mathbf{u}^p \pmod{\alpha}$. □

Theorem 16.10. *Let α be a congruence of L; then there is a down set D of P such that $\alpha = \alpha_D$. Conversely, let D be a down set of P; then α_D is a congruence of L.*

Proof. Let α be a congruence of L. Define $\mathrm{Prec}(\alpha)$ to be the set of all $(\mathbf{a}, \mathbf{b}) \in L^2$ such that $\mathbf{a} \equiv \mathbf{b} \pmod{\alpha}$ and $\mathbf{a} \prec \mathbf{b}$. Define D to be the set of all $p \in P$ such that $a_p \prec b_p$ for some $(\mathbf{a}, \mathbf{b}) \in \mathrm{Prec}(\alpha)$. By Lemma 16.9, $0 \equiv \mathbf{u}^p \pmod{\alpha}$ for all $p \in D$. By Lemma 16.7, D is a down set of P. By Lemma 16.5, $\bigvee(\mathbf{u}^p \mid p \in D) = \mathbf{u}^D$; hence, $0 \equiv \mathbf{u}^D \pmod{\alpha}$. By Lemma 16.8, $\alpha_D \subseteq \alpha$. On the other hand, since L is finite, α is the smallest equivalence relation containing $\mathrm{Prec}(\alpha)$, so we must have $\alpha \subseteq \alpha_D$. Thus $\alpha = \alpha_D$.

Conversely, let D be a down set of P. A typical α_D-class is of the form $[\mathbf{a}, \mathbf{b}]$, where for $p \in D$, $a_p = 0$ and $b_p = 1$, and otherwise $a_q = b_q$. By Lemma 16.6, $\mathbf{a} \le \mathbf{a} \vee \mathbf{u}^D = \mathbf{a} \vee_S \mathbf{u}^D = \mathbf{b}$. Let $\mathbf{c} \in L$; it suffices to show that $\mathbf{a} \vee \mathbf{c} \equiv \mathbf{b} \vee \mathbf{c} \pmod{\alpha}$ and $\mathbf{a} \wedge \mathbf{c} \equiv \mathbf{b} \wedge \mathbf{c} \pmod{\alpha}$.

For the join, we may take $\mathbf{a} \le \mathbf{c}$, so that $\mathbf{a} \vee \mathbf{c} = \mathbf{c} \le \mathbf{b} \vee \mathbf{c}$. Let us assume that $c_p < (\mathbf{b} \vee \mathbf{c})_p$; we must show that $p \in D$. If $c_p < b_p \vee c_p$, then from $a_p \le c_p$, we conclude that $a_p \ne b_p$, and so $p \in D$. Otherwise, $c_p = b_p \vee c_p < (\mathbf{b} \vee \mathbf{c})_p$ so that $c_p = v = b_p \vee c_p < 1 = (\mathbf{b} \vee \mathbf{c})_p$. Again, if $a_p \ne b_p$, then $p \in D$; so assume that $a_p = b_p \le c_p = v$. As D is a down set, we must have $a_q = b_q$ for all $q \ge p$. Now, $b_p \vee c_p = v < 1 = (\mathbf{b} \vee \mathbf{c})_p$ implies that one of Cases $(1_\vee)$–$(3_\vee)$ holds for $\{\mathbf{b}, \mathbf{c}\}$ for some $p' \ge p$. Note that $b_p \le c_p$ implies that Case $(3_\vee)$ cannot occur. If Case $(1_\vee)$ occurs, then $a_{p'} = b_{p'} = c_{p'} = v$; since p' is a $\{\mathbf{b}, \mathbf{c}\}$-fork, this forces p' to be an $\{\mathbf{a}, \mathbf{c}\}$-fork, contradicting that $\mathbf{a} \le \mathbf{c}$. If Case $(2_\vee)$ occurs, then $a_{p'} = b_{p'} \not\le c_{p'}$, again contradicting $\mathbf{a} \le \mathbf{c}$. This proves that $\mathbf{a} \vee \mathbf{c} \equiv \mathbf{b} \vee \mathbf{c} \pmod{\alpha}$.

For the meet, we proceed similarly. $\square$

16.5. The isoform property

As in Section 16.4, we assume that *each S_p is simple and $S_p \ne C_2$.*

Theorem 16.11. *L is isoform.*

Proof. Let D be a down set of P; in S every block of α_D can be written in the form $[\mathbf{a}, \mathbf{a} \vee_S \mathbf{u}^D]$ for some $\mathbf{a} \le_S (\mathbf{u}^D)'$. Hence if $\mathbf{b}, \mathbf{c} \in [\mathbf{a}, \mathbf{a} \vee_S \mathbf{u}^D]$ and $b_p \ne c_p$, then $p \in D$. The map $\phi_{\mathbf{a}} \colon [0, \mathbf{u}^D] \to [\mathbf{a}, \mathbf{a} \vee_S \mathbf{u}^D]$ defined by $\phi_{\mathbf{a}}(\mathbf{x}) = \mathbf{x} \vee_S \mathbf{u}^D$ is an isomorphism of sublattices of S. By Lemma 16.6, $\mathbf{a} \le \mathbf{a} \vee \mathbf{u}^D = \mathbf{a} \vee_S \mathbf{u}^D$; thus the blocks of α_D in L are also of the form $[\mathbf{a}, \mathbf{a} \vee_S \mathbf{u}^D]$ for some $\mathbf{a} \le (\mathbf{u}^D)'$ (which is the same as $\mathbf{a} \le_S (\mathbf{u}^D)'$). We will show that $\phi_{\mathbf{a}}$ is also an isomorphism of L-sublattices; this will prove that L is isoform. We will make use without further mention that for $\mathbf{b}, \mathbf{c} \in L$, $(\mathbf{b} \vee_S \mathbf{c})_p = b_p \vee c_p$ for all $p \in P$.

Obviously, $\phi_{\mathbf{a}}$ is a bijection. Let us assume that $0 \le \mathbf{b} \prec \mathbf{c} \le \mathbf{u}^D$. Then by Lemma 16.6 and the fact that $\phi_{\mathbf{a}}$ is an S-isomorphism, it follows that $\mathbf{b} \vee_S \mathbf{a} \prec_S \mathbf{c} \vee_S \mathbf{a}$. If we do not have $\mathbf{b} \vee_S \mathbf{a} \prec \mathbf{c} \vee_S \mathbf{a}$, then by Lemma 16.3, there are $p < p'$ with $b_p \vee a_p = v = c_p \vee a_p$ and $b_{p'} \vee a_{p'} \prec c_{p'} \vee a_{p'}$. From this latter, we get $p' \in D$; since D is a down set and $p < p'$, we have $p \in D$. But then $\mathbf{a} \le (\mathbf{u}^D)'$ implies that $a_p = a_{p'} = 0$; therefore, $v = b_p \vee a_p = b_p$ and

$v = c_p \vee a_p = c_p$, so that $b_p = c_p = v$. Since $\mathbf{b} \prec \mathbf{c}$, we have that $b_{p''} = c_{p''}$ for all $p < p''$. For $p'' = p'$, this yields $b_{p'} \vee a_{p'} = b_{p'} = c_{p'} = c_{p'} \vee a_{p'}$, contradicting that $b_{p'} \vee a_{p'} \prec c_p \vee a_p$. Thus $\mathbf{b} \vee_S \mathbf{a} \prec \mathbf{c} \vee_S \mathbf{a}$.

Now assume that $\mathbf{a} \leq \mathbf{b}' \prec \mathbf{c}' \leq \mathbf{a} \vee_S \mathbf{u}^D$; hence, $\mathbf{a} \leq_S \mathbf{b}' \prec_S \mathbf{c}' \leq_S \mathbf{a} \vee_S \mathbf{u}^D$. Since γ_a is an S-isomorphism, there are unique $0 \leq_S \mathbf{b} \prec_S \mathbf{c} \leq_S \mathbf{u}^D$ such that $\mathbf{b}' = \mathbf{b} \vee_S \mathbf{a}$ and $\mathbf{c}' = \mathbf{c} \vee_S \mathbf{a}$. Hence there is a unique $p \in D$ such that $b_p \prec c_p$, $b_p \vee a_p \prec c_p \vee a_p$, and $b_q \vee a_q = c_q \vee a_q$ for $p \neq q$. We need to show that $\mathbf{b} \leq \mathbf{c}$. So let $b_r = v = c_r$; then $r \neq p$. We have to verify that $b_q = c_q$, for $r < q$. This can fail only if $r < p$. But then $r < p \in D$, a down set; hence, $r \in D$. Since $\mathbf{a} \leq (\mathbf{u}^D)'$, it follows that $a_r = a_p = 0$. But then $b_r \vee a_r = b_r = v = c_r = c_r \vee a_r$, while $b_p \vee a_p \prec c_p \vee a_p$, contradicting that $\mathbf{b}' \leq \mathbf{c}'$. Thus $\mathbf{b} \prec \mathbf{c}$. We conclude that L is isoform. $\square$,

This completes the proof of Theorem 16.1.

16.6. Discussion

Regular lattices

Sectionally complemented lattices are regular, so we already have a congruence-preserving extension theorem for regular lattices in Chapter 14. We would like to point out that Theorem 16.1 contains this result.

Lemma 16.12. *Every finite isoform lattice is regular.*

Proof. Let L be an isoform lattice. Let α and β be congruences sharing the block A. Then A is also a block of $\alpha \wedge \beta$; in other words, we can assume that $\alpha \leq \beta$. Let B be an arbitrary α block. Since $\alpha \leq \beta$, there is a (unique) β block $\overline{B}$ containing B. Since L is isoform, $A \cong B$ and $A \cong \overline{B}$; by finiteness, $B = \overline{B}$, that is, $\alpha = \beta$. $\square$

Corollary 16.13. *The lattice L of Theorem 16.1 is regular.*

Permutable congruences

Since sectionally complemented lattices have permutable congruences, the congruence-preserving extension result for congruences permutable lattices follows from the result in Chapter 14. We would like to point out that Theorem 16.1 contains also this result. This statement follows from

Lemma 16.14. *The lattice L of Theorem 16.1 is congruence permutable.*

Proof. This is obvious since the congruences of L are congruences of S, which is a direct product of simple lattices. $\square$

Problem 2 of G. Grätzer, R. W. Quackenbush, and E. T. Schmidt [113] raised the question whether this is typical. K. Kaarli [163] provided an affirmative solution:

Theorem 16.15. *Every finite uniform lattice is congruence permutable.*

Kaarli also described a countably infinite isoform lattice that is not congruence permutable. Similarly, if an algebra has a majority term, then any two congruences whose blocks are two-element sets permute; see G. Czédli [14] and [15].

Deterministic isoform lattices

Let L be an isoform lattice, let γ be a congruence of L, and let A_γ be a congruence class of γ. Then any congruence class of γ is isomorphic to A_γ. However, if γ and δ are congruences of L, it may happen that A_γ and A_δ as lattices are isomorphic. Let us call an isoform lattice L *deterministic* if this cannot happen; in other words if $\gamma \neq \delta$ are congruences of L, then A_γ and A_δ are not isomorphic.

Lemma 16.16. *The lattice L of Theorem 16.1 can be constructed to be deterministic.*

Proof. The size $|A_\gamma|$ is the product of the $|S_\gamma|, \gamma \in \mathrm{Con}_\mathrm{M} K$. Since we can easily construct the S_γ-s so that all $|S_\gamma|$-s are distinct primes; the statement follows.   □

Naturally isoform lattices

Let L be a finite lattice. Let us call a congruence relation α of L *naturally isoform* if any two congruence classes of α are naturally isomorphic (as lattices) in the following sense: if $a \in L$ is the smallest element of the class a/α, then $x \mapsto x \vee a$ is an isomorphism between $0/\alpha$ and a/α. Let us call the lattice L *naturally isoform* if all congruences of L are naturally isoform.

The lattice L of Theorem 16.1 is not naturally isoform. There is a good reason for it.

Theorem 16.17. *Let L be a finite lattice. If L is naturally isoform, then $\mathrm{Con}\, L$ is boolean.*

Proof. If L is simple, the statement is trivial.

Let α be a nontrivial congruence relation of L. Let a be the largest element of the α-class $0/\alpha$ and let b be the smallest element of the α-class $1/\alpha$. Then obviously $a \vee b = 1$. If $a \wedge b > 0$, then $b \vee 0 = b \vee (a \wedge b)\ (= b)$; therefore, $x \mapsto x \vee b$ is not an isomorphism between $0/\alpha$ and b/α. Thus $a \wedge b = 0$.

We prove that $L \cong \mathrm{id}(a) \times \mathrm{id}(b)$. For $c \in L$, we get $c \wedge b \equiv c \wedge 1 = c$ (mod α), and so $c \wedge b \in c/\alpha$. If $d < c \wedge b$ is the smallest element of c/α, then the natural isomorphism between $0/\alpha$ and d/α would force that $c \wedge b = d \vee x$, for some $0 < x \leq a$, contradicting that $(c \wedge b) \wedge a = 0$. The natural isomorphism between $0/\alpha$ and c/α yields that $c = (c \wedge b) \vee x$ for some unique $x \leq a$. Since

$x \leq c$, clearly, $x \leq c \wedge a$. Therefore, $(c \wedge b) \vee (c \wedge a) \leq c = (c \wedge b) \vee x \leq (c \wedge b) \vee (c \wedge a)$, that is, $c = (c \wedge b) \vee (c \wedge a)$. This proves that $L \cong \mathrm{id}(a) \times \mathrm{id}(b)$.

Thus α is the kernel of the projection of L onto $\mathrm{id}(a)$ and so has a complement, the kernel of the projection of L onto $\mathrm{id}(b)$. We conclude that $\mathrm{Con}\, L$ is boolean. $\qquad\square$

A generalized construction

We defined an element v of a finite lattice A to be a separator if $0 \prec v \prec 1$. Instead, we can have the more general definition: a *separator* v is a doubly irreducible element. Then there is a unique $v_* \prec v$ and a unique $v^* \succ v$, and these take over the role of 0 and 1 in the proofs. Let us restate Theorem 16.4:

Theorem 16.18. *L is a lattice. Let $\mathbf{a} = (a_p)_{p \in P}, \mathbf{b} = (b_p)_{p \in P} \in L$. Then*

$$(\mathbf{a} \vee \mathbf{b})_p = \begin{cases} v^* & \text{if } a_p \vee b_p = v \text{ and, for some } p' \geq p, \\ & (1_\vee) \ p' \text{ is an } \{\mathbf{a}, \mathbf{b}\}\text{-fork, or} \\ & (2_\vee) \ b_p \leq a_p \text{ and } b_{p'} \not\leq a_{p'}, \text{ or} \\ & (3_\vee) \ a_p \leq b_p \text{ and } a_{p'} \not\leq b_{p'}; \\ a_p \vee b_p, & \text{otherwise;} \end{cases}$$

and

$$(\mathbf{a} \wedge \mathbf{b})_p = \begin{cases} v_* & \text{if } a_p \wedge b_p = v \text{ and for some } p' \geq p, \\ & (1_\wedge) \ p' \text{ is an } \{\mathbf{a}, \mathbf{b}\}\text{-fork, or} \\ & (2_\wedge) \ b_p \geq a_p \text{ and } b_{p'} \not\geq a_{p'}, \text{ or} \\ & (3_\wedge) \ a_p \geq b_p \text{ and } a_{p'} \not\geq b_{p'}; \\ a_p \wedge b_p, & \text{otherwise.} \end{cases}$$

Problems

Problem 16.1. Develop a theory of (deterministic) isoform lattices.

Problem 16.2. Let $f(n)$ be the smallest integer with the property that every lattice K of n elements has a congruence-preserving extension into an isoform lattice of size $f(n)$. Estimate f.

Let L be a finite lattice. A congruence α of L is *term isoform* if, for every $a \in L$, there is a unary term function $p(x)$ that is an isomorphism between $0/\alpha$ and a/α. The lattice L is term isoform if all congruences are term isoform.

Problem 16.3. Does every finite lattice have a congruence-preserving extension to an term isoform finite lattice?

Problem 16.4. Can we carry out the construction for Theorem 16.1 in case $\mathrm{Con}\, K$ is finite (as opposed to K being finite)?

Problem 16.5. Is there an analogue of Theorem 16.1 for infinite lattices?

By Lemma 16.16, every finite lattice has a congruence-preserving extension to a deterministic lattice. Can this result be extended to infinite lattices?

Problem 16.6. Does every lattice have a congruence-preserving extension to a deterministic lattice?

Chapter

17

The Congruence Lattice and the Automorphism Group

17.1. Results

The following problem was first raised in the first edition of my book [59] (Problem II.18):

Problem. Let K be a nontrivial lattice and let G be a group. Does there exist a lattice L such that the congruence lattice of L is isomorphic to the congruence lattice of K and the automorphism group of L is isomorphic to G? If K and G are finite, can L be chosen to be finite?

This problem was solved for finite lattices by V. A. Baranskiĭ [4], [5] and A. Urquhart [188]. We now state the Baranskiĭ–Urquhart theorem:

Theorem 17.1 (Independence Theorem). *Let D be a nontrivial finite distributive lattice and let G be a finite group. Then there exists a finite lattice L such that the congruence lattice of L is isomorphic to D and the automorphism group of L is isomorphic to G.*

This is a representation theorem; both published proofs rely heavily on the Dilworth Theorem (Theorem 8.1) and on the representation theorem of finite groups as automorphism groups of finite lattices (G. Birkhoff [9] and R. Frucht [52]).

The topic of this chapter is the congruence-preserving extension variant of Theorem 17.1, which we published in G. Grätzer and E. T. Schmidt [126].

© Springer International Publishing Switzerland 2016
G. Grätzer, *The Congruences of a Finite Lattice*,
DOI 10.1007/978-3-319-38798-7_17

Theorem 17.2 (Strong Independence Theorem). *Let K be a nontrivial finite lattice and let G be a finite group. Then K has a congruence-preserving extension L whose automorphism group is isomorphic to G.*

There is a stronger form of this theorem; to state it, we need the analogue of the congruence-preserving extension concept for automorphisms.

Let K be a lattice. L is an *automorphism-preserving extension* of K if L is an extension and every automorphism of K has exactly one extension to L; moreover, every automorphism of L is the extension of an automorphism of K. Of course, then the automorphism group of K is isomorphic to the automorphism group of L.

Now we state the stronger version of Theorem 17.2.

Theorem 17.3 (Strong Independence Theorem, Full Version). *Let K_C and K_A be nontrivial finite lattices. Let us assume that $K_C \cap K_A = \{0\}$. Then there exists a lattice L such that the following conditions hold:*

(1) *L is a finite, atomistic, $\{0\}$-extension of both K_A and K_C.*

(2) *L is a congruence-preserving extension of K_C.*

(3) *L is an automorphism-preserving extension of K_A.*

Of course, then the congruence lattice of L is isomorphic to the congruence lattice of K_C, and the automorphism group of L is isomorphic to the automorphism group of K_A.

Notice how the topic is shifting. In Chapters 14 and 15, we seek congruence-preserving extensions into first-order definable classes: sectionally complemented and semimodular lattices. In Chapter 16, we deal with the properties of congruences: uniform and isoform congruences. In this chapter, for the first time, we go "outside" the lattice, looking at automorphism groups. Naturally, the techniques that worked so well no longer apply, in particular, we do not use cubic extensions. Chopped lattices, however, continue to play a crucial role.

17.2.   *Proof-by-Picture*

By a *graph* $C = (V, E)$, we mean a nonempty set V of *vertices* and a set E of *edges*; every $e \in E$ is a two-element subset of V. Figure 17.1 illustrates how we draw graphs: the vertices are represented by small circles circle; the circles representing two distinct vertices x and y are connected by a line segment iff $\{x, y\}$ is an edge. An *automorphism* α of C is a bijection of V preserving the edges, that is, $\{x, y\}$ is an edge iff $\{\alpha x, \alpha y\}$ is an edge for $x \neq y \in V$. The automorphisms of C form a group, denoted by $\operatorname{Aut} C$.

Now let C be a graph (V, E), and—as in R. Frucht's [51]—we order the set $\operatorname{Frucht} C = \{0, 1\} \cup V \cup E$ by $0 < v$, $e < 1$, for all $v \in V$ and $e \in E$, and $v < e$

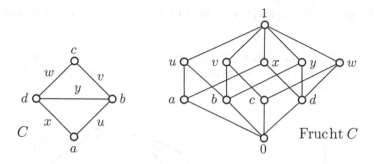

Figure 17.1: Constructing the Frucht lattice.

for $v \in V$ and $e \in E$. iff $v \in e$. The ordered set Frucht C is a lattice; it is in fact atomistic and of length 3. See Figure 17.1 for a small example. It is clear from the picture that C is coded into Frucht C, so Aut(Frucht C) = Aut C. The lattice Frucht C is almost always simple (see Section 17.5).

So we obtain the following result (R. Frucht [52]):

Theorem 17.4. *For every graph C, there is an atomistic lattice F of length 3 with* Aut $C \cong$ Aut F.

A graph $C = (V, E)$ is *rigid* iff it only has the identity map as an auto-morphism.

Rigid graphs are easy to construct. Figure 17.2 shows a rigid graph with 7 vertices, a 6-*cycle with a chord and a tail*. Replace 6 with any $n \geq 6$, and one gets a rigid graph R_n, defined on the set $\{r_0^n, r_1^n, \ldots, r_n^n\}$, with $n + 1$ elements; the edges are $\{r_0^n, r_1^n\}$ (tail), $\{r_1^n, r_2^n\}, \{r_3^n, r_2^n\}, \ldots, \{r_n^n, r_1^n\}$ (cycle), and $\{r_1^n, r_3^n\}$ (chord). These graphs have the important property that if $n \neq m$, then R_n has no embedding into R_m.

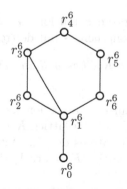

Figure 17.2: The rigid graph, R_6.

We start the *Proof-by-Picture* of Theorem 17.3 with the following modest step:

Theorem 17.5. *The lattice K_A has a finite, atomistic, simple, automorphism- and $\{0\}$-preserving extension L_A with an atom u that is fixed under all the automorphisms.*

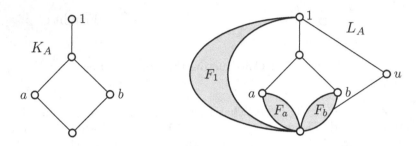

Figure 17.3: The lattice L_A for K_A.

To illustrate the construction, let us take $K_A = \mathsf{B}_2 + \mathsf{C}_1$ be the four-element boolean lattice with atoms a and b and with a new unit element, 1; see Figure 17.3. We take three copies: F_a, F_b, and F_1 of the Frucht lattice Frucht R_6 and replace the interval $[0, a]$ of K_A with F_a, the interval $[0, b]$ of K_A with F_b, and adjoin F_1 by identifying the zero of F_1 with the zero of K_A, the unit of F_1 with the unit of K_A, as depicted in Figure 17.3. In general, for any join-irreducible element x, we attach a copy F_x of F so that the zero of F_x is identified with the zero of K_A and the unit of F_x is identified with x; all the other elements of F_x are incomparable with the other elements of $[0, x]$.

Clearly, L_A is a lattice; it is atomistic because F is; it is simple because of all the complements. Since it is atomistic, automorphisms are determined by how they act on atoms. It now follows that L_A is an automorphism-preserving extension, because F is rigid.

The next step constructs an extension L_C of the lattice K_C (of Theorem 17.3) that retains the congruences but destroys the automorphisms.

Theorem 17.6. *The lattice K_C has a finite, atomistic, rigid, congruence-preserving, $\{0\}$-extension L_C.*

By Theorem 14.1, we can take a sectionally complemented, congruence-preserving, $\{0\}$-extension K_C' of the lattice K_C, with zero, 0, and unit, 1_C. Let $\{a_1, \ldots, a_n\}$ be the set of atoms of K_C'. With every atom a_i in K_C', we associate the (rigid Frucht) lattice $F_i = $ Frucht R_{i+w}, where $w = |K_C'|$, with zero, 0_i, and unit, 1_i; let $p_i = r_0^i$.

Let M_1 be the chopped lattice we obtain merging K_C' and the lattice F_1 by identifying the zero of K_C' with the zero of F_1 and by identifying the atom

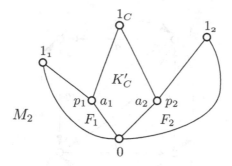

Figure 17.4: The second step in the construction of the chopped lattice M.

a_1 of K'_C with the atom p_1 of F_1; so $K'_C \cap F_1 = \{0, a_1 = p_1\}$, in M_1, as illustrated in Figure 17.4.

Now we merge F_2 and M_1, with the atoms a_2 and p_2, the same way, to obtain the chopped lattice M_2; see Figure 17.4. In n steps, we obtain the chopped lattice $M = M_n$.

The lattice L_C is defined as Id M.

We are given the lattices L_A and L_C. We assume that $L_A \cap L_C = \{0\}$. We further assume that $L_A \ncong L_C$, otherwise, we can take $L = L_A$.

Let v be any atom of L_C. Let u denote the atom of L_A that is fixed by all automorphisms. We construct the chopped lattice N by merging L_A and L_C, by identifying the zeros of L_A and L_C, and identifying u with v; see Figure 17.5.

Under some very mild conditions, the lattice $L = \mathrm{Id}\, N$ satisfies the conditions of the Strong Independence Theorem, Full Version.

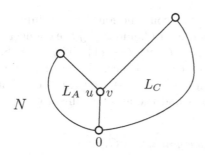

Figure 17.5: Merging the two lattices, L_A and L_C.

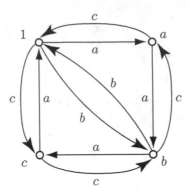

Figure 17.6: A labeled Cayley digraph.

17.3. The representation theorems

To prove the Strong Independence Theorem, we need a representation theorem for groups:

Theorem 17.7. *Every group G is isomorphic to the automorphism group of a lattice L.*

We use the term *digraph* for a directed graph (the edges have a direction) without multiple edges—for each pair of vertices v, w, there is at most one edge from v to w. The digraphs we consider here will also have *no loops*, that is, the ends of any edge will be distinct. *Labeled digraph* means that we have a set X of *labels* and a surjective mapping λ from the directed edges to X.

Step 1: *The labeled Cayley digraph,* Cayley G.

We start with the four-element commutative group $G = \{1, a, b, c\}$, with multiplication: $a^2 = c^2 = b$, $b^2 = 1$, $ab = c$, $bc = a$, and $ac = 1$. Figure 17.6 shows the associated *labeled Cayley digraph,* Cayley G, with the set of labels $X = G - \{1\}$.

We can carry out this construction for any finite group G. The graph Cayley G is defined on the set G, and $x \neq y$ are connected with an arrow from x to y labeled by z ($x, y \in G$, $z \in G - \{1\}$) iff $y = xz$. So the labeled digraph represents right multiplication, hence, the automorphisms of this structure are the left multiplications: $\lambda_u \colon x \mapsto ux$ ($u, x \in G$). The map $u \mapsto \lambda_u$ is an isomorphism between G and the automorphism group $\mathrm{Aut}(\text{Cayley}\,G)$ of the labeled digraph.

Step 2: *The Cayley graph,* Cay G.

Let $G - \{1\} = \{g_1, \ldots, g_n\}$. For $1 \leq m \leq n$, we denote by **m** the graph defined on the set $\{1, \ldots, m+2\}$, where i and j are joined by an edge for

$1 \leq i, j \leq m + 2$ iff $i \neq j$. So $\mathbf{m}$ is the full graph on $m + 2$ elements without loops or double edges.

In Cayley G, we replace the labeled directed edges by (undirected, not labeled) graphs (without loops) as follows: Let the directed edge from x to y labeled by g_m ($x, y \in G$, $1 \leq m \leq n$) be replaced by the graph defined on $\{x, y, u_{x,y}, v_{x,y}\} \cup \mathbf{m}$, where $\mathbf{m}$ is a full subgraph and there are edges between x and 1 and between y and 1, and between $u_{x,y}$ and $v_{x,y}$, as illustrated by Figure 17.7; let $S(x, y)$ denote this graph. Let $\mathrm{Cay}\,G$ denote the graph we obtain by so replacing all labeled edges.

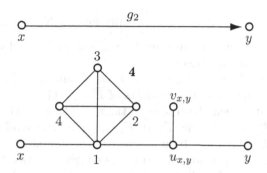

Figure 17.7: Substituting a labeled edge with a graph.

Let α be an automorphism of $\mathrm{Cay}\,G$. It is clear that a vertex which is the endpoint of exactly one edge is of the form $v_{x,y}$, for $x \neq y \in G$, so α permutes the $v_{x,y}$-s. Since $v_{x,y}$ determines $S_{x,y}$, the map α permutes the $S_{x,y}$-s. Which means that α uniquely defines an automorphism of Cayley G, and conversely.

Now by Theorem 17.4, the lattice $K_A = \mathrm{Frucht}(\mathrm{Cay}\,G)$ will do for Theorem 17.7.

The presentation in this section follows G. Grätzer and H. Lakser [95].

17.4. Formal proofs

In this section we go through the constructions and theorems of Section 17.2 and provide the details, proving the three theorems of Section 17.1.

17.4.1   An automorphism-preserving simple extension

In this section we provide a proof of Theorem 17.5.

Let $F = \mathrm{Frucht}\,R_6$, which is a finite, simple, rigid, atomistic lattice of length 3. For every join-irreducible $a \in K_A$, we take a copy F_a of the lattice F with zero 0_a and unit 1_a. We form the disjoint union

$$\bar{L} = K_A \cup \bigcup (F_a - \{0_a, 1_a\} \mid a \in K_A - \{0\})$$

and we identify 0_a with 0 and 1_a with a.

For $x, y \in \bar{L}$, we define $x \wedge y$ and $x \vee y$ as follows:

(i) Let K and all the F_a-s be sublattices of $\bar{L}$.

(ii) If $x \in F_a - K$, $y \in F_b - K$, $a \neq b$, then $x \wedge y = 0$ and $x \vee y = a \vee b$.

(iii) If $x \in F_a - K$, $y \in K - F_a$, then

$$x \wedge y = \begin{cases} x & \text{if } a \leq y, \\ 0, & \text{otherwise;} \end{cases}$$

and $x \vee y = a \vee y$. And symmetrically for $x \in K - F_a$ and $y \in F_a - K$.

It is an easy computation to show that $\bar{L}$ is a lattice containing K and all the F_a, $a \in \mathrm{J}(K)$, as sublattices.

Finally, we adjoin an element u to $\bar{L}$ to obtain L_A; the element u is a common complement to all the elements of $\bar{L} - \{0, 1\}$.

We prove that L_A satisfies the conditions of Theorem 17.5.

The lattice L_A is atomistic. Every F_a is atomistic, therefore, so is L_A.

The lattice L_A is simple. Indeed, since L_A is atomistic, every congruence relation is of the form $\mathrm{con}(0, a)$. So it is enough to prove that $\mathrm{con}(0, a)$ is the unit congruence $\mathbf{1}$ for all $0 < a$. If $a \neq u$, then $a \equiv 0 \pmod{\boldsymbol{\alpha}}$ implies that $u \equiv 1 \pmod{\boldsymbol{\alpha}}$. Let $b \in L_A - \{0, u, 1\}$; then from the last congruence we obtain that $b \equiv 0 \pmod{\boldsymbol{\alpha}}$. Since we can write 1 as the join of two such elements b, we conclude that $1 \equiv 0 \pmod{\boldsymbol{\alpha}}$, and so $\boldsymbol{\alpha} = \mathbf{1}$. A similar computation shows that $u \equiv 0 \pmod{\boldsymbol{\alpha}}$ implies that $\boldsymbol{\alpha} = \mathbf{1}$, so L_A is simple.

Now we compute that L_A is a automorphism-preserving extension of K.

First, let α be an automorphism of K. We define the extension $\hat{\alpha}$ of α to L_A. If $a, b \in K$ and $\alpha a = b$, then let $\hat{\alpha}$ map F_a onto F_b under the unique isomorphism between F_a and F_b. Finally, define $\hat{\alpha} u = u$. Then $\hat{\alpha}$ is obviously an extension of α to L_A.

Second, we prove that every automorphism β of L_A is of the form $\hat{\alpha}$ for a suitable automorphism α of K. Let $a \in \mathrm{J}(K)$. We are going to define αa.

Let p be an atom of F_a for some $a \in \mathrm{J}(K)$; then p is an atom of L_A not covered by 1. Therefore, βp is an atom of L_A not covered by 1. Now observe that every atom of L_A not covered by 1 is an atom of some F_b, with $b \in \mathrm{J}(K)$. Therefore, $\beta p \in F_b$ for some $b \in \mathrm{J}(K)$. Let q be an atom of F_a such that $\{p, q\}$ is an edge in the graph (V, E). Then $p \vee q \in F_a$ covers p and q. Therefore, $\beta(p \vee q)$ covers βp and βq. F_b is of length 3, and an atom of F_b is covered only by elements of F_b by the construction of L_A. Since βp is an atom in F_b, it follows that $\beta(p \vee q) \in F_b$, and so $\beta q \in F_b$. The graph defining F is connected, so we conclude that the image of *any* atom of F_a is in F_b. The lattice F_a is atomistic, so we obtain that βF_a is a sublattice of F_b. Since F_a and F_b are copies of the same finite rigid lattice F, it follows that β maps F_a onto F_b; in particular, a to b. We define $\alpha a = b$.

This defines α on $\mathrm{J}(K)$; adding $\alpha 0 = 0$, we now have α defined on K. It is obvious that α is an automorphism of K and $\hat{\alpha} = \beta$, by the rigidity of F.

17.4.2   A congruence-preserving rigid extension

We continue with the proof of Theorem 17.6 by verifying the properties of the lattice $L_C = \mathrm{Id}\, M$.

The lattice L_C is obviously a finite, atomistic, $\{0\}$-extension of K_C.

To show that L_C is rigid, let α be an automorphism of L_C.

We regard $M \subseteq L_C$. Observe that M and L_C have the same set of atoms.

Observation. Let $x \neq y$ be atoms. If $x \in K'_C - F_i$ and $y \in F_i - K'_C$, or if $x \in F_j - F_i$ and $y \in F_i - F_j$, where $1 \leq i \neq j \leq n$, then the interval $[0, x \vee y]$ of L_C is a four-element boolean sublattice of L_C.

Now assume that there is an i with $1 \leq i \leq n$ and an atom $r_j^i \in F_i$, with $1 \leq j \leq i + w$ such that $\alpha r_j^i \notin F_i$. We distinguish two cases.

Case 1: $\alpha r_j^i \in K'_C$. Consider the atoms r_k^i with $|j - k| > 1$. Then $r_j^i \vee r_k^i = 1_i$, so the interval $[0, r_j^i \vee r_k^i]$ is of length 3. Therefore, the interval $[\alpha 0, \alpha(r_j^i \vee r_k^i)] = [0, \alpha r_j^i \vee \alpha r_k^i]$ is also of length 3, so by the Observation above, $\alpha r_k^i \in K'_C$. So α maps all these atoms into K'_C, and therefore, α is an embedding of F_i into K'_C. This contradicts that $|F_i| > |K'_C|$.

Case 2: $\alpha r_j^i \in F_k$, with $k \neq i$. Arguing as in Case 1, we get that α is an embedding of F_i into F_k. This contradicts that the graph R_{i+w} has no embedding into the graph R_{k+w}.

This completes the verification that L_C is rigid.

Finally, the lattice L_C is a congruence-preserving extension of K_C. Indeed, the chopped lattice M is a congruence-preserving extension of K_C by a trivial induction on n, using Corollary 5.7, and L_C is a congruence-preserving extension of K_C by Theorem 5.6.

This completes the proof of Theorem 17.6.

17.4.3   Proof of the independence theorems

We are given the two finite lattices K_C and K_A with $K_C \cap K_A = \{0\}$. Let L_C be the lattice provided for K_C by Sections 17.4.1 and 17.4.2; the unit of L_C is denoted by 1_C. Let L_A be the lattice provided for K_A in Section 17.4.1; the unit of L_A is denoted by 1_A. We can assume that $L_A \cap L_C = \{0\}$. We can further assume that $L_A \ncong L_C$. Indeed, if $L_A \cong L_C$, then we can take $L = L_A$. We construct the chopped lattice N and the lattice $L = \mathrm{Id}\, N$ as in Chapter 5; see Figure 17.5.

Since L_A is simple, by Corollary 5.7, we get that N is a congruence-preserving extension of L_C, and L_C is a congruence-preserving extension of N by Theorem 5.6, so L is a congruence-preserving extension of L_C.

To verify the automorphism properties of L, we need two more statements.

Lemma 17.8. *Let U be a chopped lattice with two maximal elements m_1 and m_2. If $p = m_1 \wedge m_2$ is an atom, then $\operatorname{Id} U$ is a union of four parts:*

$$U_1 = \operatorname{id}(m_1),$$
$$U_2 = \operatorname{id}(m_2),$$
$$U_{1,2} = (\operatorname{id}(m_1) - [p, m_1]) \times (\operatorname{id}(m_2) - [p, m_2]),$$
$$U^{1,2} = [p, m_1] \times [p, m_2].$$

Proof. This is easy, if (j_1, j_2) is a compatible vector (see Section 5.2), then either $j_1 \wedge p = j_2 \wedge p = 0$ or $j_1 \wedge p = j_2 \wedge p = p$. In the first case, $(j_1, j_2) \in U_1$ if $j_2 = 0$, $(j_1, j_2) \in U_2$ if $j_1 = 0$, and $(j_1, j_2) \in U_{1,2}$ if $j_1 \neq 0$ and $j_2 \neq 0$. In the second case, $j_1 \geq p_1$ and $j_2 \geq p_2$, so $(j_1, j_2) \in U^{1,2}$. $\qquad\square$

Lemma 17.9. *Using the assumptions and notation of Lemma 17.8, let α_i be an automorphism of U_i for $i = 1, 2$. If $p = \alpha_1 p = \alpha_2 p$, then there is a unique automorphism $\hat{\alpha}$ of U that extends both α_1 and α_2.*

Proof. Let $p = \alpha_1 p = \alpha_2 p$. Then define $\hat{\alpha}$ on a compatible vector (j_1, j_2) by $\hat{\alpha}(j_1, j_2) = (\alpha_1 j_1, \alpha_2 j_2)$. It is easy to see that $\hat{\alpha}$ is an automorphism of U. Conversely, let $\hat{\alpha}$ be an automorphism of U extending both α_1 and α_2. The compatible pair (j_1, j_2) is either in $U^{1,2}$, in which case it satisfies $(j_1, j_2) = (j_1, p) \vee (p, j_2)$ or $(j_1, j_2) \notin U^{1,2}$, in which case it satisfies $(j_1, j_2) = (j_1, 0) \vee (0, j_2)$. Either one of the equations implies that $\hat{\alpha}(j_1, j_2) = (\alpha_1 j_1, \alpha_2 j_2)$. $\qquad\square$

Let us assume now that the Frucht lattice F used in the construction of L_A satisfies the condition $|F| > |L_A|, |L_C|$. To prove that L is an automorphism-preserving extension, we have to prove two claims:

Claim 1. Every automorphism of K_A has exactly one extension to L.

Proof. Let α_1 be an automorphism of K_A; let α_2 be the identity map on L_C. The atom p of $K_A \cap K_C$ is the atom u of K_A fixed by all automorphisms of K_A, so $p = \alpha_1 p = \alpha_2 p$ holds. By Lemma 17.9, α_1 (and α_2) has an extension to an automorphism $\hat{\alpha}_1$ of L. To get the uniqueness, let $\overline{\alpha}_1$ extend α_1 to L. Then $\overline{\alpha}_1 p = p$. Since $\overline{\alpha}_1$ permutes the atoms of L_A and keeps p fixed, it follows that it permutes the atoms of L other than the atoms of L_A, that is, it permutes the atoms of L_C. So $\overline{\alpha}_1$ restricted to L_C is an automorphism of L_C. Since L_C is rigid, $\overline{\alpha}_1$ restricted to L_C is the identity map. So $\overline{\alpha}_1 = \hat{\alpha}_1$ of Lemma 17.9, and it is unique. $\qquad\square$

Claim 2. Every automorphism of L is the extension of an automorphism of K_A.

Proof. Let α be an automorphism of L. Let us take the set H of all elements $h \in L$ satisfying the condition that in L, the interval $[0, h]$ contains a copy of F such that 0 is the zero of F and h is the unit of F. We claim that

$H = J(K_A)$. Indeed, If $h \in J(K_A)$, then the interval $[0, h]$ contains a copy of F by the construction of L_A; see Section 17.4.1.

If $h \in L$, then h is in one of the regions described in Lemma 17.8, namely, $U_1 = L_A$, $U_2 = L_C$, $U_{1,2}$, or $U^{1,2}$. Let $h \in H$ but $h \notin J(K_A)$. We distinguish four cases.

Case 1: $h \in U_1 = L_A$.
Then $h \in L_A$ so either $h \in K_A$ or $h \in F_a$ for some $a \in J(K_A)$ with $h < a$. If $h \in K_A$, then h is join-reducible, so the copy of F such that 0 is the zero of F and h is the unit of F, must be in K_A, contradicting the assumption: $|K_A| < |F|$. If $h \in F_a$, for some $a \in J(K_A)$ with $h < a$, then the copy of F must be disjoint to F_a; therefore, $F \subseteq K_A$, the same contradiction as in the previous case.

Case 2: $h \in U_2 = L_C$.
This contradicts the assumption: $|K_C| < |F|$.

Case 3: $h \in U_{1,2}$.
Then $[0, h]$ in L is a direct product, contradicting that F is directly irreducible.

Case 4: $h \in U^{1,2}$, but $h \notin U_1$ and $h \notin U_2$.
Then $v = (1_A, h_2)$ for some $h_2 \in [v, 1_V]$, with $h_2 \neq u$, since u is a dual atom of L_A. So the dual atoms of F in $[0, h]$ with at most one exception come from the dual atoms of $[0, h_2]$ in L_C, yielding again a contradiction with $|F| > |L_C|$.
This concludes the proof of $H = J(K_A)$.

Since α must permute H, we conclude that α permutes $J(K_A)$, and so α restricted to K_A is an automorphism of K_A. By the construction of L_A we immediately get that α restricted to L_A is an automorphism α_1 of L_A. □

These claims show that the lattice $L = \text{Id} N$ satisfies the conditions of the Strong Independence Theorem, Full Version.

17.5. Discussion

Automorphism groups

It was first proved in G. Birkhoff [9] that every finite group is isomorphic to the automorphism group of a finite (distributive) lattice. R. Frucht in [51] constructed a length 3 lattice with a given automorphism group, as described in Section 17.2.

For Frucht's construction, we need a graph with a given automorphism group. Such a graph was constructed in [51]; see also G. Sabidussi [172]. For a general introduction to this topic, see the monograph P. Pultr and V. Trnková [170].

Two groups

In the style of Part IV, in [95], G. Grätzer and H. Lakser proved a "two group" theorem:

Theorem 17.10. *Let G, G' be groups and let $\eta \colon G \to G'$ be a group homomorphism. There is a simple lattice H of length at most 6 with an ideal $H' = \mathrm{id}(i_H)$ of length at most 3 which is also simple. There are isomorphisms*

$$\tau_H \colon G \to \mathrm{Aut}\, H$$

and

$$\tau_{H'} \colon G' \to \mathrm{Aut}\, H'$$

such that, for each $g \in G$, the automorphism $\tau_H g$ of H restricts to the automorphism $\tau_{H'}\eta g$ of H'. Furthermore, if G and G' are finite, then so is the lattice H.

Simple Frucht Lattices

In [95], G. Grätzer and H. Lakser observe that the Frucht lattices are almost always simple. The Frucht lattice F constructed from the graph $C = (V, E)$ is simple iff the following condition holds: For $v \in V$, there are $a_1, a_2 \in E$ with $v \notin a_1, a_2$ and $a_1 \cap a_2 = \varnothing$.

Problems

We do not have an independence theorem for lattices with any special properties. Let us mention two possibilities:

Problem 17.1. Is there an independence theorem for sectionally complemented lattices?

Problem 17.2. Is there an independence theorem for semimodular lattices?

The combinatorial questions are all open:

Problem 17.3. What is the size of the lattice L we construct for Theorem 17.2? What is the minimum size of a lattice L—as a function of $|K|$ and $|G|$—satisfying Theorem 17.2?

Problem 17.4. What is the size of the lattice L we construct for Theorem 17.3? What is the minimum size of a lattice L—as a function of $|K_C|$ and $|K_A|$—satisfying Theorem 17.3?

Theorem 17.2 can be proved also for general lattices, see G. Grätzer and F. Wehrung [149]. Unfortunately, this proof is very long and complex.

Problem 17.5. Is there a shorter proof for the Independence Theorem for general lattices?

Magic Wands

18.1. Constructing congruence lattices

18.1.1 Bijective maps

A typical way of constructing an algebra B with a given congruence lattice C is to construct an algebra A with a much larger congruence lattice and then "collapsing" sufficiently many pairs of congruences of the form $\mathrm{con}(a, b)$ and $\mathrm{con}(c, d)$ in B, so that the congruence lattice "shrinks" to C. To do this we need a "magic wand" that will make $a \equiv b$ equivalent to $c \equiv d$. Such a magic wand may be a pair of partial operations f and g such that $f(a) = c$, $f(b) = d$, and $g(c) = a$, $g(d) = b$. This is the start of the Congruence Lattice Characterization Theorem of Universal Algebras in G. Grätzer and E. T. Schmidt (see [118], and also [54] and [57].)

If we want to construct a lattice L with a given congruence lattice C, how do we turn the action of the "magic wand" into lattice operations? To construct a simple modular lattice, E. T. Schmidt [174] started with the rational interval $K = [0, 1]$ and by a "magic wand" he required that all $[a, b]$ $(0 \leq a < b \leq 1)$ satisfy that $a \equiv b$ be equivalent to $0 \equiv 1$. The action of the magic wand was realized with the gadget $\mathsf{M}_3[K]$ (see Section 6.3).

In Sections 18.2 and 18.3 we prove that one can apply the magic wand to arbitrary lattices with zero, see G. Grätzer and E. T. Schmidt [135].

If we are considering a "magic wand" (an extension L of the lattice K) that will realize that $a \equiv b$ be equivalent to $c \equiv d$ in the lattice L, we immediately notice that we have to say something about the intervals $[a, b]$ and $[c, d]$. For instance, if $a \equiv b \pmod{\boldsymbol{\alpha} \vee \boldsymbol{\beta}}$ in K, for congruences $\boldsymbol{\alpha}$ and $\boldsymbol{\beta}$ of L, then

© Springer International Publishing Switzerland 2016   201
G. Grätzer, *The Congruences of a Finite Lattice*,
DOI 10.1007/978-3-319-38798-7_18

$c \equiv d \pmod{\overline{\alpha} \vee \overline{\beta}}$, for congruences $\overline{\alpha} = \mathrm{con}_L(\alpha)$ and $\overline{\beta} = \mathrm{con}_L(\beta)$ of L, therefore, the sequence in $[a, b]$ that forces $a \equiv b \pmod{\alpha \vee \beta}$ in K (see Theorem 1.2) must somehow be mapped in L to a sequence in $[c, d]_L$ to force $c \equiv d \pmod{\overline{\alpha} \vee \overline{\beta}}$. So for lattices, "magic wands" must act on intervals, not on pairs of elements.

To set up "magic wands"—as (convex) extensions—for lattices formally, let K be a bounded lattice, let $[a, b]$ and $[c, d]$ be intervals of K, and let $\delta \colon [a, b] \to [c, d]$ be an isomorphism between these two intervals. We can consider δ and δ^{-1} as partial unary operations. Let us call a congruence α of K a δ-*congruence* iff α satisfies the Substitution Property with respect to the partial unary operations δ and δ^{-1}. Let K_δ denote the partial algebra obtained from K by adding the partial operations δ and δ^{-1}. Thus a congruence relation of the partial algebra K_δ is the same as a δ-congruence of the lattice K.

We call an extension L of the lattice K a δ-*congruence-preserving extension* of K if a congruence of K extends to L iff it is a δ-congruence and every δ-congruence of K has *exactly one* extension to L. As a special case, we get the well-known concept of a congruence-preserving extension (in case, δ is trivial).

Let us call δ (resp., δ^{-1}) a *term function* iff there is a unary term function $p(x)$ such that $\delta x = p(x)$ for all $x \in [a, b]$ (resp., $\delta^{-1} x = p(x)$ for all $x \in [c, d]$).

In G. Grätzer and E. T. Schmidt [135], we prove the following result:

Theorem 18.1. *Let K be a bounded lattice, let $[a, b]$ and $[c, d]$ be intervals of K, and let $\varphi \colon [a, b] \to [c, d]$ be an isomorphism between these two intervals. Then K has a φ-congruence-preserving convex extension into a bounded lattice L such that both φ and φ^{-1} are term functions in L. In particular, the congruence lattice of the partial algebra K_φ is isomorphic to the congruence lattice of the bounded lattice L.*

So the lattice L constructed in this result is the magic wand for φ.

18.1.2   Surjective maps

G. Grätzer, M. Greenberg, and E. T. Schmidt [82] generalized to surjective maps the approach of Section 18.1.1.

Let K be a bounded lattice, let $[a, b]$ and $[c, d]$ be intervals of K, let φ be a homomorphism of $[a, b]$ *onto* $[c, d]$. We can consider φ as a partial unary operation. Let us call a congruence α of K a φ-*congruence* iff α satisfies the Substitution Property with respect to the partial unary operation φ, that is, $x \equiv y \pmod{\alpha}$ implies that $\varphi x \equiv \varphi y \pmod{\alpha}$ for all $x, y \in [a, b]$. Let K_φ denote the partial algebra obtained from K by adding the partial operation φ. Thus a congruence relation of the partial algebra K_φ is the same as a φ-congruence of the lattice K.

We call an extension L of the lattice K a φ-*congruence-preserving extension* of K if a congruence of K extends to L iff it is a φ-congruence and every

φ-congruence of K has *exactly one* extension to L. If L is a φ-congruence-preserving extension of K, then the congruence lattice of the partial algebra K_φ is isomorphic to the congruence lattice of the lattice L.

Theorem 18.1 generalizes as follows.

Theorem 18.2. *Let K be a bounded lattice, let $[a,b]$ and $[c,d]$ be intervals of K, and let φ be a homomorphism of $[a,b]$ onto $[c,d]$. Then K has a φ-congruence-preserving convex extension into a bounded lattice L such that φ is a term function in L. In particular, the congruence lattice of the partial algebra K_φ is isomorphic to the congruence lattice of the bounded lattice L. If K is finite, then L can be constructed as a finite lattice.*

So L realizes the "magic wand": $a \equiv b$ implies that $c \equiv d$ in L.

To illustrate the use of such a result, we reprove the Dilworth Theorem (Theorem 8.1). Let D be a finite distributive lattice and $P = J(D)$. Let K be the boolean lattice with atoms $a_i, i \in P$. For $i > j$ in P, define $\varphi_{i,j}$ as the only homomorphism of $[0, a_i]$ onto $[0, a_j]$. Apply Theorem 18.2 to these homomorphisms one at a time to construct a lattice L whose congruence lattice is isomorphic to D.

18.2.   *Proof-by-Picture* for bijective maps

In this section we present the *Proof-by-Picture* of Theorem 18.1; we build the lattice L. We use two gadgets: the boolean triple construction of Chapter 6 and its special intervals of Section 6.4.

Let the bounded lattice K, the intervals $[a,b]$ and $[c,d]$, and let the isomorphism $\varphi \colon [a,b] \to [c,d]$ be given as in Theorem 18.1. We start the construction with four gadgets, which are assumed to be pairwise disjoint, see the top left of Figure 18.1, The four gadgets.

(i) $A = \mathsf{M}_3[K]$. Let $\{0_A, p_1, p_2, p_3, 1_A\}$ be the spanning M_3 in A—as in Lemma 6.1.(iv).

(ii) $B = \mathsf{M}_3[a,b]$, with the spanning M_3, $\{0_B, q_1, q_2, q_3, 1_B\}$.

(iii) The lattice $K_{a,b}$—as in Lemma 6.19.

(iv) The lattice $K_{c,d}$—as in Lemma 6.19.

Some notation: An element of one of the four gadgets is described as a triple $(x, y, z) \in K^3$ belonging to the particular gadget. Note that a triple (x, y, z) may belong to two or more gadgets. If it is not clear the element of which gadget a triple is representing, we will make it clear with subscripts:

$$(x, y, z)_A \quad \text{for } (x, y, z) \text{ as an element of } A;$$
$$(x, y, z)_{a,b} \quad \text{for } (x, y, z) \text{ as an element of } K_{a,b};$$
$$(x, y, z)_{c,d} \quad \text{for } (x, y, z) \text{ as an element of } K_{c,d};$$
$$(x, y, z)_B \quad \text{for } (x, y, z) \text{ as an element of } B.$$

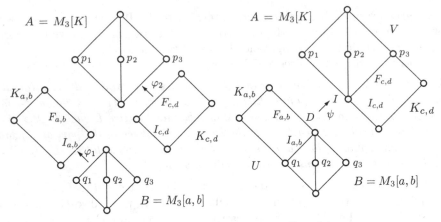

The four gadgets.   Two gluings.

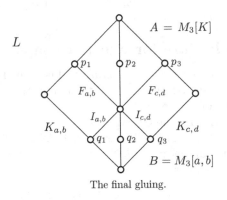

The final gluing.

Figure 18.1: Constructing the lattice L.

We start the construction by gluing together B and $K_{a,b}$ to obtain the lattice U, see the top right of Figure 18.1, Two gluings. In B, we use the filter

$$\mathrm{fil}(q_1) = \{\,(1,x,x) \mid a \leq x \leq b\,\},$$

while in $K_{a,b}$ we utilize the ideal

$$I_{a,b} = \{\,(0,x,0) \mid a \leq x \leq b\,\},$$

and we consider the isomorphism

$$\varphi_1\colon (1,x,x)_B \mapsto (0,x,0)_{a,b}, \quad x \in [a,b],$$

between the filter $\mathrm{fil}(q_1)$ of B and the ideal $I_{a,b}$ of $K_{a,b}$ to glue B and $K_{a,b}$ together to obtain the lattice U.

Similarly, we glue $K_{c,d}$ and A over the filter

$$F_{c,d} = \{\, (x, d, x \wedge d) \mid x \in K \,\}$$

of $K_{c,d}$ and the ideal

$$\mathrm{id}(p_3) = \{\, (0, 0, x) \mid x \in K \,\}$$

of A, with respect to the isomorphism

$$\varphi_2 \colon (x, d, x \wedge d)_{c,d} \mapsto (0, 0, x)_A, \quad x \in K,$$

to obtain the lattice V.

In U, we define the filter

$$F = [q_3, 1_B] \cup F_{a,b},$$

which is the union of $[q_3, 1_B]$ and $F_{a,b}$, with the unit of $[q_3, 1_B]$ identified with the zero of $F_{a,b}$.

In V, we define the ideal

$$I = I_{c,d} \cup [0_A, p_1],$$

which is the union of $I_{c,d}$ and $[0_A, p_1]$, with the unit of $I_{c,d}$ identified with the zero of $[0_A, p_1]$.

Next we set up an isomorphism $\varphi \colon F \to I$. Since

$$[q_3, 1_B] = \{\, (x, x, b) \mid a \leq x \leq b \,\}$$

and

$$I_{c,d} = \{\, (0, x, 0) \mid c \leq x \leq d \,\},$$

we define φ on $[q_3, 1_B]$ by

$$\varphi \colon (x, x, b)_B \mapsto (0, \varphi x, 0)_{c,d}, \quad x \in [a, b],$$

where $\varphi \colon [a, b] \to [c, d]$ is the isomorphism given in Theorem 18.1. This "twist" by φ—hidden in the last displayed formula—is the pivotal idea of the proof. We define φ on $F_{a,b}$ by

$$\varphi \colon (x, b, x \wedge b)_{a,b} \mapsto (x, 0, 0)_A, \quad x \in K.$$

It is clear that $\varphi \colon F \to I$ is well-defined and it is an isomorphism.

Finally, we construct the lattice L of Theorem 18.1 by gluing U over I with V over F with respect to the isomorphism $\varphi \colon F \to I$, see the bottom half of Figure 18.1: The final gluing.

By Lemma 6.1.(iii), the map $x \mapsto (x, 0, 0)_A$ is an isomorphism between the lattice K and the ideal $\mathrm{id}(p_1)$ of the lattice A; this gives us a convex

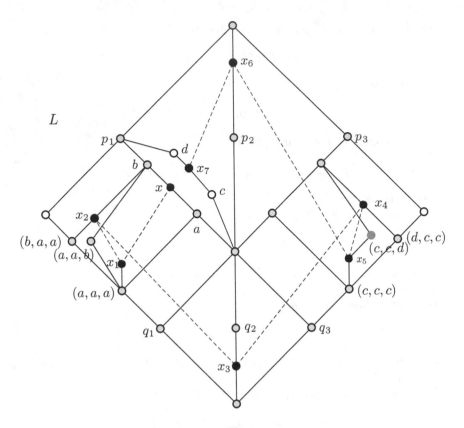

Figure 18.2: The seven steps.

embedding of K into A. We identify K with its image, and regard the lattice K as a convex sublattice of the lattice A and therefore of the lattice L. So the lattice L is a convex extension of the lattice K. We have completed the construction of the bounded lattice L of Theorem 18.1.

Now we have to show that φ and φ^{-1} are term functions in the lattice L. We define

$$p(x) = ((((((x \wedge (a,a,b)_{a,b}) \vee (b,a,a)_{a,b}) \wedge q_2) \vee (d,c,c)_{c,d}) \wedge (c,c,d)_{c,d}) \vee p_2) \wedge p_1.$$

Figure 18.2 shows that $\varphi x = p(x)$ for all $x \in [a,b]$.

The proof for φ^{-1} is similar.

18.3. Verification for bijective maps

We now verify that the lattice extension L of K satisfies the conditions stated in Theorem 18.1.

First, we will describe the congruences of L. We need some notation: For a congruence α of K,

$\qquad \alpha_A$ denotes the congruence α^3 restricted to A;

$\qquad \alpha_{a,b}$ denotes the congruence α^3 restricted to $K_{a,b}$;

$\qquad \alpha_{c,d}$ denotes the congruence α^3 restricted to $K_{c,d}$;

$\qquad \alpha_B$ denotes the congruence α^3 restricted to B.

Let us start with U. The lattices B and $K_{a,b}$ are glued together over $\mathrm{fil}(q_1)$ and $I_{a,b}$ with the isomorphism $\varphi_1\colon (1, x, x)_B \mapsto (0, x, 0)_{a,b}$, and obviously $(1, x, x) \equiv (1, y, y) \pmod{\alpha^3}$ iff $(0, x, 0) \equiv (0, y, 0) \pmod{\alpha^3}$. Hence by Lemma 2.9, the congruences of U are of the form $\alpha_B \overset{\mathrm{r}}{\circ} \alpha_{a,b}$ (the symbol $\overset{\mathrm{r}}{\circ}$ is defined in Section 2.4).

The lattice B is a congruence-preserving extension of $\mathrm{fil}(q_1)$ (formed in B) by Theorem 6.3; therefore, by Lemma 3.17 and Corollary 6.18, the lattice U is a congruence-preserving extension of $K_{a,b}$, which, in turn, is a congruence-preserving extension of $F_{a,b}$ ($\cong K$). So U is a congruence-preserving extension of $F_{a,b}$ ($\cong K$). Similarly, V is a congruence-preserving extension of A, which, in turn, is a congruence-preserving extension of $[0_A, p_1] = K$, so V is a congruence-preserving extension of $[0_A, p_1] = K$.

We glue U and V together over F and I over φ; equivalently, we identify the filter

$$F_{a,b} = \{ (x, b, x \wedge b) \mid x \in K \} \subseteq U$$

of U with

$$[0_A, p_1] = \{ (x, 0, 0) \mid x \in K \} \subseteq A,$$

and note that for any congruence α of K,

$$(x, b, x \wedge b) \equiv (y, b, y \wedge b) \pmod{\alpha^3} \qquad \text{iff} \qquad (x, 0, 0) \equiv (y, 0, 0) \pmod{\alpha^3},$$

so α^3 restricted to $F_{a,b}$ is mapped by φ to α^3 restricted to $[0_A, p_1]$—and we identify the filter

$$[q_3, 1_B] = \{ (x, x, b) \mid a \le x \le b \}$$

of B with the ideal

$$I_{c,d} = \{ (0, x, 0) \mid c \le x \le d \}$$

of $K_{c,d}$ by identifying (x, x, b) with $(0, \varphi x, 0)$ for $x \in [a, b]$. So in the lattice L,

$$(x, x, 1) \equiv (y, y, 1) \qquad \text{iff} \qquad (0, \varphi x, 0) \equiv (0, \varphi y, 0);$$

translating this back to K, we obtain that $x \equiv y \pmod{\alpha}$ iff $\varphi x \equiv \varphi y \pmod{\alpha}$. This condition is equivalent to the statement that α has the Substitution Property with respect to the partial unary operations φ and φ^{-1}.

This proves, on the one hand, that if α extends to the lattice L, then α is a φ-congruence, and, on the other hand, that a φ-congruence α extends uniquely to L, that is, the lattice L is a φ-congruence-preserving extension of the lattice K.

Second, we have to show that φ and φ^{-1} are term functions in the lattice L. We define

$$p(x) = (((((((x \wedge (a,a,b)_{a,b}) \vee (b,a,a)_{a,b}) \wedge q_2) \vee (d,c,c)_{c,d}) \wedge (c,c,d)_{c,d}) \vee p_2) \wedge p_1.$$

For $x \in [a,b]$, we want to compute $p(x)$. There are seven steps in the computation of $p(x)$ (see Figure 18.2):

$$
\begin{aligned}
x &= (x,0,0)_A = (x,b,x)_{a,b} & \text{(in } A \text{ and in } K_{a,b}),\\
x_1 &= x \wedge (a,a,b) & \text{(computed in } K_{a,b}),\\
x_2 &= x_1 \vee (b,a,a) & \text{(computed in } K_{a,b}),\\
x_3 &= x_2 \wedge q_2 & \text{(computed in } U),\\
x_4 &= x_3 \vee (d,c,c) & \text{(computed in } L),\\
x_5 &= x_4 \wedge (c,c,d) & \text{(computed in } K_{c,d}),\\
x_6 &= x_5 \vee p_2 & \text{(computed in } V),\\
x_7 &= x_6 \wedge p_1 & \text{(computed in } A).
\end{aligned}
$$

Our goal is to prove that $x_7 = \varphi x$.

By the definition of φ, when gluing U and V together, we identify $x = (x,0,0) \in A$ with $(x,b,x \wedge b) = (x,b,x) \in K_{a,b}$, so $x = (x,b,x) \in K_{a,b}$. Therefore, $x_1 = (x,b,x) \wedge (a,a,b) = (a,a,x)$, computed in $K_{a,b}$. We compute x_2 completely within $K_{a,b}$, utilizing Lemma 6.1:

$$x_2 = x_1 \vee (b,a,a) = (a,a,x) \vee (b,a,a) = \overline{(b,a,x)} = (b,x,x).$$

The element $x_3 = x_2 \wedge q_2$ is computed in U, which we obtained by gluing $K_{a,b}$ and B together with respect to the isomorphism

$$\varphi_1 \colon (1_B, x, x)_B \mapsto (0, x, 0)_{a,b}, \quad x \in [a,b],$$

between the filter $\mathrm{fil}(q_1)$ of B and the ideal $I_{a,b}$ of $K_{a,b}$. So x_3 is computed in two steps. First, in $K_{a,b}$:

$$x_2 \wedge v_{a,b} = (b,x,x) \wedge (0,b,0) = (0,x,0).$$

The image of $(0,x,0)$ under φ_1^{-1} is (b,x,x), so in B:

$$x_3 = (b,x,x) \wedge q_2 = (b,x,x) \wedge (a,b,a) = (a,x,a).$$

Now comes the crucial step. To compute $x_4 = x_3 \vee (d,c,c)$, we first compute in B:

$$x_3 \vee q_3 = (a,x,a) \vee (a,a,b) = \overline{(a,x,b)} = (x,x,b).$$

Take the image of (x, x, b) under φ and join it with (d, c, c) in $K_{c,d}$:

$$x_4 = \varphi(x, x, b) \vee (d, c, c) = (0, \varphi x, 0) \vee (d, c, c) = \overline{(d, \varphi x, c)} = (d, \varphi x, \varphi x).$$

So

$$x_5 = x_4 \wedge (c, c, d) = (d, \varphi x, \varphi x) \wedge (c, c, d) = (c, c, \varphi x),$$

computed in $K_{c,d}$.

To compute $x_6 = x_5 \vee p_2$, we note that

$$x_5 \vee v_{c,d} = (c, c, \varphi x) \vee (0, d, 0) = \overline{(c, d, \varphi x)} = (\varphi x, d, \varphi x).$$

Then we take the image of $(\varphi x, d, \varphi x)_{c,d}$ under φ_2 and join it with p_2 in A:

$$x_6 = \varphi_2(\varphi x, d, \varphi x) \vee p_2 = (0, 0, \varphi x) \vee (0, 1, 0) = \overline{(0, 1, \varphi x)} = (\varphi x, 1, \varphi x).$$

Finally, in A,

$$x_7 = x_6 \wedge p_1 = (\varphi x, 1, \varphi x) \wedge (1, 0, 0) = (\varphi x, 0, 0),$$

and φx is identified with $(\varphi x, 0, 0)$, so $x_7 = \varphi x$, as claimed.

The proof for φ^{-1} is similar, using the term function

$$q(y) = (((((y \vee p_2) \wedge (c, c, d)_{c,d}) \vee (d, c, c)_{c,d}) \wedge q_2) \vee (b, a, a)_{a,b}) \wedge (a, a, b)_{a,b}) \vee a.$$

This completes the proof of Theorem 18.1.

As you can see, the four sublattices of L isomorphic to M_3 play a crucial role in the proof; the elements forming these M_3-s are gray-filled in Figure 18.2.

18.4. 2/3-boolean triples

To prove the surjective result, we need all the gadgets from the bijective case, and a new one. The new gadget will be described in Lemma 18.14. This new gadget is based on the 2/3-boolean triple construction, which we proceed to describe.

Let $N_6 = \{o, p, q_1, q_2, r, i\}$ denote the six-element lattice depicted in Figure 18.3, with o the zero, i the unit element, p, q_1, q_2 the atoms, satisfying the relations $q_1 \vee q_2 = r$, $p \wedge q_1 = p \wedge q_2 = p \wedge r = o$, and $p \vee q_1 = p \vee q_2 = p \vee r = i$.

Following G. Grätzer, M. Greenberg, and E. T. Schmidt [82], for a bounded lattice P, we introduce 2/3-boolean triples: the element $(x, y, z) \in P^3$ is called a 2/3-*boolean triple* iff

$$y = (y \vee x) \wedge (y \vee z),$$
$$z = (z \vee x) \wedge (z \vee y).$$

We denote by $N_6[P]$ the set of all 2/3-boolean triples ordered component-wise. We prove that $N_6[P]$ is a lattice and describe the congruences of this lattice.

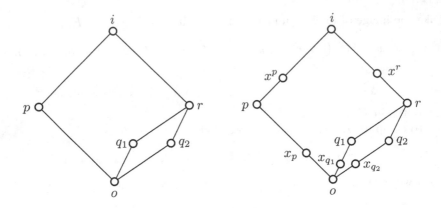

Figure 18.3: Illustrating N_6 and $N_6[P]$.

Lemma 18.3. *The subset $N_6[P]$ of P^3 is a closure system; let $\overline{(x,y,z)}$ denote the closure of $(x,y,z) \in P^3$ and call it the 2/3-boolean closure of (x,y,z). Then*

$$\overline{(x,y,z)} = (x, (y \vee x) \wedge (y \vee z), (z \vee x) \wedge (z \vee y)).$$

Proof. To simplify the notation, let $\overline{y} = (y \vee x) \wedge (y \vee z)$ and $\overline{z} = (z \vee x) \wedge (z \vee y)$. We have to verify that $\overline{(x,y,z)} = (x,\overline{y},\overline{z})$ is the closure of (x,y,z).

The triple $(x,\overline{y},\overline{z})$ is 2/3-boolean closed. Indeed, $y \leq \overline{y}$, so $x \vee y = x \vee \overline{y}$. Also, $z \leq \overline{z}$, so $\overline{y} \vee \overline{z} = y \vee z$. Therefore,

$$(\overline{y} \vee x) \wedge (\overline{y} \vee \overline{z}) = \overline{y},$$

verifying the first half of the definition of 2/3-boolean triples; the second half is proved similarly.

So $(x,y,z) \leq (x,\overline{y},\overline{z}) \in N_6[P]$. To prove that $N_6[P]$ is a closure system in P^3 and that $(x,\overline{y},\overline{z})$ is the closure of (x,y,z), it suffices to verify that if $(x_1,y_1,z_1) \in N_6[P]$ and $(x,y,z) \leq (x_1,y_1,z_1)$, then $(x,\overline{y},\overline{z}) \leq (x_1,y_1,z_1)$, which is obvious. $\qquad\square$

Corollary 18.4. *$N_6[P]$ is a lattice. Meet is componentwise and join is the closure of the componentwise join. Moreover, $N_6[P]$ has a spanning N_6 (see Figure 18.3), with the elements $o = (0,0,0)$, $p = (1,0,0)$, $q_1 = (0,1,0)$, $q_2 = (0,0,1)$, $r = (0,1,1)$, $i = (1,1,1)$.*

Lemma 18.5.

(i) *The interval $[o,p]$ of $N_6[P]$ is isomorphic to P under the isomorphism*

$$x_p = (x,0,0) \mapsto x, \quad x \in P.$$

(ii) *The interval* $[o, q_1]$ *of* $\mathsf{N}_6[P]$ *is isomorphic to* P *under the isomorphism*

$$x_{q_1} = (0, x, 0) \mapsto x, \quad x \in P.$$

(iii) *The interval* $[o, q_2]$ *of* $\mathsf{N}_6[P]$ *is isomorphic to* P *under the isomorphism*

$$x_{q_2} = (0, 0, x) \mapsto x, \quad x \in P.$$

(iv) *The interval* $[p, i]$ *of* $\mathsf{N}_6[P]$ *is isomorphic to* P *under the isomorphism*

$$x^p = (1, x, x) \mapsto x, \quad x \in P.$$

(v) *The interval* $[o, r]$ *of* $\mathsf{N}_6[P]$ *is isomorphic to* P^2 *under the isomorphism*

$$(0, x, y) \mapsto (x, y), \quad x, y \in P.$$

(vi) *The interval* $[r, i]$ *of* $\mathsf{N}_6[P]$ *is isomorphic to* P *under the isomorphism*

$$x^r = (x, 1, 1) \mapsto x, \quad x \in P.$$

Proof. By trivial computation. For instance, to prove (iv), observe that $(1, x, y)$ is closed iff $x = y$. □

For the five isomorphic copies of P in $\mathsf{N}_6[P]$, we use the notation:

$$P_p = [o, p], \quad P_{q_1} = [o, q_1], \quad P_{q_2} = [o, q_2],$$

with zero o and unit elements, p, q_1, q_2, respectively, and

$$P^p = [p, i], \quad P^r = [r, i],$$

with unit i and zero elements, p, r, respectively.

We describe the congruence structure of $\mathsf{N}_6[P]$ based on the following decomposition of elements:

Lemma 18.6. *Every* $\alpha \in \mathsf{N}_6[P]$ *has a decomposition*

$$\alpha = (\alpha \wedge p) \vee (\alpha \wedge q_1) \vee (\alpha \wedge q_2),$$

where $\alpha \wedge p \in P_p, \alpha \wedge q_1 \in P_{q_1}$, *and* $\alpha \wedge q_2 \in P_{q_2}$.

Proof. Indeed, the componentwise join of the right side equals α. □

For a congruence γ of $\mathsf{N}_6[P]$, let γ_p denote the restriction of γ to P_p, same for γ_{q_1} and γ_{q_2}. Let $\hat{\gamma}_p$ denote γ_p regarded as a congruence of P; same for $\hat{\gamma}_{q_1}$ and $\hat{\gamma}_{q_2}$. Similarly, let γ^p and γ^r denote the restriction of γ to P^p and P^r, respectively, and let $\hat{\gamma}^p$ and $\hat{\gamma}^r$ denote the corresponding congruences of P. Then we obtain:

Lemma 18.7. *Let* $\alpha, \alpha' \in \mathsf{N}_6[P]$ *and* $\gamma \in \operatorname{Con}\mathsf{N}_6[P]$. *Then*

$$\alpha \equiv \alpha' \pmod{\gamma}$$

iff

$$\alpha \wedge p \equiv \alpha' \wedge p \pmod{\gamma_p},$$
$$\alpha \wedge q_1 \equiv \alpha' \wedge q_1 \pmod{\gamma_{q_1}},$$
$$\alpha \wedge q_2 \equiv \alpha' \wedge q_2 \pmod{\gamma_{q_2}}.$$

Proof. This is clear from Lemma 18.6. □

Lemma 18.8. *Let* $\alpha = (x, y, z), \alpha' = (x', y', z') \in \mathsf{N}_6[P]$ *and* $\gamma \in \operatorname{Con}\mathsf{N}_6[P]$. *Then*

$$\alpha \equiv \alpha' \pmod{\gamma}$$

iff

$$x \equiv x' \pmod{\hat{\gamma}_p},$$
$$y \equiv y' \pmod{\hat{\gamma}_{q_1}},$$
$$z \equiv z' \pmod{\hat{\gamma}_{q_2}}.$$

Proof. This is clear from Lemma 18.7. □

Now we have the tools to describe the congruences. The next four lemmas provide the description.

Lemma 18.9. *Let* $\gamma \in \operatorname{Con}\mathsf{N}_6[P]$. *Then* $\hat{\gamma}_{q_1} = \hat{\gamma}_{q_2}$ *in* P.

Proof. Indeed, if $u \equiv u' \pmod{\hat{\gamma}_{q_1}}$, then $(0, u, 0) \equiv (0, u', 0) \pmod{\gamma_{q_1}}$, so $(0, u, 0) \equiv (0, u', 0) \pmod{\gamma}$. Therefore,

$$(0, 0, u) = ((0, u, 0) \vee (1, 0, 0)) \wedge (0, 0, 1)$$
$$\equiv ((0, u', 0) \vee (1, 0, 0)) \wedge (0, 0, 1) = (0, 0, u') \pmod{\gamma}.$$

We conclude that $(0, 0, u) \equiv (0, 0, u') \pmod{\gamma_{q_2}}$, that is, $u \equiv u' \pmod{\hat{\gamma}_{q_2}}$, proving that $\hat{\gamma}_{q_1} \leq \hat{\gamma}_{q_2}$. By symmetry, $\hat{\gamma}_{q_1} \geq \hat{\gamma}_{q_2}$, so $\hat{\gamma}_{q_1} = \hat{\gamma}_{q_2}$. □

Lemma 18.10. *For* $\gamma \in \operatorname{Con}\mathsf{N}_6[P]$, *the congruence inequality* $\hat{\gamma}_p \leq \hat{\gamma}_{q_2}$ *holds.*

Proof. Indeed, if $u \equiv u' \pmod{\hat{\gamma}_p}$, then $(u, 0, 0) \equiv (u', 0, 0) \pmod{\gamma_p}$. Therefore,

$$(0, 0, u) = (u, 1, u) \wedge (0, 0, 1) = ((u, 0, 0) \vee (0, 1, 0)) \wedge (0, 0, 1)$$
$$\equiv ((u', 0, 0) \vee (0, 1, 0)) \wedge (0, 0, 1)$$
$$= (u', 1, u') \wedge (0, 0, 1) = (0, 0, u') \pmod{\gamma},$$

that is, $u \equiv u' \pmod{\hat{\gamma}_{q_2}}$, proving that $\hat{\gamma}_p \leq \hat{\gamma}_{q_2}$. □

Lemma 18.11. *Let $\alpha \le \beta \in \operatorname{Con} P$. Then there is a unique $\gamma \in \operatorname{Con} \mathsf{N}_6[P]$, such that $\hat{\gamma}_p = \alpha$ and $\hat{\gamma}_{q_1} = \hat{\gamma}_{q_2} = \gamma$.*

Proof. The uniqueness follows from the previous lemmas. To prove the existence, for $\alpha \le \beta \in \operatorname{Con} P$, define a congruence γ on $\mathsf{N}_6[P]$ by

$$(x, y, z) \equiv (x', y', z') \pmod{\gamma}$$

iff

$$x \equiv x \pmod{\alpha}, \qquad y \equiv y' \pmod{\beta}, \qquad z \equiv z' \pmod{\beta}.$$

It is obvious that γ is an equivalence relation and it satisfies the Substitution Property for meet. To verify the Substitution Property for join, let $(x, y, z) \equiv (x', y', z') \pmod{\gamma}$ and let $(u, v, w) \in \mathsf{N}_6[P]$. Then

$$(x, y, z) \vee (u, v, w) = \overline{(x \vee u, y \vee v, z \vee w)}$$
$$= (x \vee u, (x \vee y \vee u \vee v) \wedge (y \vee z \vee v \vee w), (x \vee z \vee u \vee w) \wedge (y \vee z \vee v \vee w)).$$

Similarly,

$$(x', y', z') \vee (u, v, w)$$
$$= (x' \vee u, (x' \vee y' \vee u \vee v) \wedge (y' \vee z' \vee v \vee w), (x' \vee z' \vee u \vee w) \wedge (y' \vee z' \vee v \vee w)).$$

Since $x \equiv x' \pmod{\alpha}$, we also have

$$x \vee u \equiv x' \vee u \pmod{\alpha}.$$

From $x \equiv x' \pmod{\alpha}$ and $\alpha \le \beta$, it follows that $x \equiv x' \pmod{\beta}$. Also, $y \equiv y' \pmod{\beta}$, so $x \vee y \equiv x' \vee y' \pmod{\beta}$. Therefore, $x \vee y \vee u \vee v \equiv x' \vee y' \vee u \vee v \pmod{\beta}$. Similarly (or even simpler), $y \vee z \vee v \vee w \equiv y' \vee z' \vee v \vee w \pmod{\beta}$. Meeting the last two congruences, we obtain that

$$(x \vee y \vee u \vee v) \wedge (y \vee z \vee v \vee w) \equiv (x' \vee y' \vee u \vee v) \wedge (y' \vee z' \vee v \vee w) \pmod{\beta}.$$

Similarly,

$$(x \vee z \vee u \vee w) \wedge (y \vee z \vee v \vee w) = (x' \vee z' \vee u \vee w) \wedge (y' \vee z' \vee v \vee w) \pmod{\beta}.$$

The last three displayed equations verify that

$$(x, y, z) \vee (u, v, w) \equiv (x', y', z') \vee (u, v, w) \pmod{\beta}. \qquad \square$$

Now note that for $x, y \in P$ and congruence γ of P,

$$x_p \equiv y_p \pmod{\gamma} \qquad \text{iff} \qquad x^r \equiv y^r \pmod{\gamma}$$

and

$$x^p \equiv y^p \pmod{\gamma} \qquad \text{iff} \qquad x_{q_1} \equiv y_{q_1} \pmod{\gamma}.$$

It follows that $\hat{\gamma}^p = \hat{\gamma}_{q_1} = \hat{\gamma}_{q_2}$ and $\hat{\gamma}^r = \hat{\gamma}_p$, so Lemmas 18.9–18.11 can be restated as follows:

Corollary 18.12. *There is a one-to-one correspondence between the congruences of $N_6[P]$ and pairs of congruences $\alpha \leq \beta$ of $\operatorname{Con} P$, defined by*

$$\gamma \mapsto (\hat{\gamma}^r, \hat{\gamma}^p).$$

The inequality $\hat{\gamma}^r \leq \hat{\gamma}^p$ can be established in a stronger form by exhibiting a term function $t(x)$ on $N_6[P]$ such that $t(u^r) = u^p$ for $u \in P$.

Lemma 18.13. *There is a term function $t(x)$ on $N_6[P]$ such that $t(u^r) = u^p$ for $u \in P$.*

Proof. Define
$$t(x) = (((x \wedge p) \vee q_1) \wedge q_2) \vee p.$$

Indeed, if $u \in P$, then $u^r = (u, 1, 1)$ and so

$$
\begin{aligned}
t(u^r) &= t((u, 1, 1))\\
&= ((((u, 1, 1) \wedge (1, 0, 0)) \vee (0, 1, 0)) \wedge (0, 0, 1)) \vee (1, 0, 0)\\
&= (((u, 0, 0) \vee (0, 1, 0)) \wedge (0, 0, 1)) \vee (1, 0, 0)\\
&= (\overline{(u, 1, 0)} \wedge (0, 0, 1)) \vee (1, 0, 0)\\
&= ((u, 1, u) \wedge (0, 0, 1)) \vee (1, 0, 0)\\
&= (0, 0, u) \vee (1, 0, 0)\\
&= \overline{(1, 0, u)}\\
&= (1, u, u) = u^p,
\end{aligned}
$$

as required. $\square$

Actually, to prove Theorem 18.2, we need not the lattice $N_6[P]$, but a quotient thereof, which we now proceed to construct.

Let P and Q be nontrivial bounded lattices and let $\varphi \colon P \to Q$ be a homomorphism of P *onto* Q. By Corollary 18.12, there is a unique congruence γ of $N_6[P]$ corresponding to the congruence pair $\mathbf{0} \leq \ker(\varphi)$ of P. Define

$$B = N_6[P]/\gamma.$$

This is our new gadget.

It is useful to note that B can be represented as

$$\{\, (x, y, z) \in P \times Q \times Q \mid y = (y \vee \varphi x) \wedge (y \vee z) \text{ and } z = (z \vee \varphi x) \wedge (z \vee y) \,\}.$$

By the Second Isomorphism Theorem (Theorem 1.5), there is a one-to-one correspondence between the congruences of B and congruence pairs $\alpha \leq \beta$ of P satisfying $\ker(\varphi) \leq \beta$. For $x \in N_6[P]$, let $\overline{x}$ denote the congruence class x/β.

Using this notation, utilizing the results of this section, we state some important properties of the lattice B.

Lemma 18.14. *Let P and Q be bounded lattices and let $\varphi\colon P \to Q$ be a homomorphism of P onto Q. Then there is a lattice B, with the following properties:*

(i) *B has a spanning sublattice $\{\overline{o}, \overline{p}, \overline{q}_1, \overline{q}_2, \overline{s}, \overline{i}\}$ isomorphic to N_6.*

(ii) *The interval $[\overline{s}, \overline{i}]$ is isomorphic to P under the map $x \mapsto \overline{x^r}$, $x \in P$.*

(iii) *The interval $[\overline{p}, \overline{i}]$ is isomorphic to Q under the map $y \mapsto \overline{y^p}$, $y \in Q$, where $x \in P$ with $\varphi x = y$.*

(iv) *The congruences Σ of B are in one-to-one correspondence with pairs of congruences $(\boldsymbol{\alpha}, \boldsymbol{\beta})$, where $\boldsymbol{\alpha}$ is a congruence of P and $\boldsymbol{\beta}$ is a congruence of Q satisfying $\boldsymbol{\alpha} \le \varphi^{-1}\boldsymbol{\beta}$, where up to isomorphism, Σ restricted to $[\overline{r}, \overline{i}]$ is $\boldsymbol{\alpha}$ and Σ restricted to $[\overline{p}, \overline{i}]$ is $\boldsymbol{\beta}$.*

(v) *There is a term function $t(x)$ such that $t(\overline{u^r}) = \overline{x^p}$ for $u \in P$, where $x \in P$ with $\varphi x = u$.*

18.5. *Proof-by-Picture* for surjective maps

We proceed with the *Proof-by-Picture* of Theorem 18.2 as in Section 18.2. We can assume, without loss of generality, that $[a, b]$ and $[c, d]$ are nontrivial intervals, that is, $a < b$ and $c < d$.

We take the four gadgets: $A = \mathsf{M}_3[K]$, $K_{a,b}$, B, and $K_{c,d}$, except that now B is not $\mathsf{M}_3[a, b]$ but the lattice $B = \mathsf{N}_6[a, b]/\gamma$, see Lemma 18.14, constructed from $P = [a, b]$, $Q = [c, d]$, and the homomorphism φ from $[a, b]$ onto $[c, d]$ given in Theorem 18.2 (Figure 18.4).

We do three gluings.

First gluing. In B, we use the filter

$$\mathrm{fil}(\overline{r}) = \{\,\overline{(x, 1, 1)} \mid a \le x \le b\,\}$$

(which is isomorphic to $[a, b]$ since γ on $[r, 1]$ is $\mathbf{0}$), while in $K_{a,b}$ we utilize the ideal

$$I_{a,b} = \{\,(0, x, 0) \mid a \le x \le b\,\}$$

(which is obviously isomorphic to $[a, b]$), and we consider the isomorphism

$$\varphi_1\colon \overline{(x, 1, 1)}_B \mapsto (0, x, 0)_{K_{a,b}}, \quad x \in [a, b],$$

between the filter $\mathrm{fil}(\overline{r})$ of B and the ideal $I_{a,b}$ of $K_{a,b}$ to glue B and $K_{a,b}$ together to obtain the lattice U. (As in Section 18.2, we use the following notation: to indicate whether a triple (x, y, z) belongs to $A = \mathsf{M}_3[K]$, $K_{a,b}$, B, or $K_{c,d}$, we subscript it: with $(x, y, z)_A$, $(x, y, z)_{a,b}$, $(x, y, z)_B$, or $(x, y, z)_{c,d}$, respectively.)

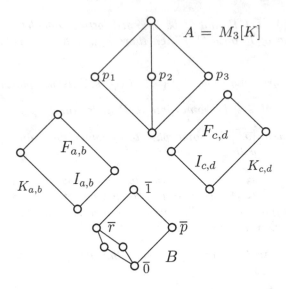

Figure 18.4: The four gadgets to construct L.

Second gluing. We glue $K_{c,d}$ and A over the filter

$$F_{c,d} = \{\,(x, c, x \wedge c) \mid x \in K\,\}$$

of $K_{c,d}$ and the ideal

$$\mathrm{id}(p_3) = \{\,(0, 0, x) \mid x \in K\,\}$$

of A, with respect to the isomorphism

$$\varphi_2 \colon (x, c, x \wedge c)_{K_{c,d}} \mapsto (0, 0, x)_A, \quad x \in K,$$

to obtain the lattice V.

Final gluing. In U, we define the filter

$$F = [\overline{p}, \overline{1}] \cup F_{a,b},$$

which is the union of $[\overline{p}, \overline{1}]$ and $F_{a,b}$, with the unit of $[\overline{p}, \overline{1}]$ identified with the zero of $F_{a,b}$.

In V, we define the ideal

$$I = I_{c,d} \cup [0_A, p_1],$$

which is the union of $I_{c,d}$ and $[0_A, p_1]$, with the unit of $I_{c,d}$ identified with the zero of $[0_A, p_1]$.

Next we set up an isomorphism $\varphi\colon F \to I$. Since

$$[\bar{p}, \bar{1}] = \{\, \overline{(1, x, x)_B} \mid a \le x \le b \,\}$$

and

$$I_{c,d} = \{\, (0, x, 0)_{c,d} \mid c \le x \le d \,\},$$

we define φ on $[\bar{p}, \bar{1}]$ by

$$\varphi\colon \overline{(1, x, x)_B} \mapsto (0, \gamma x, 0)_{c,d},$$

where $\varphi\colon [a, b] \to [c, d]$ is the homomorphism given in Theorem 18.2. We define φ on $F_{a,b}$ by

$$\varphi\colon (x, b, x \wedge b)_{a,b} \mapsto (x, 0, 0)_A, \quad x \in K.$$

It is clear that $\varphi\colon F \to I$ is well-defined and it is an isomorphism.

Finally, we construct the lattice L of Theorem 18.2 by gluing U over I with V over F with respect to the isomorphism $\varphi\colon F \to I$.

The map $x \mapsto (x, 0, 0)_A$ is an isomorphism between K and the principal ideal $\mathrm{id}(p_1)$ of A; this gives us a convex embedding of K into A. We identify K with its image, and regard K as a convex sublattice of A and therefore of L. So L is a convex extension of K. We have completed the construction of the bounded lattice L of Theorem 18.2.

18.6. Verification for surjective maps

The proof in Section 18.3 heavily depended on the fact that we glued over ideals and filters of which the building components were congruence-preserving extensions. This is no longer the case; however, a modification of Lemma 2.9 comes to the rescue.

A congruence κ of L can be described by four congruences,

κ_A, the restriction of κ to A,

$\kappa_{a,b}$, the restriction of κ to $K_{a,b}$,

$\kappa_{c,d}$, the restriction of κ to $K_{c,d}$,

κ_B, the restriction of κ to B.

These congruences satisfy a number of conditions:

(i) $(0, x, 0)_{a,b} \equiv (0, y, 0)_{a,b} \pmod{\kappa_{a,b}}$ iff $\overline{(x, 1, 1)_B} \equiv \overline{(y, 1, 1)_B} \pmod{\kappa_B}$
    for $x, y \in [a, b]$.

(ii) $(0, 0, x)_A \equiv (0, 0, y)_A \pmod{\kappa_A}$ iff

$$(x, c, x \wedge c)_{c,d} \equiv (y, c, y \wedge c)_{c,d} \pmod{\kappa_{c,d}}$$

    for $x, y \in K$.

(iii) $(x, 0, 0)_A \equiv (y, 0, 0)_A \pmod{\kappa_A}$ iff

$$(x, a, x \wedge a)_{a,b} \equiv (y, a, y \wedge a)_{a,b} \pmod{\kappa_{a,b}}$$

for $x, y \in K$.

(iv) $(0, \gamma x, 0)_{c,d} \equiv (0, \gamma y, 0)_{c,d} \pmod{\kappa_{c,d}}$ iff

$$\overline{(1, x, x)_B} \equiv \overline{(1, y, y)_B} \pmod{\kappa_B}$$

for $x, \, y \in [a, b]$.

Conversely, if we are given congruences κ_A on A, $\kappa_{a,b}$ on $K_{a,b}$, $\kappa_{c,d}$ on $K_{c,d}$, κ_B on B, then by (i), we can define a congruence κ_U on U. By (ii), we can define a congruence κ_V on V. By (iii) and (iv), we can define a congruence κ_L on L.

Now it is clear that if we start with a congruence σ of K, then we can define the congruences σ_A on A, $\sigma_{a,b}$ on $K_{a,b}$, $\sigma_{c,d}$ on $K_{c,d}$ componentwise, and σ_B on B as in Section 18.3. Conditions (i)–(iii) trivially hold (since σ_A, $\sigma_{a,b}$, and $\sigma_{c,d}$ are defined componentwise). Finally, (iv) holds if σ is a γ-congruence. So every γ-congruence of K has an extension to L.

Let σ be a congruence of K that extends to L. Since A is a congruence-preserving convex extension of $K = [0_A, p_1]$, further, $K_{a,b}$ is a congruence-preserving extension of $F_{a,b}$, and $K_{c,d}$ is a congruence-preserving extension of $F_{c,d}$, the congruence σ uniquely extends to A as σ_A, to $K_{a,b}$ as $\sigma_{a,b}$, and to $K_{c,d}$ as $\sigma_{c,d}$. Therefore, σ uniquely extends to the intervals $[\bar{r}, \bar{1}]$ and $[\bar{p}, \bar{1}]$ of B, and so by Lemma 18.14 to B. We conclude that if a congruence σ of K extends to L, then it extends uniquely.

To complete the proof, we prove that γ is a term function. Define

$$p(x) = (((((((((x \wedge (a, a, b)_{a,b}) \vee (b, a, a)_{a,b}) \wedge p) \vee q_1) \wedge q_2)$$
$$\vee (d, c, c)_{c,d}) \wedge (c, c, d)_{c,d}) \vee p_2) \wedge p_1.$$

By Lemma 18.13, $p(x)$ behaves properly in B, while outside of B, $p(x)$ is the same term function as in Section 18.3.

This completes the proof of Theorem 18.2.

18.7. Discussion

Note that Theorem 18.2 implies Theorem 18.1. However, the two theorems have different generalizations.

In [135], G. Grätzer and E. T. Schmidt proved the generalizations stated in the three following subsections.

First generalization of Theorem 18.1

Let K be a bounded lattice, let $[a_i, b_i]$, $i < \alpha$, be intervals of K (α is an initial ordinal ≥ 2), and let

$$\gamma_{i,j} \colon [a_i, b_i] \to [a_j, b_j], \quad \text{for } i, j < \alpha,$$

be an isomorphism between the intervals $[a_i, b_i]$ and $[a_j, b_j]$. For notational convenience, we write $[a, b]$ for $[a_0, b_0]$. Let

$$\gamma = \{\, \gamma_{i,j} \mid i,\ j < \alpha \,\}$$

be subject to the following conditions for $i, j < \alpha$:

(1) $\gamma_{i,i}$ is the identity map on $[a_i, b_i]$.

(2) $\gamma_{i,j}^{-1} = \gamma_{j,i}$.

(3) $\gamma_{j,k} \circ \gamma_{i,j} = \gamma_{i,k}$.

Let K_γ denote the partial algebra obtained from K by adding the partial operations $\gamma_{i,j}$ for $i, j < \alpha$. Let us call a congruence α of K a γ-congruence iff α satisfies the Substitution Property with respect to the partial unary operations $\gamma_{i,j} \in \gamma$. Thus a congruence relation of the partial algebra K_γ is the same as a γ-congruence of the lattice K. We call the lattice L a γ-congruence-preserving extension of K if a congruence of K extends to L iff it is a γ-congruence of K_γ and every γ-congruence of K has *exactly one* extension to L.

Theorem 18.15. *Let K be a bounded lattice, and let $\alpha \geq 2$ be an ordinal. Let $[a_i, b_i]$, for $i < \alpha$, be intervals of K, and let $\gamma_{i,j} \colon [a_i, b_i] \to [a_j, b_j]$ be an isomorphism between the intervals $[a_i, b_i]$ and $[a_j, b_j]$, for $i, j < \alpha$, subject to the conditions (1)–(3), where $\gamma = \{\, \gamma_{i,j} \mid i,\ j < \alpha \,\}$. Then the partial algebra K_γ has a γ-congruence-preserving convex extension into a bounded lattice L such that all $\gamma_{i,j}$, for $i, j < \alpha$, are term functions in L. In particular, the congruence lattice of the partial algebra K_γ is isomorphic to the congruence lattice of the lattice L.*

It would be nice to be able to claim that this theorem can be proved by applying Theorem 18.1 to the isomorphisms $\gamma_{i,j}$ one at a time, and then forming a direct limit. Unfortunately, the direct limit at ω produces a lattice with no zero or unit, so we cannot continue with the construction.

Second generalization of Theorem 18.1

Let K be a bounded lattice, let α be an ordinal, and for $i < \alpha$, let γ_i be an isomorphism between the interval $[a_i, b_i]$ and the interval $[c_i, d_i]$:

$$\gamma_i \colon [a_i, b_i] \to [c_i, d_i].$$

Let

$$\gamma = \{\, \gamma_i \mid i < \alpha \,\},$$

and let K_γ denote the partial algebra obtained from K by adding the partial operations γ_i for $i < \alpha$. Let us call a congruence α of the lattice K a γ-*congruence* iff α satisfies the Substitution Property with respect to the partial unary operations γ_i, for $i < \alpha$, that is, $x \equiv y \pmod{\alpha}$ implies that $x\gamma_i \equiv y\gamma_i \pmod{\alpha}$ for all $x, y \in [a_i, b_i]$ and $i < \alpha$. Thus a congruence relation of the partial algebra K_γ is the same as a γ-congruence of the lattice K. We call the lattice L a γ-*congruence-preserving extension* of the lattice K if a congruence of K extends to L iff it is a γ-congruence of K and every γ-congruence of K has *exactly one* extension to L.

Theorem 18.16. *Let K be a bounded lattice, let γ be given as above. Then the partial algebra K_γ has a γ-congruence-preserving convex extension into a lattice L such that all γ_i, for $i \in I$, are term functions in L. In particular, the congruence lattice of the partial algebra K_γ is isomorphic to the congruence lattice of the lattice L.*

A generalization of Theorem 18.2

Let K be a bounded lattice, and for $i \in I$, let γ_i be a homomorphism of the interval $[a_i, b_i]$ onto the interval $[c_i, d_i]$. Let

$$\gamma = \{\, \gamma_i \mid i \in I \,\},$$

and let K_γ denote the partial algebra obtained from K by adding the partial operations γ_i for $i \in I$. Let us call a congruence α of the lattice K a γ-*congruence* iff α satisfies the Substitution Property with respect to the partial unary operations γ_i, for $i \in I$, that is, $x \equiv y \pmod{\alpha}$ implies that $\gamma_i(x) \equiv \gamma_i \pmod{y}(\alpha)$ for all $x, y \in [a_i, b_i]$ and $i \in I$. Thus a congruence relation of the partial algebra K_γ is the same as a γ-congruence of the lattice K. We call the lattice L a γ-*congruence-preserving extension* of the lattice K if a congruence of K extends to L iff it is a γ-congruence of K and every γ-congruence of K has *exactly one* extension to L.

Theorem 18.17. *Let K be a bounded lattice, let γ be given as above. Then the partial algebra K_γ has a γ-congruence-preserving convex extension into a lattice L such that all γ_i, $i \in I$, are term functions in L. In particular, the congruence lattice of the partial algebra K_γ is isomorphic to the congruence lattice of the lattice L.*

Theorem 18.17 easily implies Theorems 18.15 and 18.16, with one important difference: In Theorem 18.15, we obtain a *bounded* lattice L.

Problem 18.1. Can we ensure that the lattice L be bounded in Theorem 18.17?

Magic wands with special properties

In Theorems 18.1 and 18.2, can we construct a lattice L with special properties, such as being semimodular? Obviously not if we insist on convex embeddings, since a convex sublattice of a semimodular lattice is semimodular again. Let Theorem 18.1* denote Theorem 18.1 with "convex" deleted, and the same for Theorem 18.2*.

Let us call be a class **C** of lattices a *CPE-class* if every finite lattice A has a finite congruence-preserving extension $B \in$ **C**.

Theorem 18.18. *Let K be a finite lattice, and let* **C** *be a CPE-class of lattices. Then in Theorems 18.1* *and 18.2*, *for a finite K, we can assume that $L \in$* **C**.

Problem 18.2. What can we say about CPE-classes?

By Theorem 14.1, sectionally complemented lattices; by Theorem 15.1, semimodular lattices; by Theorem 16.1, isoform lattices; by Theorem 17.1, lattices with a given automorphism group form such classes. So there is a sectionally complemented magic wand, a semimodular magic wand, and so on.

Problem 18.3. Is it possible to get a magic wand combining any two of these properties?

Fully invariant congruences

As usual, let us call a congruence $\boldsymbol{\alpha}$ *fully invariant* iff $a \equiv b \pmod{\boldsymbol{\alpha}}$ implies that $\delta a \equiv \delta b \pmod{\boldsymbol{\alpha}}$ for any automorphism δ.

For a lattice K, let $\mathrm{Con_{inv}}\, K$ denote the lattice of fully invariant congruences of K, and let $\mathrm{Aut}\, K$ denote the set (group) of automorphisms of K.

For a bounded lattice K, we can apply Theorem 18.17 to $I = \mathrm{Aut}\, K$; for $\delta \in \mathrm{Aut}\, K$, let $[a_\delta, b_\delta] = [c_\delta, d_\delta] = [0, 1]$, and let $\gamma_\delta = \delta$.

Theorem 18.19. *Let K be a bounded lattice. Then K has a convex extension into a lattice L such that a congruence of K extends to L iff it is fully invariant and a fully invariant congruence of K extends uniquely to L. In particular, $\mathrm{Con_{inv}}\, K$ is isomorphic to $\mathrm{Con}\, L$.*

The 1/3-boolean triple construction

The reader may ask, what is a 1/3-boolean triple construction? For a lattice P, let us call the element $(x, y, z) \in P^3$ a *1/3-boolean triple* iff

$$z = (z \vee x) \wedge (z \vee y).$$

Then instead of N_6 of Figure 18.3, we now get the lattice N_7 of Figure 18.5 (the dual of S_7), and the 1/3-boolean triples form a lattice $\mathsf{N}_7[P]$.

Problem 18.4. Is there some use of the 1/3-boolean triple construction?

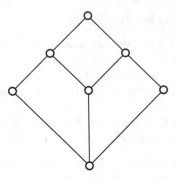

Figure 18.5: The lattice N_7.

Part V

Congruence Lattices of Two Related Lattices

Sublattices

19.1. The results

The simplest connection between two lattices K and L is the sublattice relation: $K \leq L$. How then does $\operatorname{Con} K$ relate to $\operatorname{Con} L$?

As we discussed in Section 3.3, the relation $K \leq L$ induces a map ext of $\operatorname{Con} K$ into $\operatorname{Con} L$: For a congruence relation α of K, the image $\operatorname{ext} \alpha$ is the congruence relation of L generated by α, that is, $\operatorname{ext} \alpha = \operatorname{con}_L(\alpha)$. The map ext is a $\{0\}$-separating join-homomorphism. (If we want to emphasize the embedding $\operatorname{id}_K \colon K \to L$, we write $\operatorname{ext} \operatorname{id}_K$ for ext.)

In 1974, A. P. Huhn in [158] stated the converse:

Theorem 19.1. *Let D and E be finite distributive lattices, and let $\gamma \colon D \to E$ be a $\{0\}$-separating join-homomorphism. Then there are finite lattices $K \leq L$, and isomorphisms $\alpha \colon D \to \operatorname{Con} K$ and $\beta \colon E \to \operatorname{Con} L$ satisfying*

$$\beta\gamma = (\operatorname{ext} \operatorname{id}_K)\alpha,$$

where id_K is the embedding of K into L; that is, such that the diagram

$$
\begin{array}{ccc}
D & \xrightarrow{\ \gamma\ } & E \\[4pt]
{\scriptstyle\cong}\Big\downarrow{\scriptstyle\alpha} & & {\scriptstyle\cong}\Big\downarrow{\scriptstyle\beta} \\[4pt]
\operatorname{Con} K & \xrightarrow{\ \operatorname{ext} \operatorname{id}_K\ } & \operatorname{Con} L
\end{array}
$$

is commutative.

© Springer International Publishing Switzerland 2016   225
G. Grätzer, *The Congruences of a Finite Lattice*,
DOI 10.1007/978-3-319-38798-7_19

In this chapter we prove a much stronger version of this theorem from G. Grätzer, H. Lakser, and E. T. Schmidt [103]:

Theorem 19.2. *Let K be a finite lattice, let E be a finite distributive lattice, and let $\gamma\colon \operatorname{Con} K \to E$ be a $\{0\}$-separating join-homomorphism. Then there is a finite extension L of K and an isomorphism $\beta\colon E \to \operatorname{Con} L$ with $\operatorname{ext} \operatorname{id}_K = \beta\gamma$, that is, such that the diagram*

$$
\begin{array}{ccc}
\operatorname{Con} K & \xrightarrow{\ \gamma\ } & E \\[2pt]
= \Big\downarrow & & \cong \Big\downarrow \beta \\[2pt]
\operatorname{Con} K & \xrightarrow{\ \operatorname{ext} \operatorname{id}_K\ } & \operatorname{Con} L
\end{array}
$$

is commutative.

While Theorem 19.1 claims the existence of the finite lattices $K \le L$, Theorem 19.2 claims that for every K, there is an extension L. This stronger form allows us to claim the existence of K and L in any CPE-class **C** (introduced in the subsection: Magic wands with special properties, of Section 18.7).

Theorem 19.3. *Let D and E be finite distributive lattices, and let $\gamma\colon D \to E$ be a $\{0\}$-separating join-homomorphism. Let **C** be a CPE-class of lattices. Then there are finite lattices $K, L \in \mathbf{C}$ with $K \le L$, and isomorphisms*

$$
\alpha\colon D \to \operatorname{Con} K \ \text{ and } \ \beta\colon E \to \operatorname{Con} L
$$

satisfying

$$
\gamma\beta = (\operatorname{ext} \operatorname{id}_K)\alpha,
$$

where id_K is the embedding of K into L.

So there are such lattices $K \le L$ that are *sectionally complemented*, by Theorem 14.1; *semimodular*, by Theorem 15.1; *isoform*, by Theorem 16.1; *with a given automorphism group*, by Theorem 17.1.

Theorem 19.3 trivially follows from Theorem 19.2. Indeed, let D, E, and γ be given as in Theorem 19.3. By the Dilworth Theorem, there is a finite lattice K_1 satisfying $\operatorname{Con} K_1 \cong D$ with the isomorphism φ_1. By the definition of a CPE-class, the lattice K_1 has a congruence-preserving extension $K \in \mathbf{C}$; let φ_2 be an isomorphism $\operatorname{Con} K \to \operatorname{Con} K_1$. Then $\gamma' = \varphi_2\varphi_1\gamma$ is a $\{0\}$-separating join-homomorphism of $\operatorname{Con} K$ into E. Applying Theorem 19.2 to K, E, and γ', there is a finite extension L_1 of K, an embedding $\operatorname{id}_K\colon K \to L_1$, and an isomorphism $\beta\colon E \to \operatorname{Con} L_1$ with $\operatorname{ext} \operatorname{id}_K = \gamma\beta$. We use once more that **C** is a CPE-class: there is a finite congruence-preserving extension $L \in \mathbf{C}$ of L_1. Obviously, K and L satisfy the requirements of Theorem 19.3.

So we only have to prove Theorem 19.2.

19.2.   *Proof-by-Picture*

This *Proof-by-Picture* is based on the ideas in G. Grätzer, H. Lakser, and F. Wehrung [108] (see Section 19.5 for the history of the proofs).

Let K, E, and $\gamma\colon \operatorname{Con} K \to E$ be given as in Theorem 19.2. We want to construct a finite extension L of K, an embedding $\operatorname{id}_K\colon K \to L$, and an isomorphism $\beta\colon E \to \operatorname{Con} L$ with $\operatorname{ext}\operatorname{id}_K = \beta\gamma$.

First, let $E = \mathsf{C}_2 = \{0,1\}$. Since γ is a $\{0\}$-separating join-homomorphism, it follows that $\gamma 0 = 0$, and $\gamma\alpha = 1$, for all $\alpha > \operatorname{Con} K - \{\kappa_K\}$. By Lemma 14.3, K has a finite simple extension L, with the embedding $\operatorname{id}_K\colon K \to L$. Observe that $\operatorname{ext}\mathbf{0}_K = \kappa_L$ and $\operatorname{ext}\mathbf{1}_K = \mathbf{1}_L$, since L is simple. Define $\beta\colon E \to \operatorname{Con} L$ by $\beta 0 = \kappa_L$ and $\beta 1 = \mathbf{1}_L$. So $\operatorname{ext}\operatorname{id}_K = \beta\gamma$ is obvious; it is illustrated in Figure 19.1.

Second, let $E = \mathsf{C}_2 \times \mathsf{C}_2$. We get two projection maps $\pi_i\colon E \to \mathsf{C}_2$ for $i = 1, 2$. So we can apply the $E = \mathsf{C}_2$ case—the last paragraph—to K, C_2, and $\gamma\pi_i\colon \operatorname{Con} K \to \mathsf{C}_2$, and we obtain the simple extensions L_i of K. Obviously, $L = L_1 \times L_2$ will do the job.

In general, if $E = \mathsf{B}_n$, we can take $L = L_1 \times L_2 \times \cdots \times L_n$, where the finite lattices $L_1, L_2, \ldots, L_n$ are simple extensions of K.

Now let us take the smallest non-boolean case: $E = \mathsf{C}_3 = \{0, a, 1\}$. We embed E into $B = \mathsf{B}_2 = \{0, a, b, 1\}$, and take the lattice L_B we have just constructed for B. Let $\operatorname{Con} L_B = \{\mathbf{0}, \alpha_B, \gamma_B, \mathbf{1}\}$ and choose in L_B a boolean filter $F = \{0_F, u_F, v_F, 1_B\}$ so that $\operatorname{con}(u_F, 1_B) = \alpha_B$ and $\operatorname{con}(v_F, 1_B) = \gamma_B$. Obviously, F is a congruence-determining sublattice of L_B.

By Theorem 8.9, we can represent E as the congruence lattice of a finite lattice L_E with $\operatorname{Con} L_E = \{\mathbf{0}, \alpha_E, \mathbf{1}\}$ so that L_E has a boolean ideal $I = \{0, u_E, v_E, 1_I\}$ and $\operatorname{con}(0, u_E) = \alpha_{L_E}$ and $\operatorname{con}(0, v_E) = \mathbf{1}_{L_E}$. Obviously, I is a congruence-determining sublattice of L_E.

Now we construct the lattice L by gluing L_B (with the filter F) to the lattice L_E (with the ideal I) with respect to the obvious isomorphism between F and I that maps u_F to u_I and v_F to v_I. In L, we use the notation $u = u_F = u_I$ and $v = v_F = v_I$, see Figure 19.2.

Since α_B restricted to F agrees with α_I restricted to I, it is clear that L has only one nontrivial congruence α, which agrees with α_B on L_B and with α_E on L_E. The maps work out with no difficulty (almost the same way as in Figure 19.1), verifying that L satisfies the requirements of Theorem 19.2.

Since a formal proof of Theorem 19.2 is almost the same as that of this special case, we will not present a formal version of this *Proof-by-Picture*.

19.3.   Multi-coloring

We introduced colored lattices in Section 3.2. Now we need a generalization from G. Grätzer, H. Lakser, and E. T. Schmidt [103].

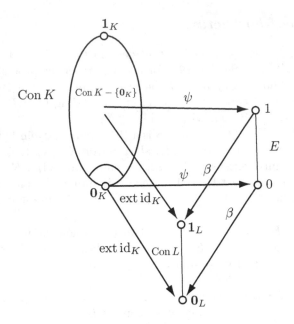

Figure 19.1: The case $E = \mathsf{C}_2$.

Let M be a finite lattice and let C be a finite set; the elements of C will be called *colors*. Recall that Prime(M) denotes the set of prime intervals of M.

A *multi-coloring* of M over C is an map μ from Prime(M) into Pow$^+ C$ (all nonempty subsets of C ordered by set-inclusion) satisfying the condition that if $\mathfrak{p}, \mathfrak{q} \in$ Prime(M) and con($\mathfrak{p}$) $\leq$ con($\mathfrak{q}$), then $\mu\mathfrak{p} \subseteq \mu\mathfrak{q}$. Coloring is the special case when all the $\mu\mathfrak{p}$ are singletons.

Equivalently, a multi-coloring is an isotone map of the ordered set Con$_J$ M into the ordered set Pow$^+ C$.

We will now show that a multi-colored lattice has a natural extension to a colored lattice.

Lemma 19.4. *Let M be a finite lattice with a multi-coloring μ over the set C. Then there exists a lattice M^* with a coloring μ^* over C such that the following conditions holds:*

(i) *M^* is the direct product of the lattices M_c, $c \in C$, where M_c is a homomorphic image of M colored by $\{c\}$.*

(ii) *There is a lattice embedding $a \mapsto a^*$ of M into M^*.*

(iii) *For every prime interval $\mathfrak{p} = [a, b]$ of M,*

$$\mu\mathfrak{p} = \{\, \mu^*(\mathfrak{q}) \mid \mathfrak{q} \in \text{Prime}(M^*) \text{ and } \mathfrak{q} \subseteq [a^*, b^*] \,\}$$

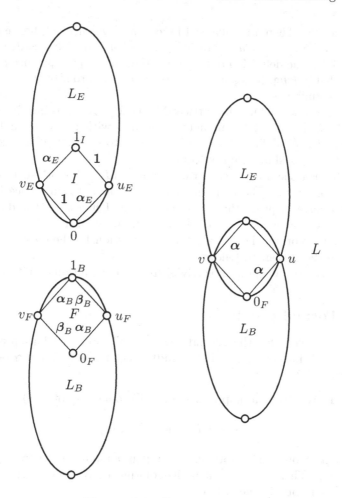

Figure 19.2: Constructing L.

and the minimal extension of $\mathrm{con}(\mathfrak{p})$ under this embedding into M^* is of the form

$$\prod(\mathrm{con}(\mathfrak{p}_c) \mid c \in C),$$

where $\mathfrak{p}_c$ is a prime interval of M_c iff $c \in \mu\mathfrak{p}$ and $\mathfrak{p}_c$ is a trivial interval otherwise (in which case, $\mathrm{con}(\mathfrak{p}_c) = \mathbf{0}_{M_c}$).

Proof. For $c \in C$, define the binary relation $\boldsymbol{\gamma}_c$ on M as follows:

$$u \equiv v \pmod{\boldsymbol{\gamma}_c} \text{ iff } c \notin \mu\mathfrak{p}, \text{ for every prime interval } \mathfrak{p} \subseteq [u \wedge v, u \vee v].$$

This relation is obviously reflexive and symmetric. To show transitivity, assume that $u \equiv v \pmod{\boldsymbol{\gamma}_c}$ and $v \equiv w \pmod{\boldsymbol{\gamma}_c}$, and let $\mathfrak{q}$ be a prime interval in $[u \wedge w, u \vee w]$. Then $\mathfrak{q}$ is collapsed by $\mathrm{con}(u, v) \vee \mathrm{con}(v, w)$, hence there is a prime interval $\mathfrak{p}$ in $[u \wedge v, u \vee v]$ or in $[v \wedge w, v \vee w]$ satisfying $\mathrm{con}(\mathfrak{q}) \leq \mathrm{con}(\mathfrak{p})$. It follows from the definition of multi-coloring that $\mu\mathfrak{q} \subseteq \mu\mathfrak{p}$; since $c \notin \mu\mathfrak{p}$, we conclude that $c \notin \mu\mathfrak{q}$, hence $u \equiv w \pmod{\gamma_c}$. The proof of the Substitution Property is similar.

For $c \in C$, we define the lattice M_c as M/γ_c. A prime interval $\mathfrak{p}$ of $M^* = \prod(M_c \mid c \in C)$ is uniquely associated with a $c \in C$ and a prime interval of M_c. We define $\mu^*(\mathfrak{p}) = c$. It is easy to see that μ^* is a coloring of M^* over C, establishing the first condition.

To establish the second condition for $a \in M$, define a^* so that its M_c-component be a/γ_c. The mapping $a \mapsto a^*$ is obviously a lattice homomorphism. We have to prove that it is one-to-one. Let $a, b \in M$ and $a \neq b$; we have to prove that $a^* \neq b^*$. Let $\mathfrak{p}$ be a prime interval in $[a \wedge b, a \vee b]$. Since μ^* is a multi-coloring, there is a $c \in \mu^*(\mathfrak{p})$. Obviously, then $a \not\equiv b \pmod{\gamma_c}$, from which the statement follows.

Finally, the third condition is trivial from the definition of M^* and μ^*. $\quad\square$

19.4.   Formal proof

Now we are ready for the second proof of Theorem 19.2 as presented in G. Grätzer, H. Lakser, and E. T. Schmidt [103]. Let K, E, and γ be given as in Theorem 19.2.

Step 1. We define a map μ of $\mathrm{Prime}(K)$ to subsets of $\mathrm{J}(E)$:

$$\mu\mathfrak{p} = \mathrm{J}(E) \cap \mathrm{id}(\gamma(\mathrm{con}(\mathfrak{p}))).$$

The map μ is obviously isotone. The join-homomorphism γ separates 0, so $\mu\mathfrak{p} \neq \varnothing$. Therefore, μ is a multi-coloring of K over $\mathrm{J}(E)$. We apply Lemma 19.4 to obtain the lattice

$$K^* = \prod(K_c \mid c \in \mathrm{J}(E)).$$

Step 2. Any finite lattice M can be embedded in a finite simple lattice, $\mathrm{Simp}\,M$, with the same zero and unit, see Lemma 14.3. Use such an extension for each K_c to obtain a simple lattice $\mathrm{Simp}\,K_c$, then define:

$$L_0 = \prod(\mathrm{Simp}\,K_c \mid c \in \mathrm{J}(E)),$$

and extend the coloring so that $\mathrm{Simp}\,K_c$ is also colored by $\{c\}$. Since L_0 is a direct product of simple lattices, it follows that $\mathrm{Con_J}\,L_0$ is an antichain; the congruence lattice of L_0 is a boolean lattice with $|\mathrm{J}(E)|$ atoms. K is a sublattice of K^* and K^* is a sublattice of L_0, so we obtain an embedding $\gamma \colon K \to L_0$.

Finally, we construct a special ideal of L_0. Let p_c be an arbitrary atom of the direct component $\text{Simp } K_c$; then the prime interval $[0, p_c]$ of L_0 has color c. The atoms p_c, for $c \in J(E)$, generate an ideal B_0 of L_0 which is a boolean lattice satisfying the following properties:

(1) any two distinct atoms have different colors;

(2) every color $c \in J(E)$ occurs in B_0.

Step 3. We continue by forming a finite atomistic lattice L_1 with $E \cong \text{Con } L_1$ under the isomorphism β_1. For L_1, we construct a chopped lattice P_1 as in Section 8.2, except that we use a uniform "tripling"—as opposed to "doubling"—of non-maximals (for every join-irreducible element p of E, we take *three* atoms p_1, p_2, and p_3, so that in P_1 they generate a sublattice isomorphic to M_3). Let L_1 be the ideal lattice of P_1. The isomorphism $J(E) \cong \text{Con}_J L_1$ is given as follows: for $p \in J(E)$, the congruence $\text{con}(0, p)$ of L_1 corresponds to p. Let β_1 denote the corresponding isomorphism $\beta_1 \colon E \to \text{Con } L_1$.

We consider on L_1 the natural coloring over $J(E)$ (a prime interval $\mathfrak{p}$ is colored by $\beta_1^{-1}(\text{con}(\mathfrak{p})) \in J(E)$). Note that L_0 and L_1 are colored over the same set, $J(E)$. Let B_1 be the ideal of L_1 generated by the atoms p_2 for $p \in J(E)$. Then the ideal B_1 is a boolean lattice satisfying the properties (1) and (2) stated in Step 2.

Step 4. We have the lattice L_0 with the ideal B_0 and L_1 with an ideal B_1. Note that B_0 and B_1 are isomorphic finite boolean lattices with the same coloring. Take the dual L_2 of L_1; in this lattice B_1 corresponds to a filter B_2. Again, note that B_0 and B_2 are isomorphic finite boolean lattices with the same coloring. Glue together L_0 and L_2 by a color preserving identification of B_0 and B_2. The resulting lattice is L. The prime intervals of L are colored by $J(E)$, and we have the isomorphism $\beta \colon E \to \text{Con } L$. Since L_0 is a sublattice of L, we may view γ as an embedding of K into L.

Step 5. Finally, we have to verify that $\text{ext } \gamma = \beta\gamma$. It is enough to prove that $(\text{ext } \gamma)(\alpha) = \beta\gamma(\alpha)$ for join-irreducible congruences α in K.

So let $\alpha = \text{con}(\mathfrak{p})$, where $\mathfrak{p} = [a, b] \in \text{Prime}(K)$. By Lemma 19.4, $\text{ext } \gamma(\text{con}(\mathfrak{p})) = \text{con}(a^*, b^*)$ collapses in K^* the prime intervals of color $\leq \gamma\alpha$; the same holds in L_0 and in L.

Computing $\beta\gamma(\alpha)$ we get the same result, hence $(\text{ext } \gamma)(\alpha) = \beta\gamma(\alpha)$, completing the proof of Theorem 19.2.

19.5.   Discussion

History

A. P. Huhn's 1983 paper [158]—and his later papers [159] and [160] (published posthumously)—attacked the Congruence Lattice Characterization Problem

(CLP) from below. He observed that a distributive algebraic lattice D with countably many compact elements is the direct limit (union) of an increasing countable family $(\,D_i \mid i < \omega\,)$ of finite distributive $\{\vee, 0\}$-subsemilattices of D. The D_i-s are, of course, finite distributive lattices. For each $i < \omega$, let us denote by $\gamma_i \colon D_i \to D_{i+1}$ the $\{\vee, 0\}$-embedding. Huhn constructs a sequence $(\,L_i \mid i < \omega\,)$ of finite lattices with lattice embeddings $\gamma_i \colon L_i \to L_{i+1}$ such that $\operatorname{ext} \gamma_i \colon \operatorname{Con} L_i \to \operatorname{Con} L_{i+1}$ represents γ_i. Then, denoting by L the direct limit of the sequence $(\,L_i \mid i < \omega\,)$, he finds the representative D of L, that is, a lattice L satisfying $\operatorname{Con} L \cong D$. The construction of the L_i and γ_i is the most complicated part of his papers. However, using our Theorem 19.2, we can proceed in a straightforward manner. We first represent D_0 by a finite lattice L_0, and, inductively, given L_i, we immediately get a finite lattice L_{i+1} and an embedding $\gamma_i \colon L_i \to L_{i+1}$ with $\operatorname{ext} \gamma_i$ representing γ_i.

P. Pudlák [168] showed that every finite subset of a distributive algebraic lattice D is contained in a finite distributive $\{\vee, 0\}$-subsemilattice S of D. Of course, S is a finite distributive lattice. P. Pudlák used this to find a new approach to E. T. Schmidt's result discussed in the Introduction.

Huhn's result also follows from Theorems 5.5 and 5.6 of M. Tischendorf's 1992 thesis [185].

In 1996, G. Grätzer, H. Lakser, and E. T. Schmidt set out to give a short proof of Huhn's result in [103]. The result published was not only a short proof but also a stronger form, with a number of applications in subsequent papers.

The *Proof-by-Picture* we present in this chapter originated in the paper [186], in which J. Tůma proved the following result.

Theorem 19.5. *Let L_0, L_1, L_2 be finite atomistic lattices and let $\eta_1 \colon L_0 \to L_1$ and $\eta_2 \colon L_0 \to L_2$ be $\{0\}$-embeddings such that $\operatorname{ext} \eta_1$ and $\operatorname{ext} \eta_2$ are injective. Let D be a finite distributive lattice, and, for $i \in \{1, 2\}$, let $\gamma_i \colon \operatorname{Con} L_i \to D$ be $\{\vee, 0\}$-embeddings such that $\gamma_1(\operatorname{ext} \eta_1) = \gamma_2(\operatorname{ext} \eta_2)$. Then there is a finite atomistic lattice L, and there are $\{0\}$-embeddings $\gamma_i \colon L_i \to L$, for $i \in \{1, 2\}$, satisfying $\gamma_1 \eta_1 = \gamma_2 \eta_2$, and there is an isomorphism $\alpha \colon \operatorname{Con} L \to D$ such that $\alpha \operatorname{ext} \gamma_i = \gamma_i$ for $i \in \{1, 2\}$.*

In G. Grätzer, H. Lakser, and F. Wehrung [108], Tůma's result was extended as follows.

Theorem 19.6. *Let L_0, L_1, L_2 be lattices and let $\eta_1 \colon L_0 \to L_1$ and $\eta_2 \colon L_0 \to L_2$ be lattice homomorphisms. Let D be a finite distributive lattice, and, for $i \in \{1, 2\}$, let $\gamma_i \colon \operatorname{Con} L_i \to D$ be complete $\bigvee$-homomorphisms such that*

$$\gamma_1(\operatorname{ext} \eta_1) = \gamma_2(\operatorname{ext} \eta_2).$$

There is then a lattice L, there are lattice homomorphisms $\gamma_i \colon L_i \to L$, for $i \in \{1, 2\}$, with $\gamma_1 \eta_1 = \gamma_2 \eta_2$, and there is an isomorphism $\alpha \colon \operatorname{Con} L \to D$ such that $\alpha \operatorname{ext} \gamma_i = \gamma_i$, for $i \in \{1, 2\}$.

If L_0, L_1, L_2 have zero and both η_1 and η_2 preserve the zero, then L can be chosen to have a zero and γ_1 and γ_2 can be chosen to preserve the zero.
If L_1 and L_2 are finite, then L can be chosen to be finite and atomistic.

This theorem is indeed an extension of Tůma's theorem—we need only observe that if the γ_i are injective, then the γ_i must be lattice embeddings. This fact follows from the elementary fact that a lattice homomorphism $\gamma \colon K \to L$ is an embedding iff ext γ *separates* 0.

Our *Proof-by-Picture* is the special case $L_0 = L_1 = L_2$ is finite and $\gamma_1 = \gamma_2$.

Applications

As outlined in the history subsection, for every distributive algebraic lattice D with countably many compact elements, we can find a lattice L representing D, where L is a ω-union of finite lattices from a CPE-class **C**. Sometimes, we get even more as in the following two results from G. Grätzer, H. Lakser, and F. Wehrung [108]:

Theorem 19.7. *Let D be a distributive algebraic lattice. If D has at most $\aleph_1$ compact elements, then there exists a locally finite, relatively complemented lattice L with zero such that $\mathrm{Con}\, L \cong D$.*

A lattice L is *congruence-finite* if $\mathrm{Con}\, L$ is finite; it is **0**-*congruence-finite* if L can be written as a union,

$$L = \bigcup(\, L_n \mid n < \omega\,),$$

where $(\, L_n \mid n < \omega\,)$ is an increasing sequence of congruence-finite sublattices of L.

We also apply Theorem 19.6 to prove the following:

Theorem 19.8. *Every **0**-congruence-finite lattice K has a **0**-congruence-finite, relatively complemented congruence-preserving extension L. Furthermore, if K has a zero, then L can be taken to have the same zero.*

Isotone maps

With restrictions we get the "dual" of Lemma 3.15.

Lemma 19.9. *Let $K \leq L$ be finite lattices. Then* re$\colon \mathrm{Con}\, L \to \mathrm{Con}\, K$ *is a $\{\wedge, 0, 1\}$-homomorphism.*

And we can obviously "dualize" Theorems 19.1 and 19.2.

So what happens if we compose an extension and a restriction? Between the first and last congruence lattices, we surely get $\{0\}$-isotone map. The converse of this was proved in G. Grätzer, H. Lakser, and E. T. Schmidt [104].

Theorem 19.10. *Let D_1 and D_2 be finite distributive lattices, and let the map $\gamma \colon D_1 \to D_2$ be $\{0\}$-isotone. Then there are finite lattices L_1, L_2, L, lattice embeddings $\gamma_1 \colon L_1 \to L$ and $\gamma_2 \colon L_2 \to L$, and isomorphisms $\alpha_i \colon D_i \to \operatorname{Con} L_i$, for $i = 1, 2$, such that*

$$\alpha_2 \gamma = (\operatorname{re} \gamma_2)(\operatorname{ext} \gamma_1)\alpha_1,$$

that is, such that the diagram

$$
\begin{array}{ccc}
D_1 & \xrightarrow{\ \gamma\ } & D_2 \\[2pt]
\cong \downarrow \alpha_1 & & \cong \downarrow \alpha_2 \\[6pt]
\operatorname{Con} L_1 \xrightarrow{\ \operatorname{ext} \gamma_1\ } & \operatorname{Con} L \xrightarrow{\ \operatorname{re} \gamma_2\ } & \operatorname{Con} L_2
\end{array}
$$

is commutative.

In G. Grätzer, H. Lakser, and E. T. Schmidt [107], there is a "concrete" version of this result:

Theorem 19.11. *Let L_1 and L_2 be arbitrary lattices with finite congruence lattices $\operatorname{Con} L_1$ and $\operatorname{Con} L_2$, respectively, and let $\gamma \colon \operatorname{Con} L_1 \to \operatorname{Con} L_2$ be an isotone map. Then there is a lattice L with a finite congruence lattice, a lattice embedding $\gamma_1 \colon L_1 \to L$, and a homomorphism $\gamma_2 \colon L \to L_2$ such that*

$$\gamma = (\operatorname{re} \gamma_2)(\operatorname{ext} \gamma_1).$$

Furthermore, γ_2 is also an embedding iff γ preserves 0. If L_1 and L_2 are finite, then L can be chosen to be finite and atomistic.

Theorem 19.6 enabled us to prove Theorem 19.11 rather easily.

Size and breadth

We now have several constructions to verify Huhn's Theorem 19.1; in all of them, the lattices K and L are very large and very "wide." We can measure the "width" of a finite lattice L as follows.

A lattice L is of *breadth* p if p is the smallest integer with the property that for every finite $X \subseteq L$, there exists a $Y \subseteq X$ such that $|Y| \leq p$ and $\bigvee X = \bigvee Y$. Note that this concept is self-dual. If a finite lattice L is of breadth p, then there is an element $a \in L$ with at least p covers. The breadth of the boolean lattice B_n is n.

A "small" version of Theorem 19.1 was proved in [106] by G. Grätzer, H. Lakser, and E. T. Schmidt.

Theorem 19.12. *Let D be a finite distributive lattice with n join-irreducible elements, let E be a finite distributive lattice with m join-irreducible elements, let $k = \max(m, n)$, and let $\mathbf{g} \colon D \to E$ be a $\{0\}$-separating join-homomorphism.*

Then there is a planar lattice K with $O(n^2)$ elements, a finite extension $L \geq K$ (with embedding $\mathrm{id}_K \colon K \to L$) of breadth 3 with $O(k^5)$ elements, and isomorphisms $\alpha \colon E \to \operatorname{Con} K$, $\beta \colon D \to \operatorname{Con} L$ with

$$\alpha\gamma = (\operatorname{ext} \mathrm{id}_K)\beta,$$

that is, such that the diagram

$$
\begin{array}{ccc}
D & \xrightarrow{\;\gamma\;} & E \\[2pt]
\cong \downarrow \beta & & \cong \downarrow \alpha \\[2pt]
\operatorname{Con} K & \xrightarrow{\operatorname{ext} \mathrm{id}_K} & \operatorname{Con} L
\end{array}
$$

is commutative.

The proof uses the planar construction of Chapter 9 and multi-coloring.

Problem 19.1. Is $O(k^5)$ optimal for the lattice L in Theorem 19.12?

In other words, can one prove (analogously to G. Grätzer, I. Rival, and N. Zaguia [114]; see Chapter 9) that the size $O(k^5)$ cannot be replaced by the size $O(k^\alpha)$ for any $\alpha < 5$?

Problem 19.2. Can one find a lower bound for $|L|$ as in G. Grätzer and D. Wang [143] and Y. Zhang [195]?

Problem 19.3. Is breadth 3 optimal for L?

This is almost certainly so since a breadth 2 lattice cannot contain B_3 as a sublattice, making it very difficult to manipulate the congruences. However, in view of the result of the next chapter (Theorem 20.3), one can ask:

Problem 19.4. Is there a planar version of Theorem 19.1?

There is an interesting way of measuring the complexity of an order. The *order dimension* of a finite ordered set $(P, \leq)$ is the smallest integer $n \geq 1$ such that $\leq$ can be represented as the intersection of the orderings of n chains defined on the set P. The order dimension of a finite lattice L is 1 iff L is a chain. The order dimension of L is 2 iff L is planar.

Problem 19.5. Prove that the lattice L we construct for Theorem 19.12 is of order dimension 3.

The construction of the lattice L for Theorem 19.12 starts with the planar lattice of Chapter 9.

Problem 19.6. Can one construct L for Theorem 19.12 starting from a different lattice K?

2-distributive lattices

A. P. Huhn introduced n-distributivity in [156] and [157]. Let $n \geq 1$ be an integer. A lattice L is n-*distributive* if for all $x, y_1, \ldots, y_{n+1} \in L$,

$$x \wedge \left(\bigvee_{i=1}^{n+1} y_i \right) = \bigvee_{i=1}^{n+1} \left(x \wedge \left(\bigvee_{\substack{j=1 \\ j \neq i}}^{n+1} y_j \right) \right).$$

In particular, a lattice L is 1-*distributive* iff it is distributive; it is 2-*distributive* iff it satisfies the identity

$$x \wedge (y_1 \vee y_2 \vee y_3) = (x \wedge (y_1 \vee y_2)) \vee (x \wedge (y_1 \vee y_3)) \vee (x \wedge (y_2 \vee y_3)).$$

We will call a lattice L *doubly* 2-*distributive* if it satisfies the 2-distributive identity and its dual. For instance, N_5 and M_3 are doubly 2-distributive lattices.

In [128], G. Grätzer and E. T. Schmidt proved the following:

Theorem 19.13. *The lattices K and L in Theorem 19.1 can be constructed as finite doubly 2-distributive lattices.*

The proof is quite complex and uses multi-coloring.

Problem 19.7. Can the construction for Theorem 19.13 be continued with the lattice L serving as the starting lattice?

If this could be done, and repeated ω-times, then we could represent distributive algebraic lattice with countably many compact elements as congruence lattices of doubly 2-distributive lattices.

Ideals

20.1. The results

The second simplest connection between two lattices K and L is that K is an ideal of L. How then does $\operatorname{Con} K$ relate to $\operatorname{Con} L$?

In Section 3.3 we discussed the connection between $\operatorname{Con} K$ and $\operatorname{Con} L$, for the lattice L and its sublattice K. In Chapter 19, we proved the corresponding representation theorem for finite lattices.

Lemma 3.14 states that if I is an ideal of the lattice L, then the restriction map re: $\operatorname{Con} L \to \operatorname{Con} I$ is a bounded homomorphism. The corresponding representation theorem for finite lattices was proved in G. Grätzer and H. Lakser [90]:

Theorem 20.1. *Let D and E be finite distributive lattices. Let φ be a bounded homomorphism of D into E. Then there exists a finite lattice L and an ideal I of L such that $D \cong \operatorname{Con} L$, $E \cong \operatorname{Con} I$, and φ is represented by re, the restriction map.*

See E. T. Schmidt [179] for an alternative proof of this result.

G. Czédli [17] represents φ with rectangular lattices; his proof is more complicated.

In the survey paper G. Grätzer and E. T. Schmidt [138], Problem 15 asks (in part), whether this result can be proved for sectionally complemented lattices.

At first glance, we may think that we can solve this problem by combining Theorem 20.1 with Theorem 14.1: *Every finite lattice has a finite, sectionally complemented, congruence-preserving extension.* We obtain a stronger

© Springer International Publishing Switzerland 2016
G. Grätzer, *The Congruences of a Finite Lattice*,
DOI 10.1007/978-3-319-38798-7_20

form of Theorem 20.1: the lattice L can be assumed to be sectionally complemented. Unfortunately, in the congruence-preserving extension of L constructed in Theorem 14.1, I is no longer an ideal, so this method fails to solve this problem.

In [97], using preorders, G. Grätzer and H. Lakser answer the question in the affirmative.

Theorem 20.2. *Let D and E be finite distributive lattices. Let φ be a bounded homomorphism of D into E. Then there exists a* sectionally complemented *finite lattice L and an ideal I of L such that $D \cong \mathrm{Con}\, L$, $E \cong \mathrm{Con}\, I$, and φ is represented by* re, *the restriction map.*

As an ideal of a sectionally complemented lattice, of course, the ideal I is also a sectionally complemented lattice.

In G. Grätzer and H. Lakser [95], we took a different approach to the congruence restriction problem—the strange story of how this came about is told in Section 20.5.

Theorem 20.3. *Let D and E be finite distributive lattices. Let φ be a bounded homomorphism of D into E. Then there exists a finite* planar *lattice L and an ideal I of L such that $D \cong \mathrm{Con}\, L$, $E \cong \mathrm{Con}\, I$, and φ is represented by* re, *the restriction map.*

This is, of course, much stronger than Theorem 20.1: we obtain planar lattices (and very small ones). However, Theorem 20.1 is the foundation on which Theorem 20.2 is built. One cannot obtain Theorem 20.2 based on Theorem 20.3.

20.2.   *Proof-by-Picture* for the main result

For the *Proof-by-Picture*, take the five-element distributive lattices D and E and the bounded homomorphism φ at the top of Figure 20.1. We choose a bounded homomorphism φ which is not an onto map. We associate with φ an "inverse map" $\varrho \colon \mathrm{J}(E) \to \mathrm{J}(D)$ (also shown in Figure 20.1), where we define ϱ on an $x \in \mathrm{J}(E)$ as the smallest element of D that is mapped to an element $\geq x$ by φ; see Section 2.5.2.

Our plan is the following: We construct a chopped lattice M as in Section 8.2—using the gadget N_6—except that we triple, not double, every join-irreducible element, starting with the order $\mathrm{J}(D) \cup \mathrm{J}(E)$ (disjoint union). Every join-irreducible element x has three copies, x_L (left), x_M (middle), and x_R (right). We further add two more copies of all these elements as in Section 8.2. So the new gadget (for $a \in \mathrm{J}(D)$) is the chopped lattice $G(a)$ in Figure 20.1, which already guarantees that $a_L \equiv 0$ is equivalent to $a_M \equiv 0$ is equivalent to $a_R \equiv 0$.

We take six gadgets: $G(x)$, for each $x \in \mathrm{J}(D) \cup \mathrm{J}(E)$—they are pairwise disjoint except they share the 0.

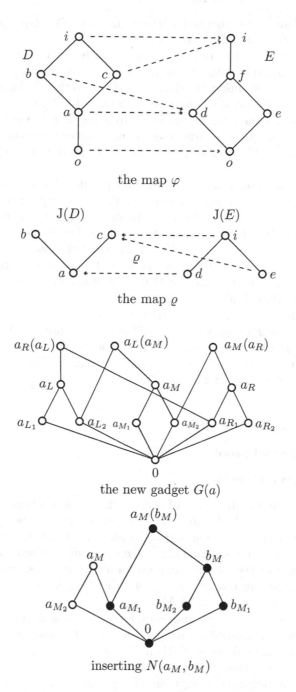

the map φ

the map ϱ

the new gadget $G(a)$

inserting $N(a_M, b_M)$

Figure 20.1: *Proof-by-Picture* for the main result.

We now have three chopped lattices: $I_C = G(d) \cup G(e) \cup G(i)$, $L'_C = G(a) \cup G(b) \cup G(c)$, and $L''_C = I_C \cup L'_C$.

We order the congruences in L'_C to agree with the ordering in $\mathrm{J}(D)$ as in Section 8.2, using the middle elements. For instance, to achieve that $b \equiv 0$ implies that $a \equiv 0$, we insert the element $b_M(a_M)$ so that $N(b_M, a_M)$ becomes a sublattice, as illustrated in Figure 20.1. So $\mathrm{Con}\, L'_C \cong D$.

Next, we order the congruences in I_C to agree with the ordering in $\mathrm{J}(D)$. Thus $\mathrm{Con}\, I_C \cong E$. At this stage, there is no congruence connection between L'_C and I_C, so $\mathrm{Con}\, L''_C$ is isomorphic to the direct product $\mathrm{Con}\, L'_C \times \mathrm{Con}\, I_C \cong D \times E$.

Finally, we want to achieve that $x \equiv 0$ is equivalent to $\varrho x \equiv 0$, for $x \in \mathrm{J}(D)$, for instance, that $e \equiv 0$ is equivalent to $c \equiv 0$. To accomplish this, we add the element $e_L(c_L)$ to L''_C so that $N(e_L, c_L)$ becomes a sublattice; therefore, in the new chopped lattice, $e_L \equiv 0$ implies that $c_L \equiv 0$. For the reverse implication, we add the element $c_R(e_R)$ so that $N(c_R, e_R)$ becomes a sublattice; now, $c_R \equiv 0$ implies that $e_R \equiv 0$. Since the gadget $G(e)$ ensures that $e_L \equiv 0$ is equivalent to $e_R \equiv 0$, we conclude that $e_L \equiv 0$ is equivalent to $c_L \equiv 0$. We proceed similarly with all $x \in \mathrm{J}(D)$.

Let L_C denote the chopped lattice we obtain after adding all these elements. By construction, $\mathrm{Con}\, L_C \cong D$ and L_C contains an ideal I_C with $\mathrm{Con}\, I_C \cong E$, and the congruences of L_C restrict to I_C as determined by ϱ.

Now we invoke Theorem 5.6, and conclude that $L = \mathrm{Id}\, L_C$ and $I = \mathrm{Id}\, I_C$ satisfy the requirements of Theorem 20.1.

20.3. The main result

20.3.1 A formal proof

Categoric preliminaries

In our proof, we will make use of several classes of algebraic structures. It will be useful to use the language of (concrete) category theory. A *concrete category* is a class $\mathbb{K}$ of algebras (partial algebras, relational systems) called *objects*, along with a designated class of *morphisms* between any two objects. It is assumed that the identity map on an object is a designated morphism, and if K, L, M are objects, $\varphi \colon K \to L$ and $\psi \colon L \to M$ are designated morphisms, then so is $\psi\varphi \colon K \to M$.

For categories $\mathbb{K}$ and $\mathbb{N}$, the map $F \colon \mathbb{K} \to \mathbb{N}$ is a *functor* if F maps objects to objects, morphisms to morphisms, and whenever $\varphi \colon K \to L$ and $\psi \colon L \to M$ are morphisms in $\mathbb{K}$, then $F(\psi)F(\varphi) = F(\psi\varphi)$. A *contravariant functor* F reverses this, that is, it satisfies $F(\varphi)F(\psi) = F(\psi\varphi)$. Also, if $\varphi \colon K \to L$ is a morphism in $\mathbb{K}$, then $F(\varphi) \colon F(K) \to F(L)$ for a functor and $F(\varphi) \colon F(L) \to F(K)$ for a contravariant functor. Moreover, $F(\mathrm{id}_K) = \mathrm{id}_{F(K)}$ for both functors and contravariant functors.

Finally, we need the concept of natural equivalence. Let $\mathbb{K}$ and $\mathbb{N}$ be categories, and let $F\colon \mathbb{K} \to \mathbb{N}$ and $G\colon \mathbb{K} \to \mathbb{N}$ be functors. Then $\eta\colon F \to G$ is a *natural transformation* if for every object A of $\mathbb{K}$, there is a morphism $\eta_A\colon F(A) \to G(A)$ such that for every morphism $\alpha\colon A \to B$ in $\mathbb{K}$, the diagram

$$
\begin{array}{ccc}
F(A) & \xrightarrow{\ F(\alpha)\ } & F(B) \\
{\scriptstyle \eta_A}\downarrow & & \downarrow{\scriptstyle \eta_B} \\
G(A) & \xrightarrow{\ G(\alpha)\ } & G(B)
\end{array}
$$

is commutative, that is, $\eta_B F(\alpha) = G(\alpha)\eta_A$.

A natural transformations $\eta\colon F \to G$ is said to be a *natural equivalence* if η_A is an isomorphism for each object A of $\mathbb{K}$.

We now introduce the various categories and some of the associated functors we will use. Note that all the objects are finite structures.

1. Let $\mathbb{D}\textsc{i}$ (for $\mathbb{D}$\textsc{istributive}) denote the category of finite distributive lattices. The morphisms are bounded homomorphisms.

2. Let $\mathbb{L}\textsc{a}$ (for $\mathbb{L}$\textsc{attice}) denote the category whose objects are the finite lattices. A morphism is an *ideal-embedding;* see Section 1.3.2.

   There is a contravariant functor $\mathrm{Con}\colon \mathbb{L}\textsc{a} \to \mathbb{D}\textsc{i}$ that associates with each lattice L its congruence lattice $\mathrm{Con}\,L$, that is, $\mathrm{Con}(L) = \mathrm{Con}\,L$. If K and L are finite lattices and $\varepsilon\colon K \to L$ is an ideal-embedding, then $\mathrm{Con}(\varepsilon)$ denotes the restriction map (denoted by re in Section 3.3 and elsewhere), that is, $\mathrm{Con}(\varepsilon)(\boldsymbol{\alpha}) = \boldsymbol{\alpha}\rceil_K$, for a congruence $\boldsymbol{\alpha}$ of L. By Lemma 3.14, $\mathrm{Con}(\varepsilon)$ is a morphism in $\mathbb{D}\textsc{i}$.

3. Let $\mathbb{O}\textsc{r}$ (for $\mathbb{O}$\textsc{r}der) denote the category whose objects are finite orders and whose morphisms are isotone maps.

   We have a contravariant functor $\mathrm{Down}\colon \mathbb{O}\textsc{r} \to \mathbb{D}\textsc{i}$; if $\varphi\colon P \to Q$ is an isotone map of orders, then the map $\mathrm{Down}\,\varphi\colon \mathrm{Down}\,Q \to \mathrm{Down}\,P$ is defined by $(\mathrm{Down}\,\varphi)(H) = \varphi^{-1}H$, for each down-set $H \subseteq Q$.

4. Let $\mathbb{C}\textsc{h}$ (for $\mathbb{C}$\textsc{h}opped) denote the category whose objects are chopped lattices. By an *embedding* ε in $\mathbb{C}\textsc{h}$, we mean a map ε which is a one-to-one meet-homomorphism that preserves $\vee$ in the strong sense: $\varepsilon x \vee \varepsilon y$ exists iff $x \vee y$ exists and then $\varepsilon(x \vee y) = \varepsilon x \vee \varepsilon y$. The morphisms of $\mathbb{C}\textsc{h}$ are ideal-embeddings $\varepsilon\colon S \to T$.

   We will show that the functor $\mathrm{Con}\colon \mathbb{C}\textsc{h} \to \mathbb{D}\textsc{i}$ that associates with each object S of $\mathbb{C}\textsc{h}$ its congruence lattice $\mathrm{Con}\,S$ is a contravariant functor. As in the case of $\mathbb{L}\textsc{a}$, if $\varepsilon\colon S \to T$ is an ideal-embedding, then $\mathrm{Con}(\varepsilon)\colon \mathrm{Con}\,T \to \mathrm{Con}\,S$ denotes the restriction map.

5. The last category we consider is $\mathbb{H}\mathbb{E}$ (for $\mathbb{H}$emi order); it is of a more technical nature. An object of $\mathbb{H}\mathbb{E}$ is a finite nonempty set Q with a binary relation ρ satisfying the following three conditions:

(a) ρ is *irreflexive*, that is, $x \, \rho \, x$ fails, for all $x \in Q$;

(b) ρ is *antisymmetric*, that is, $x \, \rho \, y$ implies that $y \, \rho \, x$ fails, for all $x, y \in Q$;

(c) ρ is *cycle-rich*, that is, for each $x \in Q$, there is a $y \in Q$ with $y \, \rho \, x$.

If Q and R are objects of $\mathbb{H}\mathbb{E}$, a morphism $\varphi \colon Q \to R$ in $\mathbb{H}\mathbb{E}$ is a one-to-one map that preserves ρ in the following strong sense:

$$\varphi x \, \rho \, \varphi y \quad \text{iff} \quad x \, \rho \, y.$$

Let Q be an object of $\mathbb{H}\mathbb{E}$; a subset H of Q is a *down-set* if $x \in H$ and $x \, \rho \, y$ imply that $y \in H$. The lattice $\operatorname{Down} Q$ of down-sets in Q is distributive with $\varnothing$ as 0 and Q as 1. If $\varphi \colon Q \to R$ is a morphism in $\mathbb{H}\mathbb{E}$, then define $(\operatorname{Down} \varphi)(H) = \varphi^{-1}(H)$, for $H \in \operatorname{Down} R$. It is easy to see that $\operatorname{Down} \varphi \colon \operatorname{Down} R \to \operatorname{Down} Q$.

So we have a contravariant functor $\operatorname{Down} \colon \mathbb{H}\mathbb{E} \to \mathbb{D}\mathbb{I}$. Note that we use the same symbol Down for the functors $\operatorname{Down} \colon \mathbb{O}\mathbb{R} \to \mathbb{D}\mathbb{I}$ and $\operatorname{Down} \colon \mathbb{H}\mathbb{E} \to \mathbb{D}\mathbb{I}$.

From $\mathbb{D}\mathbb{I}$ to $\mathbb{O}\mathbb{R}$

Let us recall and rephrase the results from Section 2.5.2.

Given an object D of $\mathbb{D}\mathbb{I}$, that is, a finite distributive lattice, we consider the order $\operatorname{J}(D)$. If $\varphi \colon D \to E$ is a morphism in $\mathbb{D}\mathbb{I}$, then $\operatorname{J}(\varphi) \colon \operatorname{J}(E) \to \operatorname{J}(D)$, with $\operatorname{J}(\varphi)(x) = \bigwedge \varphi^{-1}(x)$, is an isotone map. Thus J is a contravariant functor $\operatorname{J} \colon \mathbb{D}\mathbb{I} \to \mathbb{O}\mathbb{R}$.

Let $\operatorname{id}_{\mathbb{D}\mathbb{I}}$ be the identity map on $\mathbb{D}\mathbb{I}$, regarded as a functor $\mathbb{D}\mathbb{I} \to \mathbb{D}\mathbb{I}$. Note that $\operatorname{Down} \operatorname{J}$ is also a functor $\mathbb{D}\mathbb{I} \to \mathbb{D}\mathbb{I}$.

Lemma 20.4. *There is a natural equivalence* $\psi \colon \operatorname{id}_{\mathbb{D}\mathbb{I}} \to \operatorname{Down} \operatorname{J}$ *that associates with each object D of $\mathbb{D}\mathbb{I}$ the isomorphism* $\psi_D \colon D \to \operatorname{Down} \operatorname{J}(D)$ *defined by* $\psi_D(x) = {\downarrow} \operatorname{J}(D)$ *for $x \in D$, that is, if D and E are finite distributive lattices and $\varphi \colon D \to E$ is a bounded homomorphism, then the diagram*

$$
\begin{array}{ccc}
D & \xrightarrow{\ \psi_D\ } & \operatorname{Down} \operatorname{J}(D) \\[2pt]
{\scriptstyle \varphi}\big\downarrow & & \big\downarrow{\scriptstyle \operatorname{Down} \operatorname{J}(\varphi)} \\[2pt]
E & \xrightarrow{\ \psi_E\ } & \operatorname{Down} \operatorname{J}(E)
\end{array}
$$

commutes, in other words, $\operatorname{Down} \operatorname{J}(\varphi)\psi_D = \psi_E \varphi$, *and* ψ_D, ψ_E *are lattice isomorphisms.*

From $\mathbb{O}_{\mathbb{R}}$ to $\mathbb{H}_{\mathbb{E}}$

Let P and Q be finite orders and let $\varphi \colon Q \to P$ be an isotone map. For P, Q, and φ, we construct two $\mathbb{H}_{\mathbb{E}}$-objects, $B(Q)$ and $A(\varphi)$, and define the $\mathbb{H}_{\mathbb{E}}$-morphism $\varepsilon_\varphi \colon B(Q) \to A(\varphi)$; the notation reflects the fact that $B(Q)$ depends only on Q, while $A(\varphi)$ and ε_φ depend on φ.

Set $B(Q) = Q \times \{L, M, R\}$; for $a \in Q$, denote the ordered pairs (a, L), (a, M), (a, R) by a_L, a_M, a_R, respectively. Define ρ on $B(Q)$ by setting:

(α) $a_L \, \rho \, a_M \, \rho \, a_R \, \rho \, a_L$, for $a \in Q$;

(β) $a_M \, \rho \, b_M$, whenever $a \succ b$ in Q, where $\succ$ denotes the cover relation in Q.

Set

$$A(\varphi) = (P \times \{L, M, R\}) \cup (Q \times \{L, M, R\}),$$

where we assume that P and Q are disjoint. Define ρ on $A(\varphi)$ by setting:

(1) $a_L \, \rho \, a_M \, \rho \, a_R \, \rho \, a_L$ if $a \in P \cup Q$;

(2) $a_M \, \rho \, b_M$ if $a \succ b$, and $a, b \in P$ or $a, b \in Q$;

(3) $a_L \, \rho \, \varphi a_L$, for all $a \in Q$;

(4) $\varphi a_R \, \rho \, a_R$, for all $a \in Q$.

Define $\varepsilon_\varphi \colon B(A) \to A(\varphi)$ by setting $\varepsilon_\varphi a_K = a_K$ for $a \in Q$, $K \in \{L, M, R\}$. Then ε_φ is a $\mathbb{H}_{\mathbb{E}}$-morphism.

We define the maps $\upsilon \colon B(Q) \to Q$ and $\vartheta \colon A(\varphi) \to P$ by setting

$$\upsilon a_K = a \quad \text{for } a \in Q,\ K \in \{L, M, R\};$$
$$\vartheta a_K = \varphi a \quad \text{for } a \in Q,\ K \in \{L, M, R\};$$
$$\vartheta a_K = a \quad \text{for } a \in P,\ K \in \{L, M, R\}.$$

These determine the maps $\upsilon' \colon \operatorname{Down} Q \to \operatorname{Down} B(Q)$ with $\upsilon' H = \upsilon^{-1} H$, for $H \in \operatorname{Down} Q$, and $\vartheta' \colon \operatorname{Down} P \to \operatorname{Down} A(\varphi)$ with $\vartheta' H = \vartheta^{-1} H$ for $H \in \operatorname{Down} P$.

The following statement is easy to verify:

Lemma 20.5. *The maps $\upsilon' \colon \operatorname{Down} Q \to \operatorname{Down} B(Q)$ and $\vartheta' \colon \operatorname{Down} P \to \operatorname{Down} A(\varphi)$ are lattice isomorphisms and the diagram*

$$
\begin{array}{ccc}
\operatorname{Down} P & \xrightarrow{\ \vartheta'\ } & \operatorname{Down} A(\varphi) \\
{\scriptstyle \operatorname{Down} \varphi} \downarrow & & \downarrow {\scriptstyle \operatorname{Down} \varepsilon_\varphi} \\
\operatorname{Down} Q & \xrightarrow{\ \upsilon'\ } & \operatorname{Down} B(Q)
\end{array}
$$

commutes, that is, $(\operatorname{Down} \varepsilon_\varphi)\vartheta' = \upsilon' \operatorname{Down} \varphi$.

From $\mathbb{C}_H$ to $\mathbb{D}_I$

We need only one observation:

Lemma 20.6. Con: $\mathbb{C}_H \to \mathbb{D}_I$ *is a contravariant functor.*

Proof. Since the objects S of $\mathbb{C}_H$ are meet-semilattices, congruence relations are determined by pairs x, y with $x \leq y$. Since $\mathrm{id}(y)$ is a sublattice of S, we get, exactly as in the case of lattices, that for $x \leq y$, the congruence $x \equiv y$ (mod $\alpha \vee \beta$) holds iff there is a sequence $x = z_0 \leq z_1 \leq \cdots \leq z_n = y$, with $z_i \equiv z_{i+1}(\alpha)$ or $z_i \equiv z_{i+1}(\beta)$ for each $0 \leq i < n$.

Then we can establish that Con S is a distributive lattice and that $\mathrm{Con}(\varepsilon)$ is a $\mathbb{D}_I$-morphism for any $\mathbb{C}_H$-morphism ε, exactly as for lattices, see Theorem 3.1. $\qquad\square$

From $\mathbb{H}_E$ to $\mathbb{C}_H$

We describe a functor $S\colon \mathbb{H}_E \to \mathbb{C}_H$. Let Q be an object in $\mathbb{H}_E$ and set

$$S(Q) = \{0\} \cup \{a_1, a_2, a \mid a \in Q\} \cup \{b(a) \mid a,\ b \in Q,\ \text{with } b\,\rho\,a\,\},$$

where 0 is distinct from the other elements. Define an ordering $\leq$ on $S(Q)$:

$$0 < a_i \quad \text{for } i = 1, 2;$$
$$a_i < a \quad \text{for } i = 1, 2;$$
$$b_1 < b(a) \quad \text{if } b\,\rho\,a;$$
$$a < b(a) \quad \text{if } b\,\rho\,a.$$

The maximal elements of $S(Q)$ are then of the form $b(a)$ with $b\,\rho\,a$, and $b, a \in Q$, since for each $a \in Q$ there is a $b \in Q$ with $b\,\rho\,a$. Each $\mathrm{id}(b(a))$ is isomorphic to the lattice $\mathsf{N}_6 = N(b, a)$ depicted in Figure 8.1.

If $\varphi\colon Q \to R$ is a $\mathbb{H}_E$-morphism, then define $S(\varphi)\colon S(Q) \to S(R)$ by setting

$$S(\varphi)(0) = 0;$$
$$S(\varphi)(a_i) = \varphi a_i \quad \text{for } i = 1, 2;$$
$$S(\varphi)(a) = \varphi a;$$
$$S(\varphi)(b(a)) = \varphi b(\varphi a).$$

Note that $b\,\rho\,a$ implies that $\varphi b\,\rho\,\varphi a$, justifying the last equation.

Lemma 20.7. $S\colon \mathbb{H}_E \to \mathbb{C}_H$ *is a functor.*

Proof. In $S(Q)$,

$d(c) \wedge b(a) = 0$	if $b\,\rho\,a$, $d\,\rho\,c$ and a, b, c, d are all distinct;
$c(b) \wedge b(a) = b_1$	if $c\,\rho\,b\,\rho\,a$, and a, b, c are all distinct;
$c(a) \wedge b(a) = a$	if $c\,\rho\,a$, $b\,\rho\,a$, and a, b, c are all distinct;
$c(a) \wedge c(b) = c_1$	if $c\,\rho\,a$, $c\,\rho\,b$, and a, b, c are all distinct.

There appears to be one more case to check: $a(b) \wedge b(a)$; however, this cannot occur by property (b) in the definition of $\mathbb{H}_{\mathbb{E}}$.

Thus $S(Q)$ is a meet-semilattice and it is easy to see that $S(Q)$ is indeed a chopped lattice.

Clearly, $S(\varphi)$ is an ideal-embedding of $S(Q)$ into $S(R)$, whenever $\varphi\colon Q \to R$ is a $\mathbb{H}_{\mathbb{E}}$-morphism. $\qquad\square$

Given an object Q of $\mathbb{H}_{\mathbb{E}}$, we define a map

$$\psi_Q\colon \operatorname{Down} Q \to \operatorname{Con} S(Q)$$

by setting

$$\psi_Q(H) = \operatorname{con}(H).$$

Note that each subset of Q is a subset of $S(Q)$.

We now prepare the proof of Theorem 20.10 with two lemmas.

Given $a, b \in Q$ with $b \, \rho \, a$, the principal ideal $N(b, a)$ generated by $b(a)$ has exactly three congruence relations:

$1_{b,a}$, collapsing all of $N(b, a)$,

$0_{b,a}$, the identity relation,

$\alpha_{b,a}$, depicted in Figure 8.1 (where it is denoted by α), with congruence classes $\{b_1, b(a)\}$ and $\{0, a_1, a_2, a\}$.

We first show that ψ_Q is surjective.

Lemma 20.8. *Let Q be an object of $\mathbb{H}_{\mathbb{E}}$ and let α be a congruence relation on $S(Q)$. Then*

$$H = \{\, a \in Q \mid a \equiv 0 \, (\alpha) \,\}$$

is a down-set in Q and $\alpha = \operatorname{con}(H)$.

Proof. Let $b \in H$ and let $b \, \rho \, a$. Then $b \equiv 0 \pmod{\alpha}$, so $b_1 \equiv 0 \pmod{\alpha}$. Thus

$$\alpha\rceil_{N(b,a)} = 1_{b,a},$$

and we conclude that $a \in H$. Consequently, H is a down-set in Q.

Since $S(Q)$ is sectionally complemented, α is determined by its congruence class containing 0. However, for any congruence, $a_1 \equiv 0$ iff $a_2 \equiv 0$ iff $a \equiv 0$, and $b(a) \equiv 0$ iff $b \equiv 0$. Thus α is determined by

$$H = \{\, a \in Q \mid a \equiv 0 \, (\alpha) \,\},$$

that is, $\alpha = \operatorname{con}(H)$, concluding the proof. $\qquad\square$

We now characterize $\operatorname{con}(H)$.

Lemma 20.9. *Let Q be an object of $\mathbb{H}_E$ and let H be a down-set in Q. Let $a, b \in Q$ with $b \,\rho\, a$. Then*

$$\operatorname{con}(H)\rceil_{N(b,a)} = \begin{cases} \mathbf{1}_{b,a} & \text{if } b \in H; \\ \boldsymbol{\alpha}_{b,a} & \text{if } b \notin H \text{ and } a \in H; \\ \mathbf{0}_{b,a} & \text{if } a \notin H. \end{cases}$$

Proof. The set $\operatorname{Max} = \operatorname{Max}(S(Q))$ of maximal elements of $S(Q)$ consists of all elements of the form $b(a)$, with $b \,\rho\, a$. For each $b(a) \in \operatorname{Max}$, define the congruence relation $\boldsymbol{\alpha}_{b(a)}$ on the ideal $N(b, a)$ by setting

$$\boldsymbol{\alpha}_{b(a)} = \begin{cases} \mathbf{1}_{b,a}, & \text{if } b \in H; \\ \boldsymbol{\alpha}_{b,a}, & \text{if } b \notin H \text{ and } a \in H; \\ \mathbf{0}_{b,a}, & \text{if } a \notin H. \end{cases}$$

Then,

$$\boldsymbol{\alpha}_{b(a)}\rceil_{N(b,a) \cap N(d,c)} = \boldsymbol{\alpha}_{d(c)}\rceil_{N(b,a) \cap N(d,c)},$$

for any $b(a), d(c) \in \operatorname{Max}$. Therefore, $(\boldsymbol{\alpha}_{b(a)} \mid b \,\rho\, a)$ is a congruence-vector, introduced in Section 5.3, so by Lemma 5.5, there is a unique congruence relation $\boldsymbol{\alpha}$ on $S(Q)$ with

$$\boldsymbol{\alpha}\rceil_{N(b,a)} = \boldsymbol{\alpha}_{b(a)} \quad \text{for all } b(a) \in M.$$

By Lemma 20.9,

$$\boldsymbol{\alpha} = \operatorname{con}(H_1), \quad \text{with } H_1 = \{a \in Q \mid a \equiv 0(\boldsymbol{\alpha})\}.$$

For each $a \in Q$, there is a $b \in Q$ with $b \,\rho\, a$. Then $a \in H_1$ iff $a \equiv 0 \pmod{\boldsymbol{\alpha}}$ iff $a \equiv 0 \pmod{\boldsymbol{\alpha}_{b(a)}}$ iff $a \in H$; the last equivalence follows by the definition of $\boldsymbol{\alpha}_{b(a)}$. Thus $H = H_1$. $\square$

We are now ready for the main result of this section.

Theorem 20.10. *The map ψ establishes the natural equivalence of the functors* Down: $\mathbb{H}_E \to \mathbb{D}_I$ *and* $\operatorname{Con} S$: $\mathbb{H}_E \to \mathbb{D}_I$. *That is, for every object Q in $\mathbb{H}_E$,*

$$\psi_Q: \operatorname{Down} Q \to \operatorname{Con} S(Q)$$

is an isomorphism and, given a $\mathbb{H}_E$-morphism $\varphi: Q \to R$, the diagram in $\mathbb{D}_I$:

$$
\begin{array}{ccc}
\operatorname{Down} R & \xrightarrow{\ \psi_R\ } & \operatorname{Con} S(R) \\[4pt]
{\scriptstyle \operatorname{Down} \varphi} \downarrow & & \downarrow {\scriptstyle \operatorname{Con} S(\varphi)} \\[4pt]
\operatorname{Down} Q & \xrightarrow{\ \psi_Q\ } & \operatorname{Con} S(Q)
\end{array}
$$

commutes, that is, $\operatorname{Con} S(\varphi)\psi_R = \psi_Q \operatorname{Down} \varphi$.

Proof. We first show that, for each object Q of $\mathbb{H}_E$,

$$\psi_Q \colon \operatorname{Down} Q \to \operatorname{Con} S(Q)$$

is an isomorphism. Clearly, $H_1 \subseteq H_2$ implies that $\operatorname{con}(H_1) \subseteq \operatorname{con}(H_2)$ and, by Lemma 20.8, ψ_Q is surjective. We need to show that ψ_Q is an embedding, that is, that $\operatorname{con}(H_1) \subseteq \operatorname{con}(H_2)$ implies that $H_1 \subseteq H_2$. Let $\operatorname{con}(H_1) \subseteq \operatorname{con}(H_2)$ for $H_1, H_2 \in \operatorname{Down} Q$. Take $a \in H_1$, and let $b \in Q$ with $b \, \rho \, a$; such a b exists by property ($\mathbb{H}_E$.c). Then $a \equiv 0 \pmod{\operatorname{con}(H_2)}$ and so

$$\operatorname{con}(H_2)\!\restriction_{N(b,a)} \neq \mathbf{0}_{b,a}.$$

By Lemma 20.9, $a \in H_2$. Thus $H_1 \subseteq H_2$, concluding the proof that ψ_Q is an isomorphism.

Next we verify that the diagram is commutative. Let $H \in \operatorname{Down} R$. Then, by Lemma 20.9,

$$(\operatorname{Con} S(f)\psi_R)(H) = \operatorname{con}(H_1),$$

for some $H_1 \in \operatorname{Down} Q$. But $a \in H_1$ iff $a \equiv 0 \pmod{(\operatorname{Con} S(\varphi)\psi_R)(H)}$ iff $S(\varphi)(a) \equiv 0 \pmod{\psi_R(H)}$, that is, $\varphi a \equiv 0 \pmod{\operatorname{con}(H)}$ iff $\varphi a \in H$. Thus $H_1 = \operatorname{Down} \varphi H$. But then

$$(\operatorname{Con} S(\varphi)\psi_R)(H) = (\psi_Q \operatorname{Down} \varphi)(H),$$

and so

$$\operatorname{Con} S(\varphi)\psi_R = \psi_Q \operatorname{Down} \varphi. \qquad \square$$

From $\mathbb{C}_H$ to $\mathbb{L}_A$

The functor $\operatorname{Id} \colon \mathbb{C}_H \to \mathbb{L}_A$ associates with each object S in $\mathbb{C}_H$ the lattice $\operatorname{Id}(S)$. If $\varphi \colon S \to T$ is a $\mathbb{C}_H$-morphism, then $\operatorname{Id}(\varphi) \colon \operatorname{Id} S \to \operatorname{Id} T$ is the ideal-embedding $\operatorname{Id}(I) = \varphi I$.

Consider the functors $\operatorname{Con} \colon \mathbb{C}_H \to \mathbb{D}_I$ and $\operatorname{Con} \operatorname{Id} \colon \mathbb{C}_H \to \mathbb{D}_I$. For any chopped lattice M, Theorem 5.6 establishes that $\operatorname{Con} M$ and $\operatorname{Con} \operatorname{Id} M$ are isomorphic; we obtain an isomorphism by assigning to the congruence $\boldsymbol{\alpha}$ of M, the congruence $\sigma_S(\boldsymbol{\alpha}) = \operatorname{con}(\boldsymbol{\alpha})$, the congruence generated by $\boldsymbol{\alpha}$ in $\operatorname{Id} M$. (We regard M as a sublattice of $\operatorname{Id} M$ by identifying m with $\operatorname{id}(m)$, for $m \in M$.)

Lemma 20.11. *σ is a natural equivalence* $\operatorname{Con} \to \operatorname{Con} \operatorname{Id}$. *That is, for each object S of $\mathbb{C}_H$, the map*

$$\sigma_S \colon \operatorname{Con} S \to \operatorname{Con}(\operatorname{Id} S)$$

is an isomorphism and, given a $\mathbb{C}_H$-morphism $\varphi \colon S \to T$, the diagram

$$
\begin{array}{ccc}
\operatorname{Con} T & \xrightarrow{\ \sigma_T\ } & \operatorname{Con}(\operatorname{Id} T) \\[4pt]
{\scriptstyle \operatorname{Con}(\varphi)}\Big\downarrow & & \Big\downarrow{\scriptstyle \operatorname{Con}(\operatorname{Id}(\varphi))} \\[4pt]
\operatorname{Con} S & \xrightarrow{\ \sigma_S\ } & \operatorname{Con}(\operatorname{Id} S)
\end{array}
$$

in $\mathbb{D}\mathbb{I}$ *commutes, that is,* $\mathrm{Con}(\mathrm{Id}(\varphi))\sigma_T = \sigma_S\,\mathrm{Con}(\varphi)$.

20.3.2   The final step

By combining the natural equivalences of Lemmas 20.4, 20.11, and Theorem 20.10, and the commutative diagram of Lemma 20.5, we get our main result (for $\mathrm{Con}(\alpha)$, reference Lemma 20.6):

Theorem 20.12. *Let D and E be finite distributive lattices and let $\varphi\colon D \to E$ be a bounded homomorphism. Then there exist finite lattices I, L, an ideal-embedding α of I into L, and isomorphisms $\beta\colon D \to \mathrm{Con}\,L$, $\gamma\colon E \to \mathrm{Con}\,I$ such that the diagram*

$$
\begin{array}{ccc}
D & \xrightarrow{\ \beta\ } & \mathrm{Con}\,L \\
{\scriptstyle\varphi}\big\downarrow & & \big\downarrow{\scriptstyle\mathrm{Con}(\alpha)} \\
E & \xrightarrow{\ \gamma\ } & \mathrm{Con}\,I
\end{array}
$$

is commutative, that is,

$$
\mathrm{Con}(\alpha)\beta = \gamma\varphi.
$$

Proof. Consider the following diagram:

$$
\begin{array}{ccccccc}
D & \xrightarrow{\psi_D} & \mathrm{Down}\,\mathrm{J}(D) & \xrightarrow{\vartheta'} & \mathrm{Down}\,A(\mathrm{J}(\varphi)) & \xrightarrow{\psi_{A(\mathrm{J}(\varphi))}} & \cdots \\
{\scriptstyle\varphi}\big\downarrow & & {\scriptstyle\mathrm{Down}\,\mathrm{J}(\varphi)}\big\downarrow & & {\scriptstyle\mathrm{Down}\,\varepsilon_{\mathrm{J}(\varphi)}}\big\downarrow & & \\
E & \xrightarrow{\psi_E} & \mathrm{Down}\,\mathrm{J}(E) & \xrightarrow{v'} & \mathrm{Down}\,B(\mathrm{J}(E)) & \xrightarrow{\psi_{B(\mathrm{J}(E))}} & \cdots
\end{array}
$$

$$
\begin{array}{ccc}
\cdots \longrightarrow & \mathrm{Con}\,S(A(\mathrm{J}(\varphi))) & \xrightarrow{\sigma_{S(A(\mathrm{J}(\varphi)))}} \mathrm{Con}\,\mathrm{Id}\,S(A(\mathrm{J}(\varphi))) \\
& {\scriptstyle\mathrm{Con}\,S\varepsilon_{\mathrm{J}(\varphi)}}\big\downarrow & {\scriptstyle\mathrm{Con}(\mathrm{Id}(S(E_{\mathrm{J}(\varphi)})))}\big\downarrow \\
\cdots \longrightarrow & \mathrm{Con}\,S(B(\mathrm{J}(E))) & \xrightarrow{\sigma_{S(B(\mathrm{J}(E)))}} \mathrm{Con}\,\mathrm{Id}\,S(B(\mathrm{J}(E)))
\end{array}
$$

By Lemma 20.4, ψ_D and ψ_E are isomorphisms and the left-most square commutes. By Lemma 20.5, the maps v' and ϑ' are isomorphisms and the next square commutes. By Theorem 20.10, $\psi_{A(\mathrm{J}(\varphi))}$ and $\psi_{B(\mathrm{J}(E))}$ are isomorphisms and the corresponding square commutes. Finally, by Lemma 20.11, $\sigma_{S(A(\mathrm{J}(\varphi)))}$ and $\sigma_{S(B(\mathrm{J}(E)))}$ are isomorphisms and the right-most square commutes.

Set $L = \mathrm{Id}\,S(A(\mathrm{J}(\varphi)))$, $I = \mathrm{Id}\,S(B(\mathrm{J}(E)))$, $\alpha = \mathrm{Id}\,S(\varepsilon_{\mathrm{J}(\varphi)})$. If we set

$$
\beta = \sigma_{S(A(\mathrm{J}(\varphi)))}\psi_{A(\mathrm{J}(\varphi))}\vartheta'\psi_D
$$

and

$$
\gamma = \sigma_{S(B(\mathrm{J}(E)))}\psi_{B(\mathrm{J}(E))}v'\psi_E,
$$

the theorem follows. $\square$

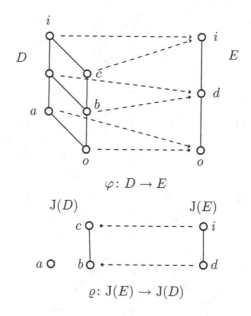

$$\varphi \colon D \to E$$

Figure 20.2: Distributive lattices for the planar example.

Observe that we can prove that the lattice L is sectionally complemented by adjusting the proof of Theorem 8.4 to the present more general setup.

20.4. *Proof-by-Picture* for planar lattices

The *Proof-by-Picture* will use the example distributive lattices D and E and the bounded homomorphism φ of Figure 20.2. The figure also shows the "inverse map" $\varrho \colon J(E) \to J(D)$, as it was done in Section 20.2; see also Section 2.5.2.

As in Section 9.2, we base the construction on the direct product of two chains C and C'. Again we use two gadgets: covering M_3-s to make a prime interval $\mathfrak{p}$ of the first chain congruence-equivalent to a prime interval $\mathfrak{q}$ of the second chain; and N_5 (rather than the $N_{5,5}$ of Section 20.2) to force that $\mathrm{con}(\mathfrak{p}) < \mathrm{con}(\mathfrak{q})$, for the prime intervals $\mathfrak{p}$ and $\mathfrak{q}$ of the first chain.

Let $C = C_7$ and $C' = C_6$. For every cover $x \prec y$ in $J(D) \cup J(E)$, we color three adjacent prime intervals of C with y, x, and y. We start with $J(E)$; it has one cover: $d \prec i$, so we color the first three prime intervals with i, d, i. Then we take the only cover of $J(E)$: $b \prec c$, and we color the next three prime intervals of C with b, b, c. We color C' by setting up a bijection between the prime intervals of C' and $J(D) \cup J(E)$. We start with $J(E)$, then with $J(D)$.

We start by adding the M_3 gadgets.

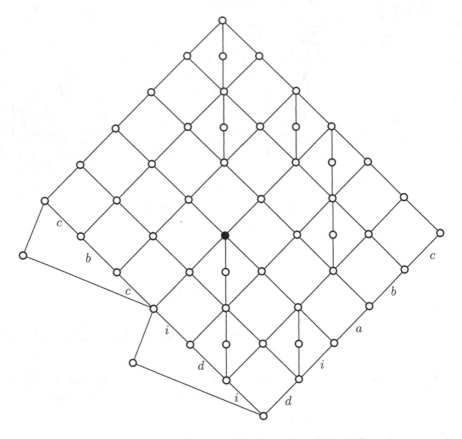

Figure 20.3: The planar lattice L and ideal I.

First, the M_3 gadget is used to "fill in" every covering square whose two sides have the same color; there are six.

Second, we "identify" by ϱ, that is, we fill in every covering square whose two sides have color x and $x\varrho$, for some $x \in J(E)$; there are two.

Next comes the N_5 gadget: for every cover $x \prec y$ in $J(D) \cup J(E)$, take the three adjacent prime intervals of C colored with y, x, and y; add an element so that they will form an N_5.

Note that the prime interval colored by a is not related by any other prime interval by the N_5 or M_3 gadget because a is not comparable to any other element of $J(D)$ and because a is not in the image of ϱ.

This completes the construction of the lattice L. The principal ideal generated by the black-filled element is I; see Figure 20.3.

It is really easy to compute that I represents E, L represents D, and the restriction map represents φ.

A formal proof Theorem 20.3 is very similar to proofs we have already presented in detail, so we leave it to the reader.

20.5. Discussion

In G. Grätzer and H. Lakser [98], we prove a very hard variant of Theorem 20.1, constructing the lattices K and L as *isoform* lattices (see Section 13.5 and Chapter 16). The proof is very technical and there is no *Proof-by-Picture*.

In general, automorphisms of a lattice do not restrict to automorphisms of an ideal. In [95], G. Grätzer and H. Lakser construct lattices where this does happen.

Theorem 20.13. *Let D and E be finite distributive lattices with more than one element, and let $\psi\colon D \to E$ be a bounded homomorphism. Let G and H be groups, and let $\eta\colon G \to H$ be a group homomorphism. Then there exist a lattice L, an ideal I in L, lattice isomorphisms*

$$\varrho_D\colon D \to \operatorname{Con} L, \qquad \varrho_E\colon E \to \operatorname{Con} I,$$

and group isomorphisms

$$\tau_G\colon G \to \operatorname{Aut} L, \qquad \tau_H\colon H \to \operatorname{Aut} I$$

such that, for each $x \in D$, the congruence relation $\varrho_E(\psi x)$ on I is the restriction to I of the congruence relation $\varrho_D(x)$ on L, and, for each $g \in G$, the automorphism $\tau_H(\eta g)$ of I is the restriction of the automorphism $\tau_G(g)$ of L.
If G and H are finite, then the lattice L can be chosen to be finite.

By identifying D with $\operatorname{Con} L$, E with $\operatorname{Con} I$, G with $\operatorname{Aut} L$, and H with $\operatorname{Aut} I$, Theorem 20.13 can be paraphrased as follows: any pair ψ, a bounded homomorphism of finite distributive lattices, and ϱ, a homomorphism of groups, can be simultaneously realized as the respective restrictions $\operatorname{Con} L \to \operatorname{Con} I$ and $\operatorname{Aut} L \to \operatorname{Aut} I$ for some lattice L and some ideal I in L.

Note that this result is a far reaching generalization of the Baranskiĭ-Urquhart Theorem discussed in Section 17.1.

In Section 19.5 we discussed the (doubly) 2-distributive lattices of A. P. Huhn.

Problem 20.1. Can Theorem 20.1 be proved for (doubly) 2-distributive lattices?

Problem 20.2. What can we say about semimodular (doubly) 2-distributive lattices?

Problem 20.3. Can Theorem 20.1 be proved for (semi) modular lattices?

Tensor Extensions

21.1. The problem

Let A and B be nontrivial finite lattices. Then $\operatorname{Con} A \times \operatorname{Con} B$ is a finite distributive lattice, so by the Dilworth Theorem (Theorem 8.1), the lattice $\operatorname{Con} A \times \operatorname{Con} B$ can be represented as $\operatorname{Con} L$, for some finite lattice L. How can we construct the lattice L from the lattices A and B? By Theorem 2.1,

$$(1) \qquad \operatorname{Con}(A \times B) \cong \operatorname{Con} A \times \operatorname{Con} B,$$

so we can take $L = A \times B$.

For a similar isomorphism, we now introduce tensor products.

For a nontrivial lattice K with zero, we denote by K^- the $\vee$-subsemilattice of K defined on $K - \{0\}$.

Let A and B be nontrivial lattices with zero. We denote by $A \otimes B$ the *tensor product* of A and B, defined as the free $\{\vee, 0\}$-semilattice generated by the set $A^- \times B^-$ and subject to the relations

$$(2) \qquad \begin{aligned} (a, b_1) \vee (a, b_2) &= (a, b_1 \vee b_2), & \text{for } a \in A^-,\ b_1, b_2 \in B^-; \\ (a_1, b) \vee (a_2, b) &= (a_1 \vee a_2, b), & \text{for } a_1, a_2 \in A^-,\ b \in B^-. \end{aligned}$$

If A and B are finite nontrivial lattices, then $A \otimes B$ is a finite lattice since it is a finite $\vee$-semilattice with zero. Since $A \otimes B$ just "happens to be a lattice", it is quite surprising that the isomorphism

$$(3) \qquad \operatorname{Con} A \otimes \operatorname{Con} B \cong \operatorname{Con}(A \otimes B),$$

© Springer International Publishing Switzerland 2016
G. Grätzer, *The Congruences of a Finite Lattice*,
DOI 10.1007/978-3-319-38798-7_21

holds, see G. Grätzer, H. Lakser, and R. W. Quackenbush [99]. The proof of this result is very complicated, and not appropriate for presentation in this book.

There is, however, a closely related result, with an approachable proof.

If L is a finite lattice and D is a finite distributive lattice, then we can define $L[D]$ as the lattice of all isotone maps from $J(D)$ to L. By E. T. Schmidt [176], we obtain that

$$\operatorname{Con} L[D] \cong (\operatorname{Con} L)[\operatorname{Con} D].$$

(See also Section 6.3 and G. Grätzer and E. T. Schmidt [121].)

So here is the question: Let A and B be nontrivial finite lattices. Then

$$(\operatorname{Con} A)[\operatorname{Con} B]$$

is a finite distributive lattice. Can we represent it as $\operatorname{Con} L$, for a finite lattice L *constructed from* A and B?

In G. Grätzer and M. Greenberg [78], we introduced a construction: the *tensor extension* $A[B]$ for nontrivial finite lattices A and B. In this chapter we will prove that

$$(4) \qquad\qquad \operatorname{Con}(A[B]) \cong (\operatorname{Con} A)[\operatorname{Con} B],$$

providing an answer to the above question.

In Section 21.7 we discuss in detail the history of this isomorphism and how it relates to tensor products. In [78], we prove that for finite lattices, a tensor extension is isomorphic to a lattice tensor product, introduced in G. Grätzer and F. Wehrung [146], so our result is the finite case of the main result of [146].

The reader should view this chapter as an introduction to a very technical chapter of congruence lattice theory, and should consult Section 21.7 for further readings. This elementary introduction is based on G. Grätzer and M. Greenberg [78] and [81].

21.2.   Three unary functions

To prepare for the introduction of tensor extensions, we define some useful functions.

Let A and B be nontrivial finite lattices.

We are going to define three unary functions, m, j, and p, on B^A, the set of all maps from B into A. First, for $a \in A$ and $\alpha \in B^A$, we define—computed

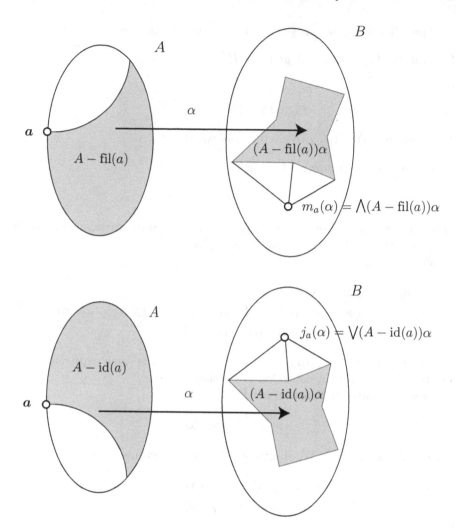

Figure 21.1: Illustrating $m_a(\alpha)$ and $j_a(\alpha)$.

in B (see Figure 21.1):

$$(5) \qquad m_a(\alpha) = \bigwedge \alpha(A - \mathrm{fil}(a)) = \bigwedge_{x \not\geq a} \alpha x,$$

$$(6) \qquad j_a(\alpha) = \bigvee \alpha(A - \mathrm{id}(a)) = \bigvee_{y \not\leq a} \alpha y,$$

$$(7) \qquad p_a(\alpha) = \bigwedge_{x \not\geq a} j_x(\alpha) = \bigwedge_{x \not\geq a} \bigvee \alpha(A - \mathrm{id}(x)) = \bigwedge_{x \not\geq a} \bigvee_{y \not\leq x} \alpha y.$$

The following statements about these functions are easy to verify:

Lemma 21.1. *For $a, b \in A$ and $\alpha \in B^A$,*

$$(8) \qquad\qquad m_a(\alpha) \wedge m_b(\alpha) = m_{a \vee b}(\alpha),$$
$$(9) \qquad\qquad j_a(\alpha) \vee j_b(\alpha) = j_{a \wedge b}(\alpha).$$

Moreover, $m_0(\alpha) = 1_B$, $j_1(\alpha) = 0_B$, and $p_0(\alpha) = 1_B$.

For $A = \mathsf{M}_3 = \{0, a, b, c, 1\}$, we get

$$m_a(\alpha) = \alpha b \wedge \alpha c \wedge \alpha 0,$$

and symmetrically for $m_b(\alpha)$ and $m_c(\alpha)$. These are the expressions that occur in the existential definition of boolean triples.
For $A = \mathsf{M}_3$,

$$\begin{aligned}
p_a(\alpha) &= j_b(\alpha) \wedge j_c(\alpha) \wedge j_0(\alpha) \\
&= (\alpha a \vee \alpha c \vee \alpha 1) \wedge (\alpha b \vee \alpha c \vee \alpha 1) \wedge (\alpha a \vee \alpha b \vee \alpha c \vee \alpha 1) \\
&= (\alpha a \vee \alpha c \vee \alpha 1) \wedge (\alpha b \vee \alpha c \vee \alpha 1),
\end{aligned}$$

and symmetrically for $p_b(\alpha)$ and $p_c(\alpha)$. These are the same expressions that occur in the fixed-point definition of boolean triples.
Now we define the function $m \colon B^A \to B^A$ by

$$(10) \qquad\qquad m(\alpha)(a) = m_a(\alpha),$$

for $a \in A$ and $\alpha \in B^A$. We similarly define the functions j and p:

$$(11) \qquad\qquad j(\alpha)(a) = j_a(\alpha),$$
$$(12) \qquad\qquad p(\alpha)(a) = p_a(\alpha).$$

21.3. Defining tensor extensions

In Section 6.1 we had three definitions of $\mathsf{M}_3[B]$:

(F), the fixed point definition;

(E), the existential definition;

the closure definition in Lemma 6.1.(ii).

The elements of $\mathsf{M}_3[B]$ are 3-tuples of elements of B. Similarly, the elements of the tensor extension $A[B]$ are defined as A-tuples of elements of B that is, elements of B^A.

Let A and B be finite nontrivial lattices. The existential definition presents the *tensor extension* of A by B:

(13) $$A[B] = \{\, m(\alpha) \mid \alpha \in B^A \,\}.$$

In other words, for $\gamma \in B^A$,

$$\gamma \in A[B] \quad \text{iff} \quad \text{there exists } \alpha \in B^A \text{ satisfying } \gamma = m(\alpha).$$

To obtain a "fixed-point description", we introduce the notation $p(\alpha) = \overline{\alpha}$ (for $\alpha \in B^A$) and call $\overline{\alpha}$ the *closure of* α (in B^A). We now verify that this is indeed a closure operation on B^A.

Lemma 21.2. *For $\alpha, \beta \in B^A$,*

(i) $\alpha \leq \overline{\alpha}$.

(ii) *If $\alpha \leq \beta$ and $\beta \in A[B]$, then $\overline{\alpha} \leq \beta$.*

(iii) $\alpha \in A[B]$ *iff $\alpha = \overline{\alpha}$.*

Proof.

(i) Let $a \in A$. Then for any $x \not\geq a$, we have $a \in A - \mathrm{id}(x)$, and so $\alpha a \leq \bigvee \alpha(A - \mathrm{id}(x))$. Therefore, by (7),

$$\alpha a \leq \bigwedge_{x \not\geq a} \bigvee \alpha(A - \mathrm{id}(x)) = p_a(\alpha) = \overline{\alpha}(a).$$

(ii) By the definition of $A[B]$, there is a $\gamma \in B^A$ such that $\beta = m(\gamma)$. Choose $a \in A$ and $x \not\geq a$. Thus for each $y \in A - \mathrm{id}(x)$, we have

$$\alpha y \leq \beta y = m_y(\gamma) = \bigwedge \gamma(A - \mathrm{fil}(y)) \leq \gamma x,$$

since $x \in A - \mathrm{fil}(y)$. Joining these inequalities for $y \in A - \mathrm{id}(x)$, we obtain that

$$\bigvee \alpha(A - \mathrm{id}(x)) \leq \gamma x.$$

Meeting these inequalities for $x \not\geq a$, we get

$$p_a(\alpha) = \bigwedge_{x \not\geq a} \bigvee \alpha(A - \mathrm{id}(x)) \leq \bigwedge_{x \not\geq a} \gamma x = m_a(\gamma) = \beta a,$$

or $\overline{\alpha}(a) \leq \beta a$, proving that $\overline{\alpha} \leq \beta$.

(iii) Let $\alpha = \overline{\alpha}$, that is, $\alpha a = p_a(\alpha)$ for all $a \in A$. Define $\beta \in B^A$ by $\beta b = \bigvee \alpha(A - \mathrm{id}(b))$ for $b \in A$. Then

$$m_a(\beta) = \bigwedge \beta(A - \mathrm{fil}(a)) = \bigwedge_{x \not\geq a} x\beta = \bigwedge_{x \not\geq a} \bigvee \alpha(A - \mathrm{id}(x)) = p_a(\alpha) = \alpha a,$$

for any $a \in A$; therefore, $\alpha \in A[B]$. Conversely, assume that $\alpha \in A[B]$. By (i), $\alpha \leq \overline{\alpha}$ holds. The reverse inequality holds by (ii) with $\beta = \alpha$, so $\alpha = \overline{\alpha}$. $\square$

We have established the following result.

Theorem 21.3. *Let A and B be nontrivial finite lattices. The tensor extension $A[B]$ is a lattice with meets computed pointwise and joins computed as the closures of the pointwise joins, that is, according to the formula*

$$\alpha \vee_{A[B]} \beta = \overline{\alpha \vee_{B^A} \beta},$$

for $\alpha, \beta \in A[B]$.

The main result on finite tensor extensions is the following:

Theorem 21.4. *Let A and B be nontrivial finite lattices. Then the following isomorphism holds:*

$$\mathrm{Con}(A[B]) \cong (\mathrm{Con}\, A)[\mathrm{Con}\, B].$$

If A is simple, then $\mathrm{Con}\, A \cong C_2$, so $(\mathrm{Con}\, A)[\mathrm{Con}\, B] \cong \mathrm{Con}\, B$. We have obtained the following generalization of Theorem 6.3.

Corollary 21.5. *Let A and B be nontrivial finite lattices. If A is simple, then $A[B]$ is a congruence-preserving extension of B.*

21.4.   Computing

21.4.1   Some special elements

Let A and B be nontrivial finite lattices.

For any $q \in B$, we define the (almost) constant function, κ_q, in B^A with value q:

$$\kappa_q(x) = \begin{cases} q & \text{for } x \in A - \{0_A\}; \\ 1_B & \text{for } x = 0_A. \end{cases}$$

Clearly $\kappa_q \in A[B]$, for any $q \in B$, since $p(\kappa_q) = \kappa_q$. We will call $\{\, \kappa_q \mid q \in B \,\}$ the *diagonal* of $A[B]$.

For $a \in A$, we define the element $\chi_a \in B^A$—the characteristic function of $\mathrm{id}(a)$—by (see Figure 21.2)

$$\chi_a(x) = \begin{cases} 1_B, & \text{if } x \in \mathrm{id}(a); \\ 0_B, & \text{otherwise.} \end{cases}$$

Lemma 21.6. *χ_a is an element of $A[B]$ for any $a \in A$.*

Proof. Define $\gamma \in B^A$ by $\gamma \mathrm{fil}(a) = 0_B$ and $\gamma(A - \mathrm{fil}(a)) = 1_B$. Clearly, $m_b(\gamma) = \chi_a(b)$, for any $b \in A$, that is, $\chi_a = m(\gamma)$, so $\chi_a \in A[B]$. $\qquad\square$

If $p \in B$ and $a, b \in A$ with $a < b$, we define $\vec{p}_{ab} = \chi_b \wedge (\chi_a \vee_{B^A} \kappa_p) \in B^A$ (see Figure 21.2).

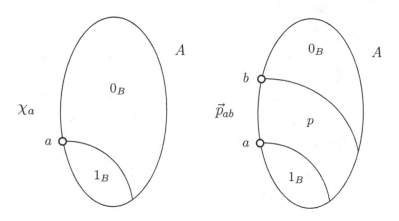

Figure 21.2: Illustrating χ_a and $\vec{p}_{ab}$.

Lemma 21.7. $\vec{p}_{ab} \in A[B]$.

Proof. Under these hypotheses, $\chi_a \vee_{B^A} \kappa_p = \chi_a \vee_{A[B]} \kappa_p \in A[B]$, therefore, $\vec{p}_{ab} \in A[B]$. $\qquad\qquad\qquad\square$

The proof of the following statement is trivial since the joins and the relevant meets in $A[B]$ are computed pointwise.

Lemma 21.8. *Let* $a, b \in A$ *with* $a < b$. *Then the map* $\gamma_{ab} \colon p \mapsto \vec{p}_{ab}$ *is an embedding of* B *into the sublattice* $[\chi_a, \chi_b]$ *of* $A[B]$; *in fact,* γ_{ab} *is a bounded embedding.*

Proof. This is obvious. The last statement adds the facts that $(\vec{0}_B)_{ab} = \chi_a$ and $(\vec{1}_B)_{ab} = \chi_b$. $\qquad\qquad\qquad\square$

We will denote by B_{ab} the image of B under the map γ_{ab}. Note that the diagonal, $\{\, \kappa_q \mid q \in B \,\}$, can also be defined as $B_{0_A 1_A}$. Moreover, if a is covered by b and b is join-irreducible, then $\mathrm{id}(b) - \mathrm{id}(a) = \{b\}$, so $B_{ab} = [\chi_a, \chi_b]$.

The next lemma will be useful in computing some joins in $A[B]$.

Lemma 21.9. *Let* $a, b, c \in A$ *with* $c = a \vee b$ *and let* $p, q \in B$ *with* $p \leq q$. *Define* $\alpha \in B^A$ *by*

$$t\alpha = \begin{cases} 1_B & \text{for } t \in \mathrm{id}(a); \\ q & \text{for } t \in \mathrm{id}(b) - \mathrm{id}(a); \\ p & \text{for } t \in \mathrm{id}(c) - (\mathrm{id}(a) \cup \mathrm{id}(b)); \\ 0_B & \text{for } t \in A - \mathrm{id}(c). \end{cases}$$

Then $\overline{\alpha} = \vec{q}_{ac}$. *(See Figure 21.3.)*

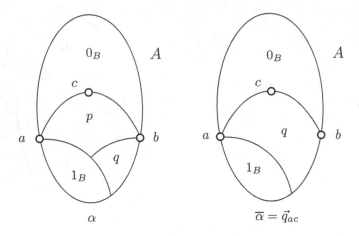

Figure 21.3: Illustrating α and $\overline{\alpha}$.

Proof. Since $\alpha \leq \vec{q}_{ac}$, it follows that $\overline{\alpha} \leq \vec{q}_{ac}$. So it is sufficient to prove that $\overline{\alpha}(t) = q$ for $t \in \mathrm{id}(c) - (\mathrm{id}(a) \cup \mathrm{id}(b))$. To see this, let $e \in \mathrm{id}(c) - (\mathrm{id}(a) \cup \mathrm{id}(b))$, that is, $e \leq c$, but $e \not\leq a$ and $e \not\leq b$. Let $f \in A - \mathrm{fil}(e)$, that is, $e \not\leq f$. Then either $a \not\leq f$ or $b \not\leq f$; otherwise, we would get $a \vee b = c \leq f$, implying that $e \leq f$, a contradiction. Therefore,

$$\bigvee \alpha(A - \mathrm{id}(f)) \geq \alpha a \wedge \alpha b = 1_B \wedge q = q,$$

from which it follows that

$$\overline{\alpha}(e) = \bigwedge_{f \in A - \mathrm{fil}(e)} \bigvee \alpha(A - \mathrm{id}(f)) \geq q,$$

for any $e \in \mathrm{id}(c) - (\mathrm{id}(a) \cup \mathrm{id}(b))$. Since $\alpha e \leq \vec{q}_{ac}$, it follows that $e\overline{\alpha} \leq \vec{q}_{ac}$, which implies that $e\overline{\alpha} \leq q$. We conclude that $e\overline{\alpha} = q$, as claimed. □

Lemma 21.10. *The map $a \mapsto \chi_a$ is an embedding of A into $A[B]$.*

Proof. This map is clearly one-to-one and meet-preserving. To compute $\chi_a \vee \chi_b$, let α be the pointwise join. Applying Lemma 21.9, we obtain that $\overline{\alpha} = \chi_{a \vee b}$, proving that $\chi_a \vee \chi_b = \chi_{a \vee b}$. □

Note that while we get many copies of B in $A[B]$, we obtain only one copy of A in $A[B]$—this is not surprising since there is, in general, only one copy of M_3 in $\mathsf{M}_3[B]$ and, further, $A[\mathsf{C}_2]$ is isomorphic to A.

21.4.2   An embedding

Let A and B be finite nontrivial lattices. Then for an ideal I of A, we have a natural embedding of $I[B]$ into $A[B]$.

Lemma 21.11. *Let I be an ideal of A. Then there exists a unique embedding*

$$f: I[B] \to A[B],$$

such that $f(\vec{p}_{ab}) = \vec{p}_{ab}$, for $p \in B$ and $a < b$ in I, where in the formula $f(\vec{p}_{ab}) = \vec{p}_{ab}$, the first $\vec{p}_{ab}$ is in $I[B]$, while the second is in $A[B]$.

Proof. For $\alpha \in I[B]$, we define α^z $(\in B^A)$, the "zero-padded" version of α, by

$$\alpha^z(x) = \begin{cases} \alpha x, & \text{if } x \in I; \\ 0_B, & \text{otherwise.} \end{cases}$$

It is obvious that $\alpha^z \in A[B]$.

Define

$$f: I[B] \to A[B],$$

by $\alpha \mapsto \alpha^z$. Clearly, f is an embedding and $f(\vec{p}_{ab}) = \vec{p}_{ab}$. By Lemma 21.22, we have the uniqueness. $\qquad\square$

21.4.3 Distributive lattices

For the lattices X and Y, let $\hom_{\{\vee,0\}}(X,Y)$ denote the set of $\{\vee,0\}$-ho-momorphism of X into Y, regarded as a $\{\vee,0\}$-semilattice. The following statement is a generalization of the fact that balanced triples are boolean for a distributive lattice B; see Section 6.3.

Lemma 21.12. *Let B be a distributive lattice. Then*

(14) $$A[B] = \hom_{\{\vee,0\}}(A, B^d).$$

Proof. By Lemma 21.1, for any lattice B, we always have

$$A[B] \subseteq \hom_{\{\vee,0\}}(A, B^d).$$

To prove the reverse containment in (14), let us assume that B is distributive. Let $\alpha \in \hom_{\{\vee,0\}}(A, B^d)$, and let p be the term introduced in (7). By Lemma 21.2, it is sufficient to show that $p(\alpha) \le \alpha$. For $a \in A$, compute:

$$
\begin{aligned}
p(\alpha)(a) &= \bigwedge_{x \nleq a} \bigvee_{y \nleq x} \alpha y \\
&= \bigvee \left(\bigwedge_{x \nleq a} \alpha \nu(x) \mid \nu \in \prod_{x \nleq a}(A - \mathrm{id}(x)) \right) \quad \text{(by the distributivity of B)} \\
&= \bigvee \left(\bigvee_{x \nleq a} \alpha \nu(x) \mid \nu \in \prod_{x \nleq a} \alpha(A - \mathrm{id}(x)) \right) \quad \text{(since $\alpha \in \hom_{\{\vee,0\}}(A, B^d)$).}
\end{aligned}
$$

To show that $p(\alpha) \leq \alpha$, that is, that $p(\alpha)a \leq \alpha a$, for all $a \in A$, it suffices to show that

$$(15) \qquad\qquad \alpha(\bigvee_{x \not\geq a} \nu(x)) \leq \alpha a,$$

for any $\nu \in \prod_{x \not\geq a}(A - \mathrm{id}(x))$. Let $\nu \in \prod_{x \not\geq a}(A - \mathrm{id}(x))$, and let $b = \bigvee_{x \not\geq a} \nu x$. We claim that $b \geq a$. Indeed, otherwise, from the definition of b, we see that $\nu b \in \mathrm{id}(b)$, contradicting the fact that $\nu \in \prod_{x \not\geq a}(A - \mathrm{id}(x))$. Since α is antitone, we obtain that $\alpha b \leq \alpha a$, completing the proof of (15). $\square$

21.5.   Congruences

In this section we compute many properties of the congruences of $A[B]$, to lay the groundwork for the proof of the isomorphism (4).

21.5.1   Congruence spreading

We show that congruence-perspectivities and congruence-projectivities in A naturally translate into a family of congruence-perspectivities and congruence-projectivities in $A[B]$.

Lemma 21.13. *Let $a, b, c, d \in A$ with $a < b$ and $c < d$. Let $p, q \in B$ with $p \leq q$. If $[a, b]$ is congruence-perspective onto $[c, d]$ in A, then $[\vec{p}_{ab}, \vec{q}_{ab}]$ is congruence-perspective onto $[\vec{p}_{cd}, \vec{q}_{cd}]$ in $A[B]$.*

Proof. For an down congruence-perspectivity of $[a, b]$ onto $[c, d]$, we only have to compute four inequalities and a componentwise meet; these are trivial. For the up congruence-perspectivity, we have to verify that $\vec{p}_{ab} \vee \vec{q}_{cd} = \vec{q}_{ab}$; we perform the componentwise join and then apply Lemma 21.9. $\square$

Since congruence-projectivity is the transitive extension of congruence-perspectivity, we obtain:

Corollary 21.14. *If the interval $[a, b]$ is congruence-projective onto the interval $[c, d]$ in A, then the interval $[\vec{p}_{ab}, \vec{q}_{ab}]$ is congruence-projective onto the interval $[\vec{p}_{cd}, \vec{q}_{cd}]$ in $A[B]$.*

If $a, b \in A$ with $a < b$ and γ is a congruence of $A[B]$, we define γ_{ab} as the congruence of B that corresponds to the restriction of γ to B_{ab} (defined in Section 21.4.1) under the isomorphism $\gamma_{ab} \colon p \mapsto \vec{p}_{ab}$ of B and B_{ab}. The following statement easily follows from Lemma 21.13:

Lemma 21.15. *Let $a, b, c \in A$ with $a \leq b \leq c$. Let γ be a congruence of $A[B]$. Then $\gamma_{ac} = \gamma_{ab} \wedge \gamma_{bc}$ in B.*

Lemma 21.16. *Let $a, b, c, d \in A$ with $a \leq b$ and $c \leq d$. If $\operatorname{con}(c, d) \leq \operatorname{con}(a, b)$, then $\gamma_{ab} \leq \gamma_{cd}$ in B.*

Proof. First, let us assume that $c \prec d$. Then it follows from $\operatorname{con}(c, d) \leq \operatorname{con}(a, b)$ that $[a, b]$ is congruence-projective onto $[c, d]$. Let $p \leq q$ in B and $p \equiv q \pmod{\gamma_{ab}}$. Then $\vec{p}_{ab} \equiv \vec{q}_{ab} \pmod{\gamma}$. By Corollary 21.14, $[\vec{p}_{ab}, \vec{q}_{ab}]$ is congruence-projective onto $[\vec{p}_{cd}, \vec{q}_{cd}]$, so $\vec{p}_{cd} \equiv \vec{q}_{cd} \pmod{\gamma}$, that is, $p \equiv q \pmod{\gamma_{cd}}$, verifying that $\gamma_{ab} \leq \gamma_{cd}$.

Second, in the general case, let $c = c_0 < c_1 < \cdots < c_n = d$ be a maximal chain in the interval $[c, d]$ in B. Then $\operatorname{con}(c_{j-1}, c_j) \leq \operatorname{con}(c, d) \leq \operatorname{con}(a, b)$ for every $0 < j \leq n$. Hence by the previous paragraph, $\gamma_{ab} \leq \gamma_{c_{j-1}c_j}$. Applying Lemma 21.15 $(n-1)$-times, we obtain that $\gamma_{ab} \leq \bigwedge_{0<j\leq n} \gamma_{c_{j-1}c_j} = \gamma_{cd}$. $\square$

Lemma 21.17. *Let A be a finite simple lattice and let z be an atom of A. Let $p, q \in B$ with $p \leq q$ and let γ be a congruence of $A[B]$. Then $\vec{p}_{0_A z} \equiv \vec{q}_{0_A z} \pmod{\gamma}$ implies that $\kappa_p \equiv \kappa_q \pmod{\gamma}$.*

Proof. Since A is simple, $\operatorname{con}(0_A, z) = \operatorname{con}(0_A, 1_A)$, so $\gamma_{0_A 1_A} = \gamma_{0_A z}$, by Lemma 21.16. Therefore, if $\vec{p}_{0_A z} \equiv \vec{q}_{0_A z} \pmod{\gamma}$, then $\vec{p} \equiv \vec{q} \pmod{\gamma_{0_A z}}$, and so $\vec{p} \equiv \vec{q} \pmod{\gamma_{0_A 1_A}}$, that is, $\vec{p}_{0_A 1_A} \equiv \vec{q}_{0_A 1_A} \pmod{\gamma}$, which completes the proof since $\kappa_p = \vec{p}_{0_A 1_A}$ and $\kappa_q = \vec{q}_{0_A 1_A}$. $\square$

For a congruence γ of $A[B]$, we define $\nabla\gamma$ to be the congruence of B induced by the restriction of γ to the diagonal of $A[B]$. Note that $\nabla\gamma = \gamma_{0_A 1_A}$. If α is a congruence of B, we define a binary relation $\Delta\alpha$ on $A[B]$ by $\alpha \equiv \beta \pmod{\Delta\alpha}$ iff $\alpha a \equiv \beta a \pmod{\alpha}$ for all $a \in A$. It is easy to verify that $\Delta\alpha$ is a congruence.

Lemma 21.18. *For any congruence α of B, $\Delta\alpha$ is a congruence of $A[B]$.*

Proof. It is obvious that $\Delta\alpha$ is an equivalence relation. The substitution property for meets is obvious; for joins it follows from our formulas for joins in $A[B]$: for $\alpha, \beta \in A[B]$, the join is the closure of the pointwise join $\alpha \vee_{B^A} \beta$, which is the same as $p(\alpha \vee_{B^A} \beta)$ (see Theorem 21.3). Since p_a is a term for every $a \in A$, it preserves the congruence α^A in B^A. The substitution property for joins follows for $\Delta\alpha$. $\square$

Lemma 21.19. *$\Delta\nabla\gamma \subseteq \gamma$ for any congruence γ of $A[B]$.*

Proof. For $\alpha, \beta \in A[B]$, let $\alpha \equiv \beta \pmod{\Delta\nabla\gamma}$. By definition, $\alpha a \equiv \beta a \pmod{\nabla\gamma}$ for any $a \in A$. Define $\gamma = j(\alpha)$ and $\delta = j(\beta)$. By (5), $\alpha a = \bigwedge \gamma(A - \operatorname{fil}(a))$ and $\beta a = \bigwedge \delta(A - \operatorname{fil}(a))$, for any $a \in A$. Also, since $\alpha a \equiv \beta a \pmod{\nabla\gamma}$, for any $a \in A$, we obtain that $\beta b \equiv b\delta \pmod{\nabla\gamma}$ for any $b \in A$.

For $b \in A$, let $\varrho_b = \chi_b \vee \kappa_{\gamma b}$. Then

$$\varrho_b(a) = \begin{cases} \beta b, & \text{if } a \notin \operatorname{id}(b); \\ 1_B, & \text{otherwise.} \end{cases}$$

Similarly, let $\sigma_b = \chi_b \vee \kappa_{\delta b}$, that is,

$$\sigma_b(a) = \begin{cases} \delta b & \text{if } a \notin \text{id}(b); \\ 1_B, & \text{otherwise.} \end{cases}$$

Since $\beta b \equiv \delta b \pmod{\nabla\gamma}$ means that $\kappa_{\gamma b} \equiv \kappa_{\delta b} \pmod{\gamma}$, for any $b \in A$, it follows that $\varrho_b \equiv \sigma_b \pmod{\gamma}$ for any b.

We claim that $\alpha = \bigwedge_{b \in A} \varrho_b$. Indeed, for all $a \in A$,

$$\bigwedge_{b \in A} \varrho_b(a) = \bigwedge_{b \not\geq a} \varrho_b(a)$$

(since the other components equal 1_B)

$$= \bigwedge_{b \notin \text{fil}(a)} \varrho_b(a) = \bigwedge_{b \notin \text{fil}(a)} \gamma b = \bigwedge(\gamma(A - \text{fil}(a)) = \alpha a.$$

Similarly, $\beta = \bigwedge_{b \in A} \sigma_b$. Since $\varrho_b \equiv \sigma_b \pmod{\gamma}$, for any $b \in A$, it follows that $\alpha \equiv \beta \pmod{\gamma}$, completing the proof. $\qquad\square$

Lemma 21.20. *Let A be a finite simple lattice. Then $\Delta\nabla\gamma = \gamma$, for any congruence γ of $A[B]$.*

Proof. Let $\alpha \equiv \beta\gamma$ with $\alpha \leq \beta$. If $d \in A$ is join-irreducible, let $c = d_*$, the unique $c \in A$ covered by d. Then, $\text{id}(d) - \text{id}(c) = \{d\}$ and

$$(\text{mod } \alpha \wedge \chi_d) \vee \chi_c = \overrightarrow{\alpha d}_{cd},$$
$$(\beta \wedge \chi_d) \vee \chi_c = \overrightarrow{\beta d}_{cd}.$$

Therefore, $\overrightarrow{\alpha d}_{bd} \equiv \overrightarrow{\beta d}_{cd} \pmod{\gamma}$, that is, $\alpha d \equiv \beta d \pmod{\gamma_{cd}}$. By Lemma 21.16, since A is simple, $\alpha d \equiv \beta d \pmod{\gamma_{0_A 1_A}}$, that is, $\overrightarrow{\alpha d}_{0_A 1_A} = \overrightarrow{\beta d}_{0_A 1_A}(\gamma)$, which is the same as $\kappa_{\alpha d} \equiv \kappa_{\beta d} \pmod{\gamma}$. By the definition of $\nabla\gamma$, we obtain that $\alpha d \equiv \beta d \pmod{\nabla\gamma}$.

If $d \in A$ is join-reducible, then $d = d_1 \vee \cdots \vee d_n$, where each d_i is join-irreducible. Then, by Lemma 21.1,

$$\alpha d = \alpha d_1 \wedge \cdots \wedge \alpha d_n \equiv \beta d_1 \wedge \cdots \wedge \beta d_n = \beta d \pmod{\nabla\gamma}.$$

So $\alpha d \equiv \beta d \pmod{\nabla\gamma}$ for any $d \in A$. Therefore, $\alpha \equiv \beta \pmod{\Delta\nabla\gamma}$, implying that $\gamma \subseteq \Delta\nabla\gamma$. The reverse containment holds by Lemma 21.19. $\qquad\square$

We have thus established the main result of this section:

Theorem 21.21. *Let A be a nontrivial finite simple lattice. Then $A[B]$ is a congruence preserving extension of B.*

Proof. Indeed, $p \mapsto \kappa_p$ is a congruence preserving embedding of B into $A[B]$. $\qquad\square$

21.5.2 Some structural observations

In this section we prove some easy structural results on tensor extension that will be needed in the subsequent sections.

Lemma 21.22. *Let B be a nontrivial bounded lattice. Then $A[B]$ is generated, as a meet-semilattice, by the set*

$$\{\, \vec{p}_{a1} \mid a \in A,\ p \in B \,\}.$$

Proof. Let $\alpha \in A[B]$. Let $\beta \in B^A$ satisfy $\alpha = m(\beta)$, and set

$$\gamma = \bigwedge_{b \in A^-} \overrightarrow{(\beta_b)}_{b1} \in A[B],$$

where $\beta_b = \beta b$. Obviously, γ is meet-generated by the set $\{\, \vec{p}_{a1} \mid a \in A^-,\ p \in B \,\}$, so it suffices to prove that $\gamma = \alpha$.

Notice that, for $a \in A$,

$$\overrightarrow{(\beta_b)}_{b1}(a) = \begin{cases} 1, & \text{if } b \geq a; \\ \beta b, & \text{otherwise.} \end{cases}$$

Therefore,

$$\gamma a = \bigwedge_{b \not\geq a} \beta_b = m(\beta)(a) = \alpha a,$$

that is, $\gamma = \alpha$. □

Lemma 21.23. *Let us assume that 1, the unit element of A, is join-reducible. Then we can represent any $\alpha \in A[B]$ in the form*

$$\alpha = \bigvee{}_{A[B]} (\alpha \wedge \chi_a \mid a \in D),$$

where D is the set of dual atoms of A.

Proof. Let $\beta = \bigvee_{A[B]} (\alpha \wedge \chi_a \mid a \in D)$. Since $\alpha \wedge \chi_a \leq \alpha$, for $a \in D$, the inequality $\beta \leq \alpha$ obviously holds. Let $c \in A - \{0, 1\}$. Then for any dual atom a containing c,

$$\beta c \geq \alpha c \wedge \chi_a(c) = \alpha c$$

holds, since $c \leq a$ implies that $\chi_a(c) = 1$. Therefore, $\beta c \geq \alpha c$ for $c \neq 1$. Since $\alpha, \beta \in \hom_{\{\vee, 0\}}(A, B^d)$ and 1 is join-reducible, $\beta 1 \geq \alpha 1$ also holds. Therefore, $\beta = \alpha$, as claimed. □

Lemma 21.24. *Let us assume that A has a unique dual atom u. Let $\alpha \leq \beta$ in $A[B]$. Then*

$$(\chi_u \vee \alpha) \wedge \beta = (\chi_u \wedge \beta) \vee \alpha.$$

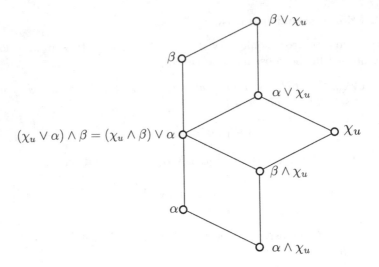

Figure 21.4: Illustrating Lemma 21.24.

See Figure 21.4.

Proof. Let $p = \alpha 1$. Then $\chi_u \vee \alpha = \vec{p}_{u1}$. Therefore,

$$((\chi_u \vee \alpha) \wedge \beta)(x) = \begin{cases} \beta x, & \text{if } x \neq 1; \\ p, & \text{if } x = 1, \end{cases}$$

since $p = \alpha 1 \leq \beta 1$. Now

$$(\chi_u \wedge \beta)(x) = \begin{cases} \beta x, & \text{if } x \neq 1; \\ 0_B, & \text{if } x = 1. \end{cases}$$

Therefore, it follows from $\alpha \leq \beta$ that

$$((\chi_u \wedge \beta) \vee_{B^A} \alpha)(x) = \begin{cases} \beta x, & \text{if } x \neq 1; \\ p, & \text{if } x = 1. \end{cases}$$

We conclude that

$$(\chi_u \wedge \beta) \vee_{B^A} \alpha = (\chi_u \vee \alpha) \wedge \beta \in A\langle<\rangle B > .$$

Therefore, $(\chi_u \wedge \beta) \vee_{B^A} \alpha$ is closed, and

$$(\chi_u \wedge \beta) \vee \alpha = (\chi_u \wedge \beta) \vee_{B^A} \alpha = (\chi_u \vee \alpha) \wedge \beta,$$

which was to be proved. $\square$

21.5.3   Lifting congruences

Let $\alpha \in \operatorname{Con} A$ and let $\alpha \colon A \to B$ be an antitone map. Define $\alpha_{\boldsymbol{\alpha}} \colon A/\boldsymbol{\alpha} \to B$ by

$$(16) \qquad \alpha_{\boldsymbol{\alpha}}(a/\boldsymbol{\alpha}) = \bigvee \alpha a/\boldsymbol{\alpha} = \alpha(\bigwedge a/\boldsymbol{\alpha}).$$

This definition makes sense since A is finite and α is antitone. It is easy to see that

$$(\alpha \wedge \beta)_{\boldsymbol{\alpha}} = \alpha_{\boldsymbol{\alpha}} \wedge \beta_{\boldsymbol{\alpha}}$$

and

$$(\alpha \vee_{B^A} \beta)_{\boldsymbol{\alpha}} = \alpha_{\boldsymbol{\alpha}} \vee_{B^{A/\boldsymbol{\alpha}}} \beta_{\boldsymbol{\alpha}},$$

for antitone maps α and β. We now show that the operations of reducing maps modulo $\boldsymbol{\alpha}$ and taking closures (see Section 21.3) commute. First, a simple observation. Since $x \le x'$ implies that $j(\alpha)(x) \ge j(\alpha)(x')$, for any $\alpha \colon A \to B$, it follows that

$$(17) \qquad p(\alpha)(a) = \bigwedge (\bigvee_{y \not\le x} \alpha y \mid x \text{ is maximal in } A - \operatorname{fil}(a)).$$

Lemma 21.25. *Let* $\boldsymbol{\alpha} \in \operatorname{Con} A$ *and let* $\alpha \colon A \to B$ *be an antitone map. Then* $p(\alpha_{\boldsymbol{\alpha}}) = p(\alpha)_{\boldsymbol{\alpha}}$.

Proof. By (16), it suffices to show that if a is minimal in $a/\boldsymbol{\alpha}$, then

$$p(\alpha_{\boldsymbol{\alpha}})(a/\boldsymbol{\alpha}) = p(\alpha)(a).$$

If a is minimal in $a/\boldsymbol{\alpha}$, then $x/\boldsymbol{\alpha} \ge a/\boldsymbol{\alpha}$ iff $x \ge a$. Therefore,

$$p(\alpha_{\boldsymbol{\alpha}})(a/\boldsymbol{\alpha}) = \bigwedge_{x \not\ge a} \bigvee_{y/\boldsymbol{\alpha} \not\le x/\boldsymbol{\alpha}} \alpha_{\boldsymbol{\alpha}}(y/\boldsymbol{\alpha})$$

$$= \bigwedge (\bigvee_{y/\boldsymbol{\alpha} \not\le x/\boldsymbol{\alpha}} \alpha_{\boldsymbol{\alpha}}(y/\boldsymbol{\alpha}) \mid x \text{ is maximal in } A - \operatorname{fil}(a)),$$

by a variant of (17). Since a is minimal in $a/\boldsymbol{\alpha}$, both $\operatorname{fil}(a)$ and $A - \operatorname{fil}(a)$ are unions of congruence classes. Therefore, if x is maximal in $A - \operatorname{fil}(a)$, then x is certainly maximal in $x/\boldsymbol{\alpha}$. But if x is maximal in $x/\boldsymbol{\alpha}$, then $y/\boldsymbol{\alpha} \le x/\boldsymbol{\alpha}$ iff $y \le x$. Thus

$$p(\alpha_{\boldsymbol{\alpha}})(a/\boldsymbol{\alpha}) = \bigwedge (\bigvee_{y \not\le x} \alpha_{\boldsymbol{\alpha}}(y/\boldsymbol{\alpha}) \mid x \text{ is maximal in } A - \operatorname{fil}(a))$$

$$= \bigwedge (\bigvee_{y \not\le x} \bigvee \alpha(y/\boldsymbol{\alpha}) \mid x \text{ is maximal in } A - \operatorname{fil}(a))$$

$$= \bigwedge (\bigvee_{y \not\le x} \alpha y \mid x \text{ is maximal in } A - \operatorname{fil}(a))$$

$$= p(\alpha)(a). \square$$

Corollary 21.26. *Let $\alpha \in \mathrm{Con}\, A$. Then there exists a unique homomorphism*

$$\gamma\colon A[B] \to (A/\alpha)[B]$$

such that $\gamma\colon \vec{p}_{ab} \mapsto \vec{p}_{a/\alpha\, b/\alpha}$.

Proof. Let $\gamma\colon \alpha \mapsto \alpha_\alpha$. This defines a lattice homomorphism by Lemma 21.25. It is easy to check that $\gamma\colon \vec{p}_{ab} \mapsto \vec{p}_{a/\alpha\, b/\alpha}$. The uniqueness follows from the fact that $A[B]$ is generated, as a lattice, by the set $\{\, \vec{p}_{ab} \mid a < b \in A \,\}$. □

For $\alpha \in \mathrm{Con}\, A$, let $\alpha[B] \in \mathrm{Con}\, A[B]$ be the kernel of the homomorphism described in Corollary 21.26. For $a < b$ in A, then it is apparent that

$$(18) \qquad \alpha[B_{ab}] = \begin{cases} 1_B & \text{if } a \equiv b\, (\alpha); \\ 0_B, & \text{otherwise.} \end{cases}$$

(B_{ab} was defined in Section 21.4.1.)

It is much easier to lift congruences of B to $A[B]$ than it is to lift congruences of A. Let $\gamma \in \mathrm{Con}\, B$ and let $\alpha\colon A \to B$. Define $\alpha^\gamma\colon A \to B/\gamma$ by composing α with the canonical projection from B to B/γ. The following statement is trivial.

Lemma 21.27. *For $\gamma \in \mathrm{Con}\, B$, there exists a unique homomorphism*

$$\lambda\colon A[B] \to A[B/\gamma]$$

such that $\lambda\colon \vec{p}_{ab} \mapsto \overrightarrow{(p/\gamma)}_{ab}$.

Let $A[\gamma] \in \mathrm{Con}\, A[B]$ be the kernel of the homomorphism described in Lemma 21.27. Then

$$(19) \qquad\qquad A[\gamma]_{ab} = \gamma,$$

for $a < b$ in A.

For $\alpha \in \mathrm{Con}\, A$ and $\gamma \in \mathrm{Con}\, B$, Corollary 21.26 and Lemma 21.27 combine in the following commutative diagram:

$$
\begin{array}{ccc}
A[B] & \longrightarrow & (A/\alpha)[B] \\
\downarrow & & \downarrow \\
A[B/\gamma] & \longrightarrow & (A/\alpha)[B/\gamma]
\end{array}
$$

Let $\alpha[\gamma] \in \mathrm{Con}\, A[B]$ be the kernel of either composite. Then by (18) and (19),

$$(20) \qquad \alpha[\gamma]_{ab} = \begin{cases} 1_B & \text{if } a \equiv b \pmod{\alpha}; \\ \gamma, & \text{otherwise,} \end{cases}$$

for $a < b$ in A.

21.5.4   The main lemma

Let I be an ideal of A, and $\gamma\colon I[B] \to A[B]$ be the embedding of Lemma 21.11. For $\gamma \in \operatorname{Con} A[B]$, let γ_I be the restriction of γ to $I[B]$, which is viewed as a sublattice of $A[B]$. Since $(\vec{p}_{ab})\gamma = \vec{p}_{ab}$, for $a < b \in I$, it follows that $(\gamma_I)_{ab} = \gamma_{ab}$.

We are now ready to prove the main lemma.

Lemma 21.28. *Let A be a finite lattice and let B be a bounded lattice. Let $\gamma, \delta \in \operatorname{Con} A[B]$ satisfy $\gamma_{ab} \leq \delta_{ab}$ for all $a < b \in A$. Then $\gamma \leq \delta$.*

Proof. We prove this statement by induction on length A. The lemma is trivial for lattices satisfying length $A \leq 2$.

Now let length $A = n$, let $\alpha \leq \beta$ in $A[B]$, and let $\alpha \equiv \beta \pmod{\gamma}$. First, we prove

Claim. *Let u be a dual atom of A. Then*

$$\alpha \wedge \chi_u \equiv \beta \wedge \chi_u \pmod{\gamma}.$$

Proof. Let $I = \operatorname{id}(u)$. Then length $I < $ length A. For all $a < b$ in I, by the induction hypothesis,

$$(\gamma_I)_{ab} = \gamma_{ab} \leq \delta_{ab} = (\delta_I)_{ab},$$

therefore, $\gamma_I \leq \delta_I$. Note that $\alpha \wedge \chi_u, \beta \wedge \chi_u \in I[B]$ since, in the terminology of Lemma 21.11, $\alpha \wedge \chi_u$ and $\beta \wedge \chi_u$ are "zero padded." The congruence $\alpha \equiv \beta \pmod{\gamma}$ implies that $\alpha \wedge \chi_u \equiv \beta \wedge \chi_u \pmod{\delta}$, so that $\alpha \wedge \chi_u \equiv \beta \wedge \chi_u \pmod{\gamma_I}$. Since $\gamma_I \leq \delta_I$, we obtain that $\alpha \wedge \chi_u \equiv \beta \wedge \chi_u \pmod{\delta_I}$, that is, $\alpha \wedge \chi_u \equiv \beta \wedge \chi_u \pmod{\delta}$. $\qquad\square$

To prove the lemma, we consider two cases.

Case 1: 1_A *is join-reducible.*

By Lemma 21.23, $\alpha = \bigvee(\alpha \wedge \chi_a \mid a \in D)$ and $\beta = \bigvee(\beta \wedge \chi_a \mid a \in D)$, where D is the set of dual atoms of A. By the Claim, $\alpha \wedge \chi_a \equiv \beta \wedge \chi_a \pmod{\gamma}$ for $a \in D$. Therefore, $\alpha \equiv \beta \pmod{\delta}$.

Case 2: 1_A *is join-irreducible.*

A has a unique dual atom u. Let $I = \operatorname{id}(u)$ and define

$$\alpha_1 = \alpha \vee \chi_u, \qquad \beta_1 = \beta \vee \chi_u,$$
$$\alpha_2 = \alpha \wedge \chi_u, \qquad \beta_2 = \beta \wedge \chi_u.$$

Let $p = \alpha 1$ and $q = \beta 1$. Noting that $\alpha_1 = \vec{p}_{u1}$ and $\beta_1 = \vec{q}_{u1}$, it follows from $\gamma_{u1} \leq \delta_{u1}$ that $\alpha_1 \equiv \beta_1 \pmod{\delta}$. By the Claim, $\alpha_2 \equiv \beta_2 \pmod{\delta}$. By Lemma 21.24, $\alpha_1 \wedge \beta = \beta_2 \vee \alpha$. Thus

$$\alpha = \alpha \vee \alpha_2 \equiv \alpha \vee \beta_2 = \alpha_1 \wedge \beta \equiv \beta_1 \wedge \beta = \beta \pmod{\delta}.$$

This completes the induction. $\qquad\square$

21.6.   The congruence isomorphism

In this section we prove Theorem 21.4 in a form that describes the isomorphism.

For $\Gamma \in \operatorname{Con} A[B]$, let $\mathcal{F}(\Gamma) \in (\operatorname{Con} B)^{\operatorname{Con} A}$ be defined by

$$
(21) \qquad \mathcal{F}(\Gamma)\colon \bigvee_{i \le n} \alpha_A(a_i, b_i) \to \bigwedge_{i \le n} \Gamma_{a_i b_i}.
$$

It follows from Lemma 21.16 that $\mathcal{F}$ is a well defined. By definition, $\mathcal{F}(\Gamma)$ is a $\{\vee, 0\}$-homomorphism from $\operatorname{Con} A$ into $(\operatorname{Con} B)^d$. Since $\operatorname{Con} B$ is distributive, by Lemma 21.12, $\mathcal{F}(\Gamma) \in (\operatorname{Con} A)[\operatorname{Con} B]$. Therefore,

$$
\mathcal{F}\colon \operatorname{Con} A[B] \to (\operatorname{Con} A)[\operatorname{Con} B].
$$

We show that $\mathcal{F}$ is an isomorphism. It is clear that $\mathcal{F}$ is an isotone map. It is a consequence of Lemma 21.28 that $\mathcal{F}$ is injective with an isotone inverse.

To complete the proof, we must show that $\mathcal{F}$ is onto.

Let $\eta \in (\operatorname{Con} A)[\operatorname{Con} B]$. Then there is $\xi\colon \operatorname{Con} A \to \operatorname{Con} B$ such that $\eta = m(\xi)$. In particular,

$$
(22) \qquad \eta\operatorname{con}(a, b) = m(\xi)(\operatorname{con}(a, b)) = \bigwedge_{\alpha \not\le \alpha_A(a,b)} \xi\alpha.
$$

Define $\Gamma \in \operatorname{Con} A[B]$ by

$$
\Gamma = \bigwedge_{\alpha \in \operatorname{Con} A} \xi\operatorname{con}(\alpha).
$$

Let $a < b$ in A. Compute:

$$
\begin{aligned}
\mathcal{F}(\Gamma)(\operatorname{con}(a, b)) &= \Gamma_{ab} \\
&= \Big(\bigwedge_{\alpha \in \operatorname{Con} A} \alpha[\xi\alpha] \Big)_{ab} \\
&= \bigwedge_{\alpha \in \operatorname{Con} A} (\alpha[\xi\alpha])_{ab} \\
&= \bigwedge_{\alpha \not\le \operatorname{con}(a,b)} \xi\alpha \qquad\qquad \text{(by (20))} \\
&= \eta\operatorname{con}(a, b).
\end{aligned}
$$

Since the $\operatorname{con}(a, b)$ $\{\vee, 0\}$-generate $\operatorname{Con} A$, for $a < b \in A$, and $\mathcal{F}$ is a $\{\vee, 0\}$-homomorphism, it follows that $\mathcal{F}(\Gamma) = \eta$, so $\mathcal{F}$ is surjective. We have proved the following theorem:

Theorem 21.29. *Let A and B be nontrivial finite lattices. Then the map*

$$
\mathcal{F}\colon \operatorname{Con} A[B] \to (\operatorname{Con} A)[\operatorname{Con} B]
$$

is an isomorphism.

21.7. Discussion

This topic started with the 1968 paper J. Anderson and J. Kimura [2] and the 1974 paper E. T. Schmidt [174] (see also E. T. Schmidt [176] and G. Grätzer and E. T. Schmidt [121]), already discussed in Section 6.3. J. Anderson and J. Kimura [2] introduced tensor products (see also G. A. Fraser [46] and Z. Shmuley [182]) as in Section 21.1.

In general, for the nontrivial lattices A and B with zero, $A \otimes B$ is not a lattice. (For instance, $M_3 \otimes F(3)$ is not a lattice by G. Grätzer and F. Wehrung [147].) Obviously it is if A and B are finite.

For A and B finite, the lattice $A[B]$ has a natural meet-embedding into $A \otimes B$. This embedding is an isomorphism if either A or B is distributive. In general, however, $|A[B]| < |A \otimes B|$. R. W. Quackenbush computed some small examples: $|M_3[M_3]| = 44$ and $|M_3 \otimes M_3| = 50$; $|N_5[N_5]| = 42$ and $|N_5 \otimes N_5| = 43$; $|N_5 \otimes M_3| = |N_5[M_3]| = 41$.

Problem 21.1. When is $|A[B]| = |A \otimes B|$? When is $|A[B]| = |A \otimes B| - 1$? When is $|A[B]| = |A \otimes B| - n$, where n is small compared to $|A|$ and $|B|$?

Problem 21.2. What does the defect $|A \otimes B| - |A[B]|$ signify? Given $|A|$ and $|B|$, what is the maximum value of the defect?

Since $A \otimes B$ and $A[B]$ are not the same, in general, the main result of this chapter sheds no light on the old result:

$$(23) \qquad \operatorname{Con} A \otimes \operatorname{Con} B \cong \operatorname{Con}(A \otimes B),$$

of G. Grätzer, H. Lakser, and R. W. Quackenbush [99].

The isomorphisms obtained in Theorem 21.29 and in equation (23) are special cases of the main result of G. Grätzer and F. Wehrung [148], using results from G. Grätzer and F. Wehrung [146] and G. Grätzer and M. Greenberg [80].

Problem 21.3. Can we utilize Theorem 21.29 to obtain a proof of the isomorphism (3)?

G. Grätzer and F. Wehrung [146] and [148] present tensor product-like constructions for general lattices that guarantee that the resulting structure is a lattice: *lattice tensor products* in [146] and *capped tensor products* in [148]. For both constructions the analogues of (3) are proved. The reader is referred to G. Grätzer and F. Wehrung [150] for a survey. For some more recent results, see B. Chornomaz [12] and F. Wehrung [191].

It is proved in G. Grätzer and M. Greenberg [78] that for nontrivial finite lattices A and B, the tensor extension $A[B]$ is isomorphic to the lattice tensor product of A and B. (In particular, this implies the isomorphism $A[B] \cong B[A]$.) So Theorem 21.29 should be viewed as an elementary proof for the finite case of the general result in G. Grätzer and F. Wehrung [148].

Problem 21.4. Find a direct proof of the isomorphism $A[B] \cong B[A]$ for nontrivial finite lattices A and B?

In the Introduction, we described the topic of this chapter as follows: Let ⊛ be a construction for finite distributive lattices (that is, if D and E are finite distributive lattices, then $D \circledast E$ is a finite distributive lattice). Find a construction ⊙ of finite lattices (that is, if K and L are finite lattices, then $K \odot L$ is a finite lattice) satisfying $\operatorname{Con}(K \odot L) \cong \operatorname{Con} K \circledast \operatorname{Con} L$.

There are three relevant results, for $A \times B$, $A \otimes B$, and for $A[B]$, and in all three cases we get ⊛ = ⊙. There are trivial results, for instance, glued sums, $A \dotplus B$ and $B \dotplus A$, in which case ⊙ is the direct product.

Problem 21.5. Are there nontrivial lattice constructions $A \circledast B$ for finite lattices A and B satisfying

$$\operatorname{Con} A \circledast \operatorname{Con} B \cong \operatorname{Con}(A \circledast B),$$

other than $A \times B$, $A \otimes B$, and $A[B]$?

Problem 21.6. Are there nontrivial constructions $A \circledast B$ for finite distributive lattices A and B and constructions $A \odot B$ for finite lattices A and B satisfying

$$\operatorname{Con} A \circledast \operatorname{Con} B \cong \operatorname{Con}(A \odot B),$$

other than those listed in Problem 21.5?

Part VI

The Ordered Set of
Principal Congruences

Representation Theorems

22.1. Representing the ordered set of principal congruences

This section is based on my paper [65].

As we noted in Lemma 3.6, the ordered set $\operatorname{Princ} K$ is bounded for a bounded lattice K. We now state the converse.

Theorem 22.1. *Let P be an ordered set with zero and unit. Then there is a bounded lattice* $\operatorname{Lat} P$ *such that*

$$P \cong \operatorname{Princ}(\operatorname{Lat} P).$$

If P is finite, we can construct $\operatorname{Lat} P$ *as a finite lattice.*

22.2. *Proof-by-Picture*

For a bounded ordered set P, let P^- denote the ordered set P with the bounds removed.

Let P be the ordered set in Theorem 22.1. Let 0 and 1 denote the zero and unit of P, respectively. We denote by $P^{\|}$ those elements of P^- that are not comparable to any other element of P^-.

© Springer International Publishing Switzerland 2016

G. Grätzer, *The Congruences of a Finite Lattice*,

DOI 10.1007/978-3-319-38798-7_22

The lattice Frame P

We first construct the lattice Frame P consisting of the elements o, i and the elements a_p, b_p for every $p \in P$, where $a_p \neq b_p$ for every $p \in P^-$ and $a_0 = b_0$, $a_1 = b_1$. These elements are ordered and the lattice operations are formed as in Figure 22.1. For $p \in P$, the p-edge of Frame P is a pair of elements (a_p, b_p).

Let X be a set. Define Frame$_X P$, a lattice extension of Frame P, defined on the set Frame $P \cup X$ (disjoint union), where every $x \in X$ is a complement of every $y \in (\text{Frame } P)^-$.

The lattice $S(p, q)$

For $p < q \in P^-$, we define the lattice $S(p, q)$, see Figure 22.2.

A congruence $\alpha > 0$ of a bounded lattice L is *Bound Isolating* (BI, for short), if $\{0\}$ and $\{1\}$ are congruence blocks of α. The lattice $S(p, q)$ has two BI congruences:

$$\text{con}(a_p, b_p) < \text{con}(a_q, b_q),$$

see Figure 22.3.

The lattice $S(p, q)$ has two BI congruences:

$$\text{con}(a_p, b_p) < \text{con}(a_q, b_q),$$

see Figure 22.3. In addition it has the two trivial congruences $\mathbf{0}$ and $\mathbf{1}$, so Con $S(p, q)$ is the four element chain.

The lattice Lat P

We are going to construct the lattice Lat P (of Theorem 22.1) as an extension of the lattice Frame P. This was discussed in LTS1 (G. Grätzer [68]), so here we only sketch the construction and the proof.

We insert the set

$$\{c_{p,q}, d_{p,q}, e_{p,q}, f_{p,q}, g_{p,q}\}$$

to the sublattice

$$\{o, a_p, b_p, a_q, b_q, i\}$$

of Frame P for $p < q \in P^-$, as illustrated in Figure 22.2, to form the sublattice $S(p, q)$. We call such an $S(p, q)$ a *flap*. The flaps $A \neq B$ are *disjoint* if $A \cap B = \{o, i\}$; otherwise, they are *adjacent*. If A are B are adjacent, then $A \cap B = \{o, a_r, b_r, i\}$ for some $r \in P$.

For $p \in P^\parallel$, let $C_p = \{o < a_p < b_p < i\}$ be a four-element chain.

Now we make the set Lat P into a lattice as expected, so that all flaps are sublattices.

The nine rules in [65] describing the joins in Lat P formally say that for $x \parallel y \in \text{Lat } P$, either x and y are complementary or x and y both are in the same flap and we form the join in the flap, or x is in a flap A, y is in a flap

B, where A and B are adjacent. In the last case, x and y both are in one of the lattices L_C, L_V, and L_H depicted in Figures 22.4–22.6, respectively, and the join is formed in the appropriate lattice. For a detailed discussion, see my paper [65] or Section 10.6 in LTS1.

For $p \in P^-$, the principal congruence $\mathrm{con}(a_p, b_p)$ is a BI congruence and every principal BI congruence is of this form. There are two principal congruences that are not BI, namely, $\mathbf{0}$ and $\mathbf{1}$.

For $p \in P^\|$, let $\varepsilon_p = \mathrm{con}(a_p, b_p)$, viewed as a partition. Let $H \subseteq P^\|$, and let ε_H denote the equivalence relation

$$\varepsilon_H = \bigvee (\,\varepsilon_p \mid p \in H\,)$$

on $\mathrm{Lat}\, P$.

Let β be a BI congruence of the lattice $\mathrm{Lat}\, P$. We associate with β a subset of the ordered set P^-:

$$\mathrm{Base}(\beta) = \{\, p \in P^- \mid a_p \equiv b_p \pmod{\beta}\,\}.$$

Then $\mathrm{Base}(\beta)$ is a down set of P^-.

The following lemma is easy to compute.

Lemma 22.2. *Let H be a down set of P^-. Then the binary relation:*

$$\beta_H = \varepsilon_H \cup \bigcup (\,\mathrm{con}_{S(p,q)}(a_q, b_q) \mid q \in H\,) \cup \bigcup (\,\mathrm{con}_{S(p,q)}(a_p, b_p) \mid p \in H\,).$$

is a BI congruence on $\mathrm{Lat}\, P$ and the correspondence

$$\gamma \colon \beta \mapsto \mathrm{Base}(\beta)$$

is an order preserving bijection between the ordered set of BI congruences of $\mathrm{Lat}\, P$ and the ordered set of down sets of P^-.

We extend γ by $\mathbf{0} \mapsto \{0\}$ and $\mathbf{1} \mapsto P$. Then γ is an isomorphism between $\mathrm{Con}(\mathrm{Lat}\, P)$ and $\mathrm{Down}^- P$, the ordered set of nonempty down sets of P verifying Theorem 22.1.

22.3. An independence theorem

For a bounded lattice L, G. Czédli [28] proved that the two related structures $\mathrm{Princ}\, L$ and $\mathrm{Aut}\, L$ are independent.

Theorem 22.3. *Let P is a bounded ordered set with at least two elements and let G be a group. Then there exists a bounded lattice L such that $\mathrm{Princ}\, L$ is order isomorphic to P and $\mathrm{Aut}\, L$ is group isomorphic to G.*

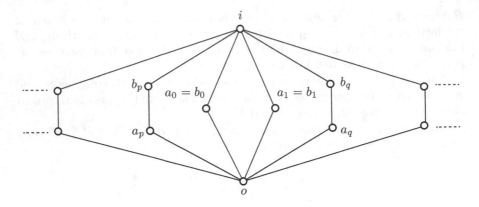

Figure 22.1: The lattice Frame P.

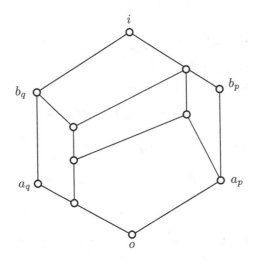

Figure 22.2: The lattice $S(p, q)$ for $p < q \in P$.

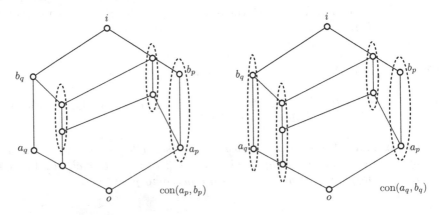

Figure 22.3: The BI congruences of $S(p, q)$.

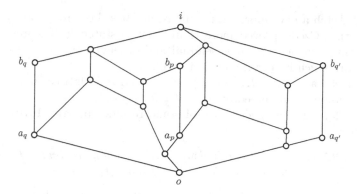

Figure 22.4: The lattice L_C for $q < p < q'$.

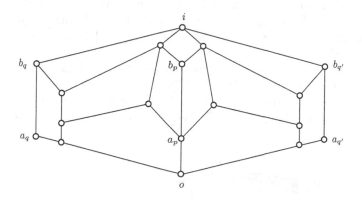

Figure 22.5: The lattice L_V for $p < q$ and $p < q'$ with $q \neq q'$.

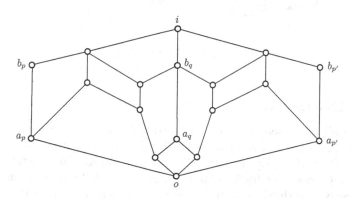

Figure 22.6: The lattice L_H for $p < q$ and $p' < q$ with $p \neq p'$.

I will sketch a new approach to this result, based on my paper [76].

Note that Czédli proved more than what is stated in Theorem 22.3 by constructing a lattice L that is selfdual and of length 16. To prove this, he had to work much harder than I do in [76].

Let us call a family $\{ K_i \mid i \in I \}$ of bounded rigid lattices *mutually rigid* if K_i has no bounded embedding into K_j for $i, j \in I$.

The following is a special case of some results from the 1970s; see also G. Czédli [29].

Theorem 22.4. *For every cardinal* $\mathfrak{m}$, *there exists a mutually rigid and simple family* $\{ K_i \mid i \in I \}$ *of bounded lattices satisfying* $|I| = \mathfrak{m}$.

Much more sophisticated results were published in the early seventies dealing with endomorphism semigroups and full embeddings of categories, see for example, G. Grätzer and J. Sichler [142], Z. Hedrlin and J. Sichler [154].

Let P and G be given as in Theorem 22.3. The lattice K we have just constructed satisfies that $\operatorname{Princ} K \cong P$. However, $\operatorname{Aut} K \cong \operatorname{Aut} P$.

Let $I = P^-$. For every $[o, a_p]$, we insert the lattice K_p into K provided by Theorem 22.4, identifying the bounds, obtaining the lattice $\overline{K}$. We then have $\operatorname{Princ} \overline{K} \cong P$ and $\operatorname{Aut} \overline{K}$ is rigid.

Now take the Frucht lattice, $\operatorname{Frucht} G$ (see Section 17.5); it is a lattice of length 3, satisfying $\operatorname{Aut}(\operatorname{Frucht} G) \cong G$.

We obtain the lattice L by forming the disjoint union of $\overline{K}$ and $\operatorname{Frucht} G$ and identifying the bounds, see Figure 22.7. The diagram does not show the lattices K_p.

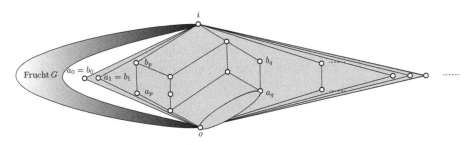

Figure 22.7: The final step.

22.4.   Discussion

We can easily represent the bounded ordered set C_2^2, see the first diagram of Figure 22.8, as $\operatorname{Princ} K$ of a lattice K. Indeed, take the lattice $K = \mathsf{C}_2^2$ and then $\operatorname{Princ} K = \mathsf{C}_2^2$, in fact, $\operatorname{Princ} K = \operatorname{Con} K$. In general, however, $\operatorname{Princ} K = \mathsf{C}_2^2$ does not imply that $\operatorname{Princ} K = \operatorname{Con} K$. For an example, take the lattice K, the third diagram in Figure 22.8. Then we have $\operatorname{Princ} K = \mathsf{C}_2^2$

but Princ $K = \operatorname{Con} K$, the second diagram in Figure 22.8. So we can ask for the characterization of Princ K as a bounded ordered subset of $\operatorname{Con} K$ for a finite lattice K. Clearly, $\operatorname{Con}_J K \subseteq$ Princ K.

For a finite lattice L, define $J^+(L) = \{0, 1\} \cup J(L)$.

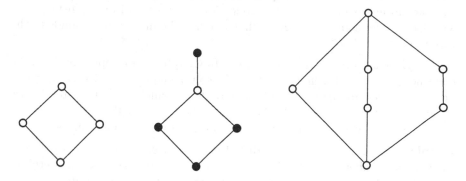

Figure 22.8: A Princ K with a Con K and a lattice K.

Problem 22.1. Let D be a finite distributive lattice. Let Q be a subset of D satisfying $J^+(D) \subseteq Q \subseteq D$. When is there a finite lattice K such that $\operatorname{Con} K$ is isomorphic to D and under this isomorphism Princ K corresponds to Q?

An affirmative solution for the special case $J^+(D) = Q$ is presented in the proof of Theorem 22.1.

The most natural problem concerning Theorem 22.1 is the following.

Problem 22.2. Can we characterize the ordered set Princ L for a lattice L as a directed ordered set with zero?

G. Czédli [18] solved this problem in the affirmative for countable lattices.

Even more interesting would be to characterize the pair $P =$ Princ L in $S = \operatorname{Con}_c L$ (the join-semilattice of compact congruences of L) by the properties that P is a directed ordered set with zero that join-generates S. We rephrase this to make this independent of the congruence lattice characterization problem, see LTF and LTS1.

Problem 22.3. Let S be a representable join-semilattice. Let $P \subseteq S$ be a directed ordered set with zero and let P join-generate S. Under what conditions is there a lattice K such that $\operatorname{Con}_c K$ is isomorphic to S and under this isomorphism Princ K corresponds to P?

For a lattice L, let us define a *valuation* v on $\operatorname{Con}_c L$ as follows.

For a compact congruence $\boldsymbol{\alpha}$ of L, let $v(\boldsymbol{\alpha})$ be the smallest integer n such that the congruence $\boldsymbol{\alpha}$ is the join of n principal congruences.

A valuation v has some obvious properties, for instance, $v(0) = 0$ and $v(\alpha \vee \beta) \leq v(\alpha) + v(\beta)$. Note the connection with $\operatorname{Princ} L$:

$$\operatorname{Princ} L = \{\, \alpha \in \operatorname{Con_c} L \mid v(\alpha) \leq 1 \,\}.$$

Problem 22.4. Let S be a representable join-semilattice. Let v map S to the natural numbers. Under what conditions is there an isomorphism γ of S with $\operatorname{Con_c} K$ for some lattice K so that under γ the map v corresponds to the valuation on $\operatorname{Con_c} K$?

Let D be a finite distributive lattice. In Chapter 8, we represent D as the congruence lattice of a finite (sectionally complemented) lattice K in which *all congruences are principal* (that is, $\operatorname{Con} K = \operatorname{Princ} K$). In Chapter 10, we construct a planar semimodular lattice K with $\operatorname{Con} K = \operatorname{Princ} K \cong D$.

In the finite variant of Problem 22.4, we need an additional property.

Problem 22.5. Let S be a finite distributive lattice. Let v be a map of D to the natural numbers satisfying $v(0) = 0$, $v(1) = 1$, and $v(a \vee b) \leq v(a) + v(b)$ for $a, b \in D$. Is there an isomorphism γ of D with $\operatorname{Con} K$ for some finite lattice K such that under γ the map v corresponds to the valuation on $\operatorname{Con} K$?

Problem 22.6. Let K be a bounded lattice. Does there exist a complete lattice L such that $\operatorname{Con} K \cong \operatorname{Con} L$?

What can we say about the (planar) semimodular case?

Problem 22.7. Characterize $\operatorname{Princ} L$ for

(i) a finite semimodular lattice L;

(ii) a planar semimodular lattice L;

(iii) an SPS lattice L.

Problem 22.1 becomes much harder for planar semimodular lattices.

Problem 22.8. Let D be a finite distributive lattice. Let Q be a subset of D satisfying $J^+(D) \subseteq Q \subseteq D$. When is there

(i) a finite semimodular lattice L;

(ii) a finite planar semimodular lattice L;

(iii) an SPS lattice L

such that $\operatorname{Con} L$ is isomorphic to D and under this isomorphism $\operatorname{Princ} L$ corresponds to Q?

We state an interesting special case.

Problem 22.9. Let D be a finite distributive lattice. When is there a finite planar semimodular lattice L such that $\operatorname{Con} L$ is isomorphic to D and under this isomorphism $\operatorname{Princ} L$ corresponds to $J^+(D)$?

Contrast this with the note following Problem 22.1.

Isotone Maps

In Chapter 22, we characterized the ordered set $\operatorname{Princ} L$ of principal congruences of a bounded lattice L as a bounded ordered set, see Theorem 22.1.

If K and L are bounded lattices and φ is a bounded homomorphism of K into L, then there is a natural isotone bounded map $\operatorname{Princ} \varphi$ from $\operatorname{Princ} K$ into $\operatorname{Princ} L$.

We prove the converse: For the bounded ordered sets P and Q and an isotone bounded map ψ of P into Q, we represent P and Q as $\operatorname{Princ} K$ and $\operatorname{Princ} L$ for bounded lattices K and L with a bounded homomorphism φ of K into L, so that ψ is represented as $\operatorname{Princ} \varphi$.

23.1. Two isotone maps

23.1.1 Sublattices

G. Czédli [26] and [27] extended Theorem 22.1, the representation theorem of the ordered set of principal congruences, to a bounded lattice and a bounded sublattice.

Let K be a bounded sublattice of a bounded lattice L. Then the map, see Section 3.3,

(1) $$\operatorname{ext}(K, L) \colon \operatorname{con}_K(x, y) \mapsto \operatorname{con}_L(x, y),$$

for $x, y \in K$, is an bounded isotone map of $\operatorname{Princ} K$ into $\operatorname{Princ} L$. Observe that the bounded map $\operatorname{ext}(K, L)$ is $\{0\}$-*separating* , that is, $\mathbf{0}_K$ is the only element mapped by $\operatorname{ext}(K, L)$ to $\mathbf{0}_L$.

Now we can state Czédli's result.

Theorem 23.1. *Let P and Q be bounded ordered sets and let ψ be an isotone $\{0\}$-separating bounded map from P into Q. Then there exist a bounded lattice L and a bounded sublattice K of L representing P, Q, and ψ as $\operatorname{Princ} K$, $\operatorname{Princ} L$, and $\operatorname{ext}(K, L)$, up to isomorphism, as in the following commutative diagram:*

$$
\begin{array}{ccc}
P & \xrightarrow{\ \psi\ } & Q \\
\cong \downarrow & & \cong \downarrow \\
\operatorname{Princ} K & \xrightarrow{\ \operatorname{ext}(K,L)\ } & \operatorname{Princ} L
\end{array}
$$

Note that if $K = L$, then $\operatorname{ext}(K, L)$ is the identity map on $\operatorname{Princ} K = \operatorname{Princ} L$. Therefore, Theorem 22.1 follows from Theorem 23.1 with $P = Q$ and ψ the identity map.

23.1.2   Bounded homomorphisms

Next we take up the analogous problem with a bounded homomorphism.

We start with the following observation, which easily follows from Theorem 3.1 and Lemma 3.2.

Lemma 23.2. *Let K and L be bounded lattices and let φ be a bounded homomorphism of K into L. Define*

(2) $$\operatorname{Princ} \varphi \colon \operatorname{con}_K(a, b) \mapsto \operatorname{con}_L(\varphi a, \varphi b)$$

for $a, b \in K$. Then $\operatorname{Princ} \varphi$ is a bounded isotone map of $\operatorname{Princ} K$ into $\operatorname{Princ} L$.

Now we state the main result of this chapter, see G. Grätzer [75].

Theorem 23.3. *Let P and Q be bounded ordered sets and let ψ be an isotone bounded map from P into Q. Then there exist bounded lattices K, L, and a bounded homomorphism φ of K into L, so that P, Q, and ψ are represented by $\operatorname{Princ} K, \operatorname{Princ} L$, and $\operatorname{Princ} \varphi$, up to isomorphism, as in the following commutative diagram:*

$$
\begin{array}{ccc}
P & \xrightarrow{\ \psi\ } & Q \\
\cong \downarrow & & \cong \downarrow \\
\operatorname{Princ} K & \xrightarrow{\ \operatorname{Princ} \varphi\ } & \operatorname{Princ} L
\end{array}
$$

We will consider *lattice-triples*: $\mathcal{L} = (K, L, \varphi)$, where K and L are bounded lattices and φ is a bounded homomorphism of K into L. Similarly, we consider *order-triples* $\mathcal{P} = (P, Q, \psi)$, where P and Q are bounded ordered sets and ψ

is an isotone bounded map of P into Q. By Lemma 23.2, a lattice-triple $\mathcal{L}$ defines an order-triple $\mathcal{P}$ in the natural way: $P = \operatorname{Princ} K$, $Q = \operatorname{Princ} L$, and $\psi = \operatorname{Princ} \varphi$. A *representable* order-triple $\mathcal{P}$ arises from a lattice-triple $\mathcal{L}$ in this way.

Now we restate Theorem 23.3.

Theorem 23.4. *Every order-triple is representable.*

23.2.   Sublattices, sketching the proof

The present approach to Czédli's result is based on my paper [77].

Let P, Q, and ψ be given as in Theorem 23.1. We form the bounded ordered set $R = P \cup Q$, a disjoint union with $0_P, 0_Q$ and $1_P, 1_Q$ identified, see Figure 23.1. So R is a bounded ordered set containing P and Q as bounded ordered subsets. Observe that Frame P is a bounded sublattice of Frame R.

For $p < q$ in P^- and for $p < q$ in Q^-, we insert $S(p,q)$, see Figure 22.2, into Frame R so that $\operatorname{con}(a_p, b_p) < \operatorname{con}(a_q, b_q)$ will hold. Also, for $p \in P^-$, we insert $S(p, \psi p)$ as a sublattice; note that $\psi p \in Q^-$.

Let L^+ denote the ordered set we obtain. We slim L^+ down to the ordered set L by deleting all the elements of the form $x_{p,\psi p}$ for $p \in P^-$. Since $x_{p,\psi p}$ is not join-reducible, the ordered set L is a lattice (but it is neither a sublattice nor a quotient of L^+). The joins and meets of any two elements u and v in L are the same as in L^+, except for a meet: $u \wedge v = x_{p,\psi p}$ for some $p \in P^-$ and $u \parallel v$; in this case, $u \wedge v = (x_{p,\psi p})_*$, the unique element covered by $x_{p,\psi p}$ in L^+.

We view K as the bounded sublattice of L built on Frame P.

Observe that $\operatorname{con}(a_p, b_p) = \operatorname{con}(\psi p, \psi q)$, since $[a_p, b_p]$ is (three step) projective to $[a_{\psi p}, b_{\psi q}]$, so all principal congruences of L are of the form $\operatorname{con}(a_q, b_q)$ for $q \in Q^-$. It follows that $\operatorname{Princ} L$ is isomorphic to Q.

Now we can prove Theorem 23.1 as we sketched Theorem 22.1 in Section 22.2 and fully proved in LTS1.

23.3.   Isotone surjective maps

Reading Czédli's papers, [26] and [27], the question came up, what about surjective maps? The general problem is still open, but I got the following result, see G. Grätzer [75].

We need some notation. For an order-triple $\mathcal{P} = (P, Q, \psi)$, we define

$$\operatorname{Top} \mathcal{P} = \{\, x \in P \mid \psi x > 0_Q \,\} \cup \{0_P\}$$

and let $\operatorname{Top} \psi$ be the restriction of ψ to $\operatorname{Top} \mathcal{P}$. We also need the "bottom" of $\mathcal{P}$:

$$\operatorname{Btm} \mathcal{P} = \{\, x \in P \mid \psi x = 0_Q \,\}.$$

Note that

$$\text{Top}\,\mathcal{P} \cup \text{Btm}\,\mathcal{P} = P,$$
$$\text{Top}\,\mathcal{P} \cap \text{Btm}\,\mathcal{P} = \{0_P\}.$$

Recall that the $\text{Lat}\,P$ and $\text{Lat}_X\,P$ constructions were defined in the Section 22.2.

Lemma 23.5. *Let $\mathcal{P} = (P, Q, \psi)$ be an order-triple. If ψ is surjective and $\text{Top}\,\psi = \psi\!\restriction\text{Top}\,\mathcal{P}$ (ψ restricted to $\text{Top}\,\mathcal{P}$) is an isomorphism between $\text{Top}\,\mathcal{P}$ and Q, then $\mathcal{P}$ has a representation $\mathcal{L} = (K, L, \varphi)$, where $\varphi = \text{Lat}\,\psi$ is surjective and $K = \text{Lat}\,P$, $L = \text{Lat}_X\,Q$ with $X = \text{Btm}\,\mathcal{P}$.*

Proof. Let $\mathcal{P} = (P, Q, \psi)$ be a order-triple with ψ surjective and let $\text{Top}\,\psi$ be an isomorphism between $\text{Top}\,\mathcal{P}$ and Q.

It follows from Lemma 22.2 that $P \cong \text{Princ}(\text{Lat}\,P)$.

Now we describe a lattice-triple $\mathcal{L} = (K, L, \text{Lat}\,\psi)$. We define $K = \text{Lat}\,P$. Since $\text{Btm}\,\mathcal{P}$ is a down-set, it follows from the above statement that there is a congruence $\boldsymbol{\alpha}$ of K with $\text{Btm}\,\mathcal{P} = \text{Base}\,\boldsymbol{\alpha}$. Define the bounded lattice $L = K/\boldsymbol{\alpha}$, and let φ be the natural bounded homomorphism of K onto L. By definition, the lattice-triple $\mathcal{L} = (K, L, \varphi)$ represents $\mathcal{P}$ and φ is surjective, so we can define $\text{Lat}\,\psi = \varphi$. Note that the lattice L can be represented in the form $\text{Lat}_X\,Q$, where $X = \text{Btm}\,\mathcal{P}$, concluding the proof of the lemma. $\qquad\square$

23.4.   Proving the Representation Theorem

Now we prove Theorem 23.3 in the form of Theorem 23.4 as in G. Grätzer [75].

Let $\mathcal{P} = (P, Q, \psi)$ be a an order-triple. We consider the bounded ordered set $R = \text{Top}\,\mathcal{P}$ and define the isotone map $\alpha\colon P \to R$ as follows:

$$(3) \qquad\qquad \alpha x = \begin{cases} x & \text{for } x \in \text{Top}\,\mathcal{P}; \\ 0_P = 0_R, & \text{otherwise.} \end{cases}$$

We also define the bounded isotone map $\beta\colon R \to Q$ as the restriction of ψ to $R = \text{Top}\,\mathcal{P}$, see Figure 23.1. Note that $\beta\alpha = \psi$.

This defines the order-triples $\mathcal{P}_\alpha = (P, R, \alpha)$ and $\mathcal{P}_\beta = (R, Q, \beta)$.

It follows from Lemma 23.5 that $\mathcal{P}_\alpha$ has a representation $\mathcal{L}_\alpha = (K, M, \text{Lat}\,\alpha)$ with $K = \text{Lat}\,P$ and $M = \text{Lat}(\text{Top}\,\mathcal{P})$, and $\text{Lat}\,\alpha$ is surjective.

Now we apply Theorem 23.1. In $\mathcal{P}_\beta$, the map β is $\{0\}$-separating, therefore, we can apply Theorem 23.1 to R, Q, and β. So we obtain

(a) a lattice-triple, $\mathcal{L}_\alpha = (K, M, \text{Lat}\,\alpha)$ representing $\mathcal{P}_\alpha$,

(b) a lattice-triple $(M', L, \text{Lat}\,\beta)$ representing $\mathcal{P}_\beta$,

see Figure 23.2.

We conclude that $\mathcal{L} = (K, L, \text{Lat}\,\alpha\,\text{Lat}\,\beta)$ is a lattice-triple representing $\mathcal{P}$, verifying the representation theorem.

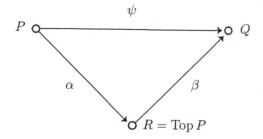

Figure 23.1: The order-triples $\mathcal{P}_\alpha = (P, R, \mathrm{Lat}\,\alpha)$ and $\mathcal{P}_\beta = (R, Q, \mathrm{Lat}\,\beta)$.

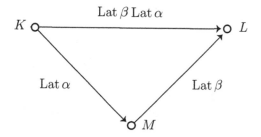

Figure 23.2: The lattice-triples $\mathcal{L}_\alpha = (K, M, \mathrm{Lat}\,\alpha)$ and $\mathcal{L}_\beta = (M, L, \mathrm{Lat}\,\beta)$.

23.5. Discussion

Problem 23.1. Can we extend Theorem 23.4 to the countable case?

Let us call a triple *surjective* if the map in it is surjective.

Problem 23.2. Characterize the surjective order-triples that have representations with surjective lattice-triples?

Part VII

Congruence Structure

Prime Intervals and Congruences

24.1. Introduction

As we have discussed in Section 3.2, to investigate the congruences of a finite lattice L, we should focus on the prime intervals. The congruence-projectivity relation $\Rightarrow$ is a preordering on $\mathrm{Prime}(L)$. The equivalence classes under $\Leftrightarrow$ form an ordered set that is isomorphic to $\mathrm{Con}_J L$.

Recall that the spreading of a congruence from a prime interval to another prime interval involves intervals of arbitrary size, as illustrated by Figure 3.2. We would like to describe such a spreading with prime intervals only.

In Section 24.2, we introduce the concepts of prime-perspectivity and prime-projectivity and prove Lemma 24.1, the Prime-Projectivity Lemma, see G. Grätzer [69]. It verifies that, indeed, we can describe the spreading of a congruence from a prime interval to a prime interval involving only prime intervals.

In Section 24.3, for SPS (slim, planar, and semimodular) lattices, we introduce the simple but powerful concept of a *swing*, illustrated in Figure 24.4. We prove the Swing Lemma, a very strong form of the Prime-Projectivity Lemma, stating that in an SPS lattice, a congruence spreads from a prime interval to another by an up perspectivity and a sequence of down perspectivities and swings.

© Springer International Publishing Switzerland 2016
G. Grätzer, *The Congruences of a Finite Lattice*,
DOI 10.1007/978-3-319-38798-7_24

In Section 4.3, we described how to construct all planar semimodular lattices from planar distributive lattices using the fork construction; in Section 24.5, we examine in detail the congruences of a fork extension.

24.2. The Prime-projectivity Lemma

Let $\mathfrak{p}$ and $\mathfrak{q}$ be prime intervals in a finite lattice L. Let us call $\mathfrak{p}$ *prime-perspective down* to $\mathfrak{q}$, in formula, $\mathfrak{p} \xrightarrow{\text{p-dn}} \mathfrak{q}$, if

$$0_\mathfrak{p} \vee 1_\mathfrak{q} = 1_\mathfrak{p},$$
$$0_\mathfrak{p} \wedge 1_\mathfrak{q} = 0_\mathfrak{q}.$$

So if $\mathfrak{p} \xrightarrow{\text{p-dn}} \mathfrak{q}$, then $\mathfrak{p}$ and $\mathfrak{q}$ generate an N_5, as in the second diagram of Figure 24.1, or they are perspective (in the usual sense).

We define *prime-perspective up*, $\mathfrak{p} \xrightarrow{\text{p-up}} \mathfrak{q}$, dually. Let *prime-perspective*, $\mathfrak{p} \xrightarrow{\text{p}} \mathfrak{q}$, mean that $\mathfrak{p} \xrightarrow{\text{p-up}} \mathfrak{q}$ or $\mathfrak{p} \xrightarrow{\text{p-dn}} \mathfrak{q}$ and let *prime-projective*, $\mathfrak{p} \xRightarrow{\text{p}} \mathfrak{q}$, be the transitive extension of $\xrightarrow{\text{p}}$. See Figure 24.2 for an illustration.

We will say that a prime-perspectivity $\mathfrak{p} \xrightarrow{\text{p-dn}} \mathfrak{q}$ that is not a perspectivity is *established* by an N_5 sublattice.

Observe that the relation $\mathfrak{p} \xrightarrow{\text{p}} \mathfrak{q}$ holds iff the elements $\{0_\mathfrak{p}, 0_\mathfrak{q}, 1_\mathfrak{p}, 1_\mathfrak{q}\}$ satisfy a simple (universal) condition.

Now we state our result: we only have to go through prime intervals to spread a congruence from a prime interval to another.

Lemma 24.1 (Prime-Projectivity Lemma). *Let L be a finite lattice and let $\mathfrak{p}$ and $\mathfrak{q}$ be distinct prime intervals in L. Then $\mathfrak{q}$ is collapsed by $\mathrm{con}(\mathfrak{p})$ iff $\mathfrak{p} \xRightarrow{\text{p}} \mathfrak{q}$, that is, iff there exists a sequence of pairwise distinct prime intervals $\mathfrak{p} = \mathfrak{r}_0, \mathfrak{r}_1, \ldots, \mathfrak{r}_n = \mathfrak{q}$ satisfying*

$$(1) \qquad \mathfrak{p} = \mathfrak{r}_0 \xrightarrow{\text{p}} \mathfrak{r}_1 \xrightarrow{\text{p}} \cdots \xrightarrow{\text{p}} \mathfrak{r}_n = \mathfrak{q}.$$

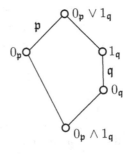

Figure 24.1: Introducing prime-perspectivity, $\mathfrak{p} \xrightarrow{\text{p-dn}} \mathfrak{q}$.

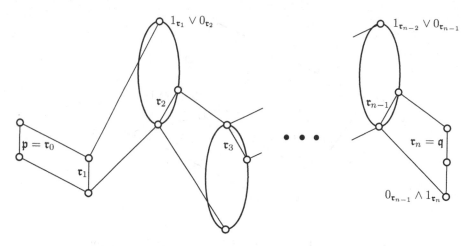

Figure 24.2: Prime-projectivity: $\mathfrak{p} \xRightarrow{\text{p}} \mathfrak{q}$.

To prove the Prime-projectivity Lemma, let $\mathfrak{p}$ and $\mathfrak{q}$ be prime intervals in a finite lattice L with $\text{con}(\mathfrak{p}) \geq \text{con}(\mathfrak{q})$. By Lemma 3.8, there is a sequence of congruence-perspectivities

$$(2) \qquad \mathfrak{p} = I_0 \twoheadrightarrow I_1 \twoheadrightarrow \cdots \twoheadrightarrow I_m = \mathfrak{q}.$$

To get from (2) to (1), by induction on m, it is sufficient to prove the following statement.

Lemma 24.2. *Let L be a finite lattice and let $I \twoheadrightarrow J$ be intervals of L. Let $\mathfrak{b}$ be a prime interval in J. Then there exists a prime interval $\mathfrak{a} \subseteq I$ satisfying $\mathfrak{a} \xRightarrow{\text{p}} \mathfrak{b}$.*

Proof. By duality, we can assume that $I \xrightarrow{\text{up}} J$. We prove the statement by induction on $\text{length}(I)$, the length of I (that is, the length of the longest chain in I). Note that if I' is an interval properly contained in I, then $\text{length}(I') < \text{length}(I)$.

For the induction base, let I be prime. Then take $\mathfrak{a} = I$.

For the induction step, we can assume that I is not prime and that the statement is proved for intervals shorter than I.

Without loss of generality, we can assume that

(i) $J \not\subseteq I$—otherwise, take $\mathfrak{a} = \mathfrak{b} \subseteq I$;

(ii) $0_{\mathfrak{b}} = 0_J$—otherwise, replace J with $[0_{\mathfrak{b}}, 1_J]$;

(iii) $0_I = 1_I \wedge 0_{\mathfrak{b}}$—otherwise, replace I with $[1_I \wedge 0_J, 1_I] \subseteq I$.

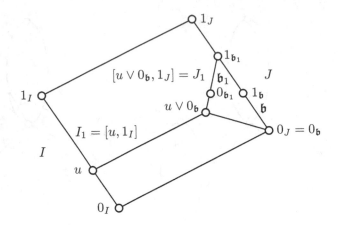

Figure 24.3: Proving the Prime-projectivity Lemma.

Since I is not a prime interval, there is an element $u \in I$ satisfying $0_I \prec u < 1_I$. If $1_\mathfrak{b} \leq u \vee 0_\mathfrak{b}$, then take $\mathfrak{a} = [0_I, u]$; clearly, $\mathfrak{a} \xRightarrow{\text{p-up}} \mathfrak{b}$.

Therefore, we can additionally assume that

(iv) $u \vee 0_\mathfrak{b} \parallel 1_\mathfrak{b}$,

see Figure 24.3. Then

$$u \vee 1_\mathfrak{b} = (u \vee 0_\mathfrak{b}) \vee 1_\mathfrak{b} > u \vee 0_\mathfrak{b}.$$

So we can define a prime interval $\mathfrak{b}_1$ with

$$u \vee 0_\mathfrak{b} \leq 0_{\mathfrak{b}_1} \prec 1_{\mathfrak{b}_1} = u \vee 1_\mathfrak{b}.$$

Observe that

(3) $\mathfrak{b}_1 \overset{\text{dn}}{\sim} \mathfrak{b}$.

Applying the induction hypotheses to $I_1 = [u, 1_I]$, $J_1 = [0_{\mathfrak{b}_1}, 1_J]$, and $\mathfrak{b}_1$, we obtain a prime interval $\mathfrak{a} \subseteq I_1 \subseteq I$ satisfying $\mathfrak{a} \xRightarrow{\text{p}} \mathfrak{b}_1$. Combining $\mathfrak{a} \xRightarrow{\text{p}} \mathfrak{b}_1$ and (3), we obtain that $\mathfrak{a} \xRightarrow{\text{p}} \mathfrak{b}$, as required.   □

The proof actually verifies more than stated. Every step of the induction adds no more than one prime interval to the sequence.

24.3.   The Swing Lemma

As introduced in Section 4.2, a lattice L is an *SPS lattice*, if it is slim, planar, and semimodular. In this section, we state the Swing Lemma, a strong form of the Prime-projectivity Lemma for SPS lattices, see my paper [70].

For the prime intervals $\mathfrak{p}, \mathfrak{q}$ of an SPS lattice L, we define a binary relation: $\mathfrak{p}$ *swings* to $\mathfrak{q}$, in formula, $\mathfrak{p} \curvearrowright \mathfrak{q}$, if $1_\mathfrak{p} = 1_\mathfrak{q}$, this element covers at least three elements, and $0_\mathfrak{q}$ is neither the left-most nor the right-most element covered by $1_\mathfrak{p} = 1_\mathfrak{q}$. We call the element $1_\mathfrak{p} = 1_\mathfrak{q}$ the *hinge* of the swing. If $0_\mathfrak{p}$ is either the left-most or the right-most element covered by $1_\mathfrak{p} = 1_\mathfrak{q}$, then we call the swing *external*, in formula, $\mathfrak{p} \overset{ex}{\curvearrowright} \mathfrak{q}$. Otherwise, the swing is *internal*, in formula, $\mathfrak{p} \overset{in}{\curvearrowright} \mathfrak{q}$.

See Figure 24.4 for two examples; in the first, the hinge covers three elements and $\mathfrak{p} \overset{ex}{\curvearrowright} \mathfrak{q}$; in the second, the hinge covers five elements and $\mathfrak{p} \overset{in}{\curvearrowright} \mathfrak{q}$.

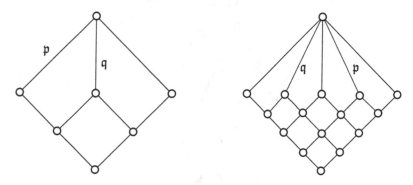

Figure 24.4: Swings, $\mathfrak{p} \curvearrowright \mathfrak{q}$.

Lemma 24.3 (Swing Lemma). *Let L be an SPS lattice and let $\mathfrak{p}$ and $\mathfrak{q}$ be distinct prime intervals in L. Then $\mathfrak{q}$ is collapsed by $\mathrm{con}(\mathfrak{p})$ iff there exists a prime interval $\mathfrak{r}$ and sequence of pairwise distinct prime intervals*

$$(4) \qquad \mathfrak{r} = \mathfrak{r}_0, \mathfrak{r}_1, \dots, \mathfrak{r}_n = \mathfrak{q}$$

such that $\mathfrak{p}$ is up perspective to $\mathfrak{r}$, and $\mathfrak{r}_i$ is down perspective to or swings to $\mathfrak{r}_{i+1}$ for $i = 0, \dots, n-1$. In addition, the sequence (4) also satisfies

$$(5) \qquad 1_{\mathfrak{r}_0} \geq 1_{\mathfrak{r}_1} \geq \cdots \geq 1_{\mathfrak{r}_n}.$$

The Swing Lemma is easy to visualize. Perspectivity up is "climbing", perspectivity down is "sliding". So we get from $\mathfrak{p}$ to $\mathfrak{q}$ by climbing once and then alternating sliding and swinging. In Figure 24.5, we climb up from $\mathfrak{p}$ to $\mathfrak{r} = \mathfrak{r}_0$, swing from $\mathfrak{r}_0$ to $\mathfrak{r}_1$, slide down from $\mathfrak{r}_1$ to $\mathfrak{r}_2$, swing from $\mathfrak{r}_2$ to $\mathfrak{r}_3$, and finally slide down from $\mathfrak{r}_3$ to $\mathfrak{r}_4$.

The following lemma is a crucial step in the proof of the Swing Lemma.

Lemma 24.4. *Let L be an SPS lattice. Let $N = \{o, u, i, v, w\}$ be an N_5 sublattice of L, with $o < u < i$ and $o < v < w < i$. Let us assume that $[v, w]$ is a prime interval. Let $u \leq x \prec i$. Then $y = x \wedge w < v$.*

Proof. There are three mutually exclusive possibilities: $y \geq v$, $y \parallel v$, and $y < v$.

Since $i = u \vee v$, we cannot have $y \geq v$, because it would imply that $x = u \vee v$.

We want to prove that $y < v$. So by way of contradiction, let us assume

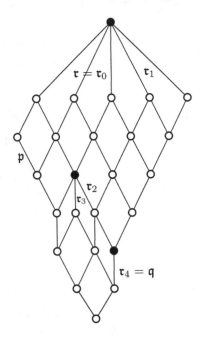

Figure 24.5: Illustrating the Swing Lemma.

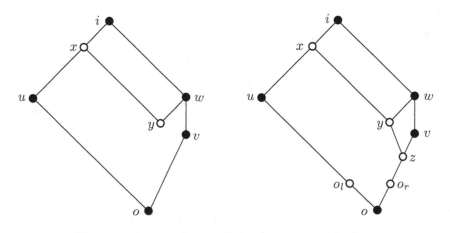

Figure 24.6: The elements for Lemma 24.4.

that

(6)
$$y \parallel v,$$

see the first diagram of Figure 24.6. Define the elements o_l and o_r satisfying

(7)
$$o \prec o_l \le u \text{ and } o \prec o_r \le v,$$

see the second diagram of Figure 24.6. Since $u \wedge v = 0$, it follows that $o_l \wedge o_r = 0$. Note that

(8)
$$o_l \nleq y.$$

Indeed, if $o_l \le y$, then $o_l \le y \le w$, and so $o_l \le u \wedge w = o$, contradicting (7). We can further assume that

(9)
$$o_r < y.$$

Since $o_r = y$ contradicts (6), if (9) fails, then $o_r \nleq y$. Thus $o_l \wedge o_r = o_l \wedge y = o_r \wedge y$, contradicting Lemma 4.3(ii), and thereby verifying (9).

So we have (6)–(9). It follows that $o_r < v$; indeed if $o_r = v$, then $v < y$, contradicting (6). Let $z = y \wedge v$. Since $o_r < v$, and by (7), $o_r < y$; therefore, $o_r \le z$ and $z < v$ by (6). Also, $z \prec z \vee o_l \ne y$ by semimodularity and (8). Then $(z \vee o_l) \wedge y = (z \vee o_l) \wedge v = y \wedge v$, contradicting Lemma 4.3(ii). □

The next statement is a very special case of the Swing Lemma; it is also a crucial step in its proof. We are considering the following condition for a slim patch lattice K:

(SL) Let $\mathfrak{q}$ be a prime interval of K on the lower right boundary of K, that is, let $1_{\mathfrak{q}} \le c_r$. Then there exists a sequence of prime intervals $\mathfrak{p}_l = \mathfrak{r}_0, \mathfrak{r}_1, \ldots, \mathfrak{r}_n = \mathfrak{q}$ such that $\mathfrak{r}_i$ is down-perspective to or swings to $\mathfrak{r}_{i+1}$ for $i = 0, \ldots, n-1$.

Lemma 24.5. *Let K be a slim patch lattice and let $S = \{o, a_l, a_r, t\}$ be a covering square of K, with a_l to the left of a_r. If (SL) hold in K, then (SL) also holds in $K[S]$.*

Proof. Note that $\mathfrak{p}_l$ and $\mathfrak{p}_r$ are also the two prime intervals of $K[S]$ on the top boundaries of $K[S]$; the elements c_l and c_r are also remain the same..

To verify (SL) for $K[S]$, let $\mathfrak{q}$ be a prime interval of $K[S]$ on the lower right boundary, that is, $1_{\mathfrak{q}} \le c_r$.

If $K = B_2$, then (SL) is trivial because $K[S] = S_7$. So we can assume that $K \ne B_2$.

There are two cases to consider.

Case 1: $\mathfrak{q} \subseteq K$. Since $\mathfrak{q}$ is prime in $K[S]$ and $\mathfrak{q} \subseteq K$, it follows that $\mathfrak{q}$ is prime in K. So we can apply (SL) to $\mathfrak{q}$ in K, to obtain a shortest sequence of prime intervals in K and a sequence of binary relations

(10) $$\mathfrak{p}_1 = \mathfrak{r}_0 \; \varrho_1 \; \mathfrak{r}_1 \; \varrho_2 \; \mathfrak{r}_2 \ldots \varrho_n \; \mathfrak{r}_n = \mathfrak{q},$$

where each relation ϱ_i is $\overset{\text{dn}}{\sim}$ or $\curvearrowleft$. If all the $\mathfrak{r}_i$ are prime intervals in $K[S]$, then the sequence (10) verifies (SL) in $K[S]$ for $\mathfrak{q}$. So let some $\mathfrak{r}_i$ not be prime in $K[S]$; we choose the $\mathfrak{r}_j$ with the largest j so that $\mathfrak{r}_j$ is not a prime. Since no element of $F[S]$ (see Figure 4.4) can be on the upper left boundary of L, we conclude that $j \neq 0$. Since $\mathfrak{q}$ is prime in K, it follows that $j \neq n$. Therefore,

(11) $$0 < j < n$$

and the intervals $\mathfrak{r}_{j+1}, \ldots, \mathfrak{r}_n = \mathfrak{q}$ are prime in $K[S]$, while the interval $\mathfrak{r}_j$ is not.

There are two possibilities: $\mathfrak{r}_j \overset{\text{dn}}{\sim} \mathfrak{r}_{j+1}$ or $\mathfrak{r}_j \curvearrowleft \mathfrak{r}_{j+1}$ in K—note that $j+1 \leq n$ by (11). If $\mathfrak{r}_j \overset{\text{dn}}{\sim} \mathfrak{r}_{j+1}$ in K, then $\mathfrak{r}_{j+1} \overset{\text{up}}{\sim} \mathfrak{r}_j$ in K and in $K[S]$. Since $\mathfrak{r}_{j+1}$ is prime in $K[S]$ but $\mathfrak{r}_j$ is not, this conflicts with the semimodularity of $K[S]$. We conclude that $\mathfrak{r}_j \curvearrowleft \mathfrak{r}_{j+1}$. Let $\mathfrak{r}_j \curvearrowleft \mathfrak{r}_{j+1}$ be established by an S_7 generated by $0_{\mathfrak{r}_j}, 0_{\mathfrak{r}_{j+1}}, w$, where w is the right-most element covered by $1_{\mathfrak{r}_j}$ if $\mathfrak{r}_j$ is to the left of $\mathfrak{r}_{j+1}$ and the left-most element covered by $1_{\mathfrak{r}_j}$, otherwise. Note that $\{0_{\mathfrak{r}_j}, 0_{\mathfrak{r}_{j+1}}, w\}$ is a three-element set since $\mathfrak{r}_j \curvearrowleft \mathfrak{r}_{j+1}$.

Since $\mathfrak{r}_j$ is not prime in $K[S]$, it follows that $0_{\mathfrak{r}_j} < z < 1_{\mathfrak{r}_j}$ for some $z \in F[S]$. We cannot have $z = m$, because in $K[S]$, z is contained in an interval $[0_{\mathfrak{r}_j}, 1_{\mathfrak{r}_j}]$ that is prime in K, while m is not contained in an interval that is prime in K. We conclude that

(12) $$z = z_{r,p}, \text{ for some } 1 \leq p \leq n_r,$$

or symmetrically. It follows that $[o, a_r] \overset{\text{dn}}{\sim} \mathfrak{r}_j$ in K and so

(13) $$[b_r, a_r] \overset{\text{dn}}{\sim} [z, 1_{\mathfrak{r}_j}].$$

Since $\mathfrak{r}_j \curvearrowleft \mathfrak{r}_{j+1}$ in K, it follows that $1_{\mathfrak{r}_j}$ covers at least three elements in K, and so $1_{\mathfrak{r}_j}$ covers at least three elements in $K[S]$. Therefore, in $K[S]$,

(14) $$[z, 1_{\mathfrak{r}_j}] \curvearrowleft \mathfrak{r}_{j+1}.$$

Since $[o, a_r] \overset{\text{dn}}{\sim} \mathfrak{r}_j$ and $\mathfrak{r}_{j-1} \overset{\text{dn}}{\sim} \mathfrak{r}_j$, we can apply Lemma 4.4 to conclude that either

(15) $$[o, a_r] \overset{\text{dn}}{\sim} \mathfrak{r}_{j-1} \overset{\text{dn}}{\sim} \mathfrak{r}_j$$

or

(16) $$\mathfrak{r}_{j-1} \overset{\text{dn}}{\sim} [o, a_r] \overset{\text{dn}}{\sim} \mathfrak{r}_j.$$

If (15) holds, then $\mathfrak{p}_1 \neq \mathfrak{r}_{j-2}$, since $[o, a_r] \overset{\mathrm{dn}}{\sim} \mathfrak{r}_{j-1}$ and $[o, a_r]$ is not on the left boundary of $K[S]$. So we have the prime interval $\mathfrak{r}_{j-2}$ satisfying that $\mathfrak{r}_{j-2} \curvearrowleft \mathfrak{r}_{j-1}$, so $\mathfrak{r}_{j-2} \curvearrowright \mathfrak{r}_{j-1}$ cannot hold. It follows that (16) holds.

By (11)–(14), and (16), the sequence of prime intervals with the binary relations

$$\mathfrak{p}_1 = \mathfrak{r}_0 \, \varrho_1 \, \mathfrak{r}_1, \ldots, \varrho_{j-1} \, \mathfrak{r}_{j-1} \overset{\mathrm{dn}}{\sim} [a_1, t] \curvearrowleft [m, t] \overset{\mathrm{dn}}{\sim} [b_r \wedge 1_{\mathfrak{r}_{j-1}}, 1_{\mathfrak{r}_{j-1}}]$$
$$\curvearrowleft \mathfrak{r}_{j+1} \, \varrho_{j+2} \cdots \varrho_n \, \mathfrak{r}_n = \mathfrak{q}$$

establishes (SL) for $K[S]$, see Figure 24.7.

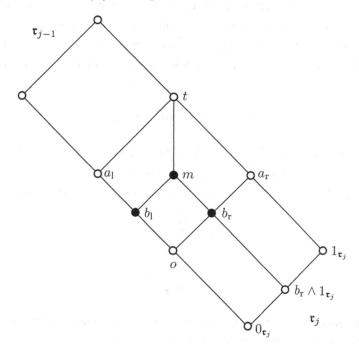

Figure 24.7: Case 1 of Lemma 24.5.

Case 2: $\mathfrak{q} \nsubseteq K$. Since $\mathfrak{q} \nsubseteq K$ is a prime interval on the lower right boundary of $K[S]$, it follows that

$$\mathfrak{q} = [y_{r,n_r}, x_{r,n_r}]_{K[S]} = \{y_{r,n_r}, z, x_{r,n_r}\},$$

where $z = z_{r,n_r}$ using the notation of Figure 4.4, or symmetrically. Let $\mathfrak{q}_{\mathrm{up}} = [z_{r,n_r}, x_{r,n_r}]$ and $\mathfrak{q}_{\mathrm{dn}} = [y_{r,n_r}, z_{r,n_r}]$; they are prime intervals in $K[S]$ and $\mathfrak{q} = \mathfrak{q}_{\mathrm{up}}$ or $\mathfrak{q} = \mathfrak{q}_{\mathrm{dn}}$.

To verify Case 2, we have to prove (SL) in $K[S]$ for $\mathfrak{q} = \mathfrak{q}_{\mathrm{up}}$ and $\mathfrak{q} = \mathfrak{q}_{\mathrm{dn}}$.

Let $\mathfrak{q}' = [0_{\mathfrak{q}_{\mathrm{dn}}}, 1_{\mathfrak{q}_{\mathrm{up}}}] = [y_{r,n_r}, x_{r,n_r}]$; it is a prime interval of K on the lower right boundary of K. By applying (SL) to K and $\mathfrak{q}'$, we obtain a shortest sequence of prime intervals in K and a sequence of binary relations

$$(17) \qquad\qquad \mathfrak{p}_l = \mathfrak{r}_0 \; \varrho_1 \; \mathfrak{r}_1 \; \varrho_2 \; \mathfrak{r}_2 \dots \varrho_n \; \mathfrak{r}_n = \mathfrak{q}',$$

where each relation ϱ_i is $\overset{\text{dn}}{\sim}$ or $\backsim$. Utilizing Lemma 4.6, the lower right boundary of K is an interval, so the last step from $\mathfrak{r}_{n-1}$ to $\mathfrak{r}_n = \mathfrak{q}'$ cannot be a swing (if it were, $1_\mathfrak{q}$ would cover at least three elements; it covers exactly one), so $\mathfrak{r}_{n-1} \overset{\text{dn}}{\sim} \mathfrak{r}_n = \mathfrak{q}'$ holds in K.

We have two subcases to consider.

Case 2a: $n = 1$, that is, $\mathfrak{p}_l = \mathfrak{r}_{n-1}$, see Figure 24.8. We cannot have $\mathfrak{p}_l \backsim \mathfrak{q}'$ because $\mathfrak{q}'$ is on the lower right boundary of K; therefore, $\mathfrak{p}_l \overset{\text{dn}}{\sim} \mathfrak{q}'$. We also have $[a_1, t] \overset{\text{dn}}{\sim} \mathfrak{q}'$, so by Lemma 4.4, we obtain that $\mathfrak{p}_l \sim [a_1, t]$. Since $\mathfrak{p}_l$ is the top left prime interval of K, it follows that $\mathfrak{p}_l \overset{\text{dn}}{\sim} [a_1, t]$. Then in $K[S]$, see Figure 4.4,

$$(18) \qquad\qquad \mathfrak{p}_l \overset{\text{dn}}{\sim} [a_1, t] \backsim [m, t] \overset{\text{dn}}{\sim} \mathfrak{q}_{\text{up}}$$

and of course, $\mathfrak{p}_l \overset{\text{dn}}{\sim} \mathfrak{q}_{\text{dn}}$. This completes the verification of (SL) for $\mathfrak{p}_l$ and $\mathfrak{q}$.

Case 2b: $n > 1$, and so, $\mathfrak{p}_l \neq \mathfrak{r}_{n-1}$. We conclude that $\mathfrak{r}_{n-1} \overset{\text{dn}}{\sim} \mathfrak{r}_n = \mathfrak{q}'$. Since $[o, a_r] \overset{\text{dn}}{\sim} \mathfrak{r}_n = \mathfrak{q}'$ also holds, we use Lemma 4.4 to obtain that

$$(19) \qquad\qquad \mathfrak{r}_{n-1} \overset{\text{dn}}{\sim} [o, a_r]$$

or

$$(20) \qquad\qquad [o, a_r] \overset{\text{dn}}{\sim} \mathfrak{r}_{n-1}.$$

But (20) would imply that $0_{\mathfrak{r}_{n-1}}$ is meet-reducible, contradicting that $0_{\mathfrak{r}_{n-1}}$ is not the left-most or right-most element covered by $1_{\mathfrak{r}_{n-2}} = 1_{\mathfrak{r}_{n-1}}$. We conclude that (19) holds.

Then $\mathfrak{r}_{n-2} \backsim \mathfrak{r}_{n-1}$ and $1_{\mathfrak{r}_{n-2}} = 1_{\mathfrak{r}_{n-1}}$ by the definition of the swing relation. The element $1_{\mathfrak{r}_{n-2}} = 1_{\mathfrak{r}_{n-1}}$ covers at least three elements and $0_{\mathfrak{r}_{n-1}}$ is not the left-most or right-most element covered by $1_{\mathfrak{r}_{n-2}} = 1_{\mathfrak{r}_{n-1}}$. We can also assume that $0_{\mathfrak{r}_{n-21}}$ is to the right of $0_{\mathfrak{r}_{n-2}}$ and the down-perceptivity $\mathfrak{r}_{n-1} \overset{\text{dn}}{\sim} \mathfrak{r}_n$ is also to the right, as in Figure 24.9. Then $\mathfrak{r}_{n-1} \overset{\text{dn}}{\sim} \mathfrak{q}_{\text{dn}}$ in $K[S]$. So the sequence

$$\mathfrak{p}_l = \mathfrak{r}_0, \mathfrak{r}_1, \dots, \mathfrak{r}_{n-1}, \mathfrak{q}_{\text{dn}}$$

verifies (SL) for $\mathfrak{p}_l$ and $\mathfrak{q}_{\text{dn}}$. $\qquad\qquad\qquad\qquad\qquad\qquad\qquad$ $\square$

The following statement almost yields the Swing Lemma.

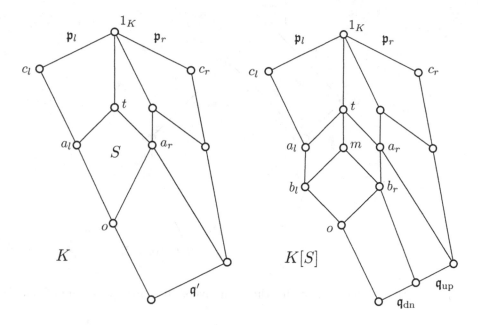

Figure 24.8: Case 2a of Lemma 24.5,

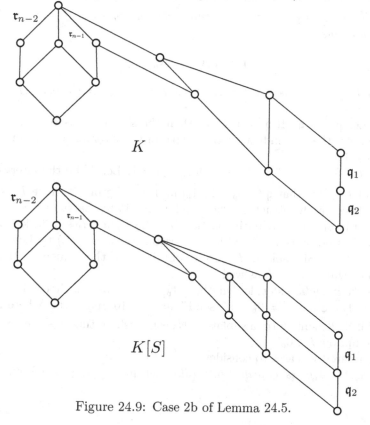

Figure 24.9: Case 2b of Lemma 24.5.

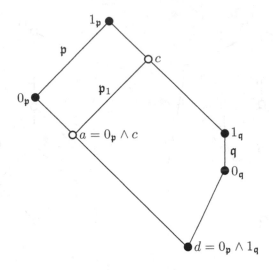

Figure 24.10: The elements for the inductive step, Case 1 in Lemma 24.6.

Lemma 24.6. *Let L be an SPS lattice and let $\mathfrak{p}$ and $\mathfrak{q}$ be distinct prime intervals in L. If $\mathfrak{p} \xrightarrow{p\text{-}dn} \mathfrak{q}$, then there exists a sequence of pairwise distinct prime intervals*

$$(21) \qquad\qquad \mathfrak{p} = \mathfrak{r}_0, \mathfrak{r}_1, \ldots, \mathfrak{r}_n = \mathfrak{q}$$

such that $\mathfrak{r}_i$ is down-perspective to or swings to $\mathfrak{r}_{i+1}$ for $i = 0, \ldots, n-1$.

Proof. Let $\mathfrak{p} \xrightarrow{p\text{-}dn} \mathfrak{q}$. If $\mathfrak{p} \overset{dn}{\sim} \mathfrak{q}$ holds, then the statement is trivial. If $\mathfrak{p} \overset{dn}{\sim} \mathfrak{q}$ fails to hold, then we induct on the length of the interval $[1_{\mathfrak{q}}, 1_{\mathfrak{p}}]$, in formula, $\text{length}[1_{\mathfrak{q}}, 1_{\mathfrak{p}}]$.

For the induction base, let $\text{length}[1_{\mathfrak{q}}, 1_{\mathfrak{p}}] = 1$. Let L' be the interval $[0_{\mathfrak{p}} \wedge 1_{\mathfrak{q}}, 1_{\mathfrak{p}}]$ of L. Note that $\mathfrak{q} \subseteq [0_{\mathfrak{p}} \wedge 1_{\mathfrak{q}}, 1_{\mathfrak{p}}]$ and $\mathfrak{p} \xrightarrow{p\text{-}dn} \mathfrak{q}$ in L'. Since L' is a slim patch lattice, by the Structure Theorem for Slim Patch Lattices, we can obtain L' from the planar distributive lattice $D = \mathsf{B}_2$ by a series of fork insertions. Since D has property (SL) and fork insertions preserve (SL) by Lemma 24.5, it follows that (SL) hold in L'. So we obtain in L' the sequence (21), which of course, will serve in L as well.

For the induction step, let $\text{length}[1_{\mathfrak{q}}, 1_{\mathfrak{p}}] > 1$. So we can choose $1_{\mathfrak{q}} < c \prec 1_{\mathfrak{p}}$. Let $a = 0_{\mathfrak{p}} \wedge c$ and $d = 0_{\mathfrak{p}} \wedge 1_{\mathfrak{q}}$, see Figures 24.10 and 24.11, where the five black filled elements form a sublattice N_5 establishing that $\mathfrak{p} \xrightarrow{P} \mathfrak{q}$. Note that by assumption $d < 0_{\mathfrak{q}}$.

There are two cases to consider.

Case 1: $[a, c]$ is a prime interval. Let $\mathfrak{p}_1 = [a, c]$, see Figure 24.10.

We claim that $\mathfrak{p}_1 \overset{\mathrm{P}}{\longrightarrow} \mathfrak{q}$. Indeed, $1_{\mathfrak{p}_1} = c > 1_{\mathfrak{q}}$ and

(22) $0_{\mathfrak{p}_1} \wedge 1_{\mathfrak{q}} = a \wedge 1_{\mathfrak{q}} = (0_{\mathfrak{p}} \wedge a) \wedge 1_{\mathfrak{q}} = a \wedge (0_{\mathfrak{p}} \wedge 1_{\mathfrak{q}}) = a \wedge d = d < 0_{\mathfrak{q}}.$

If $a = a \vee 0_{\mathfrak{q}}$, then $0_{\mathfrak{p}} \vee 0_{\mathfrak{q}} = 0_{\mathfrak{p}}$, in conflict with the assumption that $\mathfrak{p} \overset{\mathrm{p\text{-}dn}}{\longrightarrow} \mathfrak{q}$. So $a < a \vee 0_{\mathfrak{q}} \leq c$; since $[a, c]$ is assumed to be a prime interval, it follows that $a \vee 0_{\mathfrak{q}} = c$. Along with (22), this verifies that $\mathfrak{p}_1 \overset{\mathrm{P}}{\longrightarrow} \mathfrak{q}$. Since

$$\mathrm{length}[1_{\mathfrak{q}}, 1_{\mathfrak{p}_1}] < \mathrm{length}[1_{\mathfrak{q}}, 1_{\mathfrak{p}}],$$

by the inductive hypothesis, we conclude that $\mathfrak{p}_1 \overset{\mathrm{P}}{\longrightarrow} \mathfrak{q}$. Combining this relation with $\mathfrak{p} \overset{\mathrm{dn}}{\sim} \mathfrak{p}_1$, we obtain (21), completing the proof for Case 1.

   Case 2: $[a, c]$ is not a prime interval. Let $e = a \vee 1_{\mathfrak{q}} \leq c$. Choose an element b so that $a < b \prec c$, see Figure 24.11, and let $\mathfrak{p}_1 = [b, e]$. Then

(23) $\mathfrak{p} \overset{\mathrm{p\text{-}dn}}{\longrightarrow} \mathfrak{p}_1$

established by the $\mathsf{N}_5 = \{a, 0_{\mathfrak{p}}, 1_{\mathfrak{p}}, b, e\}$. We apply Lemma 24.4 with $\mathfrak{p}_1 = [x, i]$,

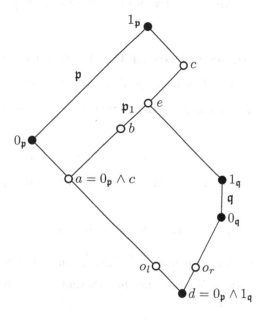

Figure 24.11: The elements for the inductive step, Case 2.

$a = u$, $\mathfrak{q} = [v, w]$, and $o = d$. Then we conclude that $0_{\mathfrak{p}_1} \wedge 1_{\mathfrak{q}} = b \wedge 1_{\mathfrak{q}} < 0_{\mathfrak{q}}$, therefore,

(24) $\mathfrak{p}_1 \overset{\mathrm{p\text{-}dn}}{\longrightarrow} \mathfrak{q}.$

Now (23) and (24) imply that $\mathfrak{p} \overset{\mathrm{p\text{-}dn}}{\longrightarrow} \mathfrak{q}$, which we are required to prove.   □

Now we are ready to prove the Swing Lemma. Let L be an SPS lattice and let $\mathfrak{p}$ and $\mathfrak{q}$ be distinct prime intervals in L so that $\mathfrak{q}$ is collapsed by $\mathrm{con}(\mathfrak{p})$. By the Prime-projectivity Lemma, there exists a sequence of pairwise distinct prime intervals $\mathfrak{p} = \mathfrak{u}_0, \mathfrak{u}_1, \ldots, \mathfrak{u}_n = \mathfrak{q}$ satisfying

$$(25) \qquad \mathfrak{p} = \mathfrak{u}_0 \xrightarrow{\;P\;} \mathfrak{u}_1 \xrightarrow{\;P\;} \cdots \xrightarrow{\;P\;} \mathfrak{u}_n = \mathfrak{q}.$$

If $\mathfrak{u}_{i-1} \xrightarrow{\text{p-up}} \mathfrak{u}_i$ for $i = 1, \ldots, n$, then $\mathfrak{u}_{i-1} \overset{\text{up}}{\sim} \mathfrak{u}_i$ by semimodularity. If $\mathfrak{u}_{i-1} \xrightarrow{\text{p-dn}} \mathfrak{u}_i$ for $i = 1, \ldots, n$, then by Lemma 24.6, we get a sequence of down perspectivities and swings. So (25) turns into a sequence of up perspectivities, down perspectivities, and swings. By Lemma 4.3(i) (or Lemma 4.4), a down perceptivity cannot be followed by an up perceptivity. A swing cannot be followed by an up perceptivity. So if there is an up perceptivity, it must be the first binary relation. Since two down perspectivities can be replaced by one and two swings can be replaced by one, we conclude that the sequence of binary relations start with at most one up perceptivity, followed by an alternating sequence of down perspectivities and swings, as claimed by the Swing Lemma.

24.4. Some consequences of the Swing Lemma

We now make a number of elementary observations about the Swing Lemma.

Observation 1. As in the proof of Lemma 24.5, we associate with the sequence

$$(26) \qquad \mathfrak{r} = \mathfrak{r}_0, \mathfrak{r}_1, \ldots, \mathfrak{r}_n = \mathfrak{q}$$

of prime intervals (as in (4)), a sequence of binary relations $\varrho_1, \ldots, \varrho_{n-1}$ such that

$$(27) \qquad \mathfrak{r} = \mathfrak{r}_0 \, \varrho_1 \, \mathfrak{r}_1 \, \varrho_2 \cdots \varrho_n \, \mathfrak{r}_n = \mathfrak{q},$$

where each binary relation is one of $\overset{\text{dn}}{\sim}, \overset{\text{ex}}{\leftthreetimes}, \overset{\text{in}}{\leftthreetimes}$ and (and in the subsequent discussions) the relations $\overset{\text{dn}}{\sim}$ and $\overset{\text{in}}{\leftthreetimes}$ are *proper*, that is, they relate two distinct prime intervals.

Observation 2. We can assume that down perspectivities and swings alternate.

Indeed, the relations: $\overset{\text{dn}}{\sim}$ and $\overset{\text{in}}{\leftthreetimes}$ are transitive, so

$$\overset{\text{dn}}{\sim} \circ \overset{\text{dn}}{\sim} = \overset{\text{dn}}{\sim},$$

$$\overset{\text{in}}{\leftthreetimes} \circ \overset{\text{in}}{\leftthreetimes} = \overset{\text{in}}{\leftthreetimes}.$$

Observation 3. If $\varrho_i = \overset{\text{dn}}{\sim}$, for $i < n$, then $\varrho_{i+1} = \overset{\text{ex}}{\curvearrowright}$.

Observation 4. ϱ_1 may be an interior swing. All the other swings in the sequence (27) are exterior swings.

The last two observations follow from the fact that there is no down perspectivity to an interior prime interval of a multifork in an SPS lattice by Lemma 4.4.

If $\mathfrak{p} \overset{\text{in}}{\curvearrowright} \mathfrak{q}$ (as in the second diagram of Figure 24.4), then $\operatorname{con}(\mathfrak{p}) = \operatorname{con}(\mathfrak{q})$; nevertheless, interior swings play an important role, see the example in Figure 24.5.

In view of these observations, we derive some simple consequences of the Swing Lemma.

Corollary 24.7. *Let L be an SPS lattice. If $\mathfrak{q}$ is an exterior and $\mathfrak{p}$ is an interior prime interval of a multifork, then $\operatorname{con}(\mathfrak{q}) > \operatorname{con}(\mathfrak{p})$.*

Proof. We know that $\operatorname{con}(\mathfrak{q}) \geq \operatorname{con}(\mathfrak{p})$. Let us assume that $\operatorname{con}(\mathfrak{q}) = \operatorname{con}(\mathfrak{p})$. Then $\operatorname{con}(\mathfrak{p}) \geq \operatorname{con}(\mathfrak{q})$ and by Observation 1 there are sequences (26) and (27). We must have $\mathfrak{p} = \mathfrak{r}$, because $\mathfrak{p}$ is an interior prime interval. If the first step is a swing, it is to another interior prime interval. So the next step is a down perspectivity. By (5), none of the $\mathfrak{r}_i$ can reach the height of $\mathfrak{q}$ for $i = 2, \ldots, n$. This proves the statement. □

Corollary 24.8. *Let $\mathfrak{p}$ and $\mathfrak{q}$ be prime intervals in an SPS lattice L satisfying $\operatorname{con}(\mathfrak{p}) = \operatorname{con}(\mathfrak{q})$. Then there is a prime interval $\mathfrak{r}$ such that one of the following two conditions hold (see Figure 24.12):*

(i) *$\mathfrak{p}$ is up perspective to $\mathfrak{q}$ and $\mathfrak{q}$ is down perspective to $\mathfrak{r}$; in formula,*

$$\mathfrak{p} \overset{up}{\sim} \mathfrak{r} \overset{dn}{\sim} \mathfrak{q}.$$

(ii) *$\mathfrak{p}$ has an interior swing to $\mathfrak{r}$ and $\mathfrak{r}$ is down perspective to $\mathfrak{q}$; in formula,*

$$\mathfrak{p} \overset{in}{\curvearrowright} \mathfrak{r} \overset{dn}{\sim} \mathfrak{q}.$$

Proof. If there are no swings in (26), we get (i).

For the sequence (27), by Corollary 24.7, there can be no external swings. By Observation 3, a perspectivity cannot be followed by an interior swing. So we are left with (ii). □

Corollary 24.9. *Let L be an SPS lattice. If $\mathfrak{s}$ is an exterior prime interval and $\mathfrak{t}$ is an interior prime interval of a multifork, then $\operatorname{con}(\mathfrak{s}) \succ \operatorname{con}(\mathfrak{t})$ in the ordered set, $\operatorname{Con}_J L$, of join-irreducible congruences of L.*

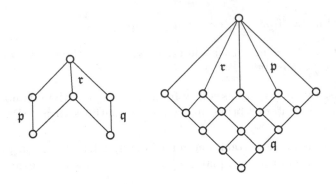

Figure 24.12: $\mathrm{con}(\mathfrak{p}) = \mathrm{con}(\mathfrak{q})$.

Proof. Let $\mathfrak{s}'$ denote the other external prime interval. If $\mathfrak{t}$ is a prime interval with $\mathrm{con}(\mathfrak{t}) > \mathrm{con}(\mathfrak{p})$, then we can take a sequence as in (27). We can assume that $\mathfrak{t} = \mathfrak{r}$. Working our way back from $\mathfrak{r}_n = \mathfrak{p}$, the last step cannot be a down perspectivity, because $\mathfrak{r}_n = \mathfrak{p}$ is an interior prime interval. So it must be a swing. If it is an external swing, we get $\mathrm{con}(\mathfrak{t}) \geq \mathrm{con}(\mathfrak{q})$ or $\mathrm{con}(\mathfrak{t}) \geq \mathrm{con}(\mathfrak{q}')$. This proves the statement. □

Let us call a sequence (26) *reduced*, if down perspectivities and swings alternate.

Corollary 24.10. *In the Swing Lemma, we can always choose a reduced sequence of prime intervals.*

For the prime intervals $\mathfrak{p}$ and $\mathfrak{q}$, let us call a reduced sequence of prime intervals (26) *neutral*, if $\mathrm{con}(\mathfrak{p}) = \mathrm{con}(\mathfrak{q})$.

Lemma 24.11. *A reduced sequence of prime intervals (26) is neutral iff it is one of the following:*

(i) $\mathfrak{p} \overset{\mathrm{up}}{\sim} \mathfrak{r} \overset{\mathrm{dn}}{\sim} \mathfrak{q}$ *(and the shorter versions, $\mathfrak{p} \overset{\mathrm{up}}{\sim} \mathfrak{q}$, $\mathfrak{p} \overset{\mathrm{dn}}{\sim} \mathfrak{q}$);*

(ii) $\mathfrak{r}_0 \overset{\mathrm{in}}{\leftthreetimes} \mathfrak{r}_1 \overset{\mathrm{dn}}{\sim} \mathfrak{r}_2$ *(and the shorter versions, $\mathfrak{r}_0 \overset{\mathrm{in}}{\leftthreetimes} \mathfrak{r}_1$, $\mathfrak{r}_0 \overset{\mathrm{dn}}{\sim} \mathfrak{r}_1$).*

Proof. Indeed, because by Corollary 24.9, a reduced neutral sequence of prime intervals (26) cannot contain an external swing. □

Corollary 24.12. *Let L be an SPS lattice and let $\mathfrak{p}$ and $\mathfrak{q}$ be prime intervals in L. Then $\mathrm{con}(\mathfrak{p}) \succ \mathrm{con}(\mathfrak{q})$ in $\mathrm{J}(\mathrm{Con}\,L)$ iff we can form a reduced sequence of prime intervals (26) as follows: the sequence $\mathfrak{p}_1 \overset{\mathrm{ex}}{\leftthreetimes} \mathfrak{q}_1$, preceded by a neutral sequence between $\mathfrak{p}$ and $\mathfrak{p}_1$ or by $\mathfrak{p} = \mathfrak{p}_1$, and followed by a neutral sequence between $\mathfrak{q}_1$ and $\mathfrak{q}$ or by $\mathfrak{q}_1 = \mathfrak{q}$.*

Proof. By Corollary 24.9, a reduced sequence of prime intervals (26) between $\mathfrak{p}$ and $\mathfrak{q}$ with $\mathrm{con}(\mathfrak{p}) \succ \mathrm{con}(\mathfrak{q})$ in $J(\mathrm{Con}\,L)$ must contain exactly one external swing. Before and after the swing the sequence must be neutral.   □

We define the *height* of a prime interval $\mathfrak{u}$ as $\mathrm{height}(1_\mathfrak{u})$; let $\mathrm{height}(\mathfrak{u})$ denote the height of $\mathfrak{u}$. For a join-irreducible congruence α of L, the *peak* of α is a prime interval $\mathfrak{p}$, so that $\alpha = \mathrm{con}(\mathfrak{p})$ and $\mathrm{height}(\mathfrak{p})$ is maximal.

Lemma 24.13. *Let L be an SPS lattice. Then every join-irreducible congruence α of L has a unique peak up to internal swings.*

Proof. Let $\mathfrak{u}$ and $\mathfrak{v}$ both be peaks of α. Then by Lemma 24.3, there is a sequence (26) from $\mathfrak{u}$ to $\mathfrak{v}$ satisfying (5). In particular, $\mathfrak{u}$ is up perspective to $\mathfrak{r}$. Since $\mathfrak{u}$ is a peak, it follows that $\mathfrak{u} = \mathfrak{r}$. By (5), $\mathrm{height}(\mathfrak{u}) \geq \mathrm{height}(\mathfrak{v})$. Symmetrically, $\mathrm{height}(\mathfrak{v}) \geq \mathrm{height}(\mathfrak{u})$, so that $\mathrm{height}(\mathfrak{u}) = \mathrm{height}(\mathfrak{v})$, that is, $1_\mathfrak{u} = 1_\mathfrak{v}$. By Lemma 24.3, we must have that $\mathfrak{u}$ swings to $\mathfrak{v}$. Symmetrically, $\mathfrak{v}$ swings to $\mathfrak{u}$, so the swing must be internal.   □

24.5. Fork congruences

In this section, we examine the extendibility of congruences to a fork extension.

Theorem 24.14. *Let L be an SPS lattice and α be a congruence of L. Let $S = \{o, a_l, a_r, t\}$ be a covering square of L.*

(i) *If $\alpha\rceil S = 1_S$, then α extends to $L[S]$.*

(ii) *If $\alpha\rceil S = 0_S$, then α extends to $L[S]$.*

(iii) *If $\alpha\rceil S$ is not trivial, then α may or may not extend to $L[S]$.*

Every prime interval of $L[S]$ is perspective to a prime interval of L, except for $[m, t]$ (using the notation of Figure 4.4) and the prime intervals perspective to it, so the only candidate for a new join-irreducible congruence in $L[S]$ is the congruence $\gamma(S) = \mathrm{con}_{L[S]}(m, t)$.

Let L and S be as in Theorem 24.14. We call the covering square of $S = \{o, a_l, a_r, t\}$ a *tight square*, if t covers exactly two elements, namely, a_l and a_r, in L; otherwise, S is a *wide square*.

Theorem 24.15. *Let L be an SPS lattice. If S is a wide square, then $\gamma(S) = \mathrm{con}_{L[S]}(m, t)$ is generated by a congruence of L.*

Theorem 24.16. *Let L be an SPS lattice. Let $S = \{o, a_l, a_r, t\}$ be a tight square. Then $L[S]$ has exactly one join-irreducible congruence, namely,*

$$\gamma(S) = \mathrm{con}_{L[S]}(m, t)$$

that is not generated by a congruence of L.

We name a few join-irreducible congruences of L and $L[S]$ that will play an important role.

Join-irreducible congruences in L:

(28) $$\alpha_l(S) = \mathrm{con}_L(a_l, t),$$

(29) $$\alpha_r(S) = \mathrm{con}_L(a_r, t).$$

Join-irreducible congruences in $L[S]$, see Figure 4.4:

(30) $$\overline{\alpha}_l(S) = \mathrm{con}_{L[S]}(a_l, t),$$

(31) $$\overline{\alpha}_r(S) = \mathrm{con}_{L[S]}(a_r, t),$$

(32) $$\gamma(S) = \mathrm{con}_{[L[S]}(m, t).$$

Proof of Theorem 24.14.(i). Let α be a congruence of L satisfying $\alpha\rceil S = 1_S$. We define the partition:

$$\pi = \{\, [u, v]_{L[S]} \mid u, v \in L \text{ and } [u, v]_L \text{ is a congruence class of } \alpha \,\}.$$

To verify that π is indeed a partition of $L[S]$, let

$$A = \bigcup(\, [u, v]_{L[S]} \mid u, v \in L \text{ and } [u, v]_L \text{ is a congruence class of } \alpha \,).$$

Clearly, $L \subseteq A$. By assumption, $\alpha\rceil S = 1$, that is, $[o, t]$ is in a congruence class of α, so there is a congruence class $[u, v]_L$ of L containing o and t. Hence $m \in A$. Since $o \equiv a_l \pmod{\alpha}$, and so $y_{l_i} \equiv x_{l_i} \pmod{\alpha}$ for $i = 1, \ldots, n_l$, therefore, there is a congruence class $[u_i, v_i]_L$ containing $x_{l,i}$ and $y_{l,i}$ for $i = 1, \ldots, n_l$. Hence $z_{l,i} \in [u_i, v_i]_{L[S]} \subseteq A$ and symmetrically. This proves that $A = L$.

Next we observe that x belongs to a π-class iff so do x^+ and x^-. This implies that the sets in π are pairwise disjoint.

Finally, we verify the substitution properties. Let $a, b, c \in L[S]$ and $a \equiv b \pmod{\pi}$. Then there exist $u, v \in L$ with $u \equiv v \pmod{\alpha}$ such that $a, b \in [u, v]_{L[S]}$. There is also an interval $[u', v']_{L[S]} \in \pi$ with $c \in [u', v']_{L[S]}$. Since $u \equiv v \pmod{\alpha}$ and $u' \equiv v' \pmod{\alpha}$, it follows that $u \vee u' \equiv v \vee v' \pmod{\alpha}$, so there is a congruence class $[u'', v'']$ of α containing $u \vee u'$ and $v \vee v'$. So $u \vee u', v \vee v' \in [u'', v'']_{L[S]}$, verifying the substitution property for joins. The dual proof verifies the substitution property for meets.

So π is a congruence of $L[S]$. Clearly, $\pi = \overline{\alpha}$. The uniqueness statement is obvious. $\square$

Proof of Theorem 24.14.(ii). Now let α be a congruence of L satisfying $\alpha\rceil S = 0_S$. We are going to define a congruence β of $L[S]$ extending α.

For $i = 1, \ldots, n_l$, define $\underline{i}$ as the smallest element in $\{1, \ldots, n_l\}$ satisfying $x_{l,i} \equiv x_{l,\underline{i}} \pmod{\alpha}$; let $\overline{i}$ be the largest one. Clearly, $\underline{i} \leq i \leq \overline{i}$. Similarly, by a slight abuse of notation, we define $\overline{i}$ and $\underline{i}$ on the right.

We define β as the partition:

(33) $\{\, [u,v]_{L[S]} \mid [u,v]_L$ is a congruence class of α with $u < v\,\}$
    $\cup \{m\} \cup \{\, [z_{l,\underline{i}}, z_{l,\overline{i}}] \mid i = 1, \dots, n_l \,\} \cup \{\, [z_{r,\underline{i}}, z_{r,\overline{i}}] \mid i = 1, \dots, n_r \,\}.$

To see that β is a partition, observe that it covers $L[S]$. Note that if $u, v \in L$ with $u < v$ such that $[u,v]_L$ is a congruence class of α, then $[u,v]_L = [u,v]_{L[S]}$ unless $u < z_{l,i} < v$ for some $i = 1, \dots, n_l$ or symmetrically. But in this case we would have that $y_{l,i} \equiv x_{l,i} \pmod{\alpha}$, implying that $o = y_{l,1} \equiv x_{l,1} \equiv a_l \pmod{\alpha}$, contrary to the assumption, $\alpha]S = \mathbf{0}_S$. Clearly, two distinct $[z_{l,\underline{i}}, z_{l,\overline{i}}]$ classes cannot intersect by the definition of $\underline{i}$ and $\overline{i}$.

So if two classes intersect, they must be of the form $[u,v]_{L[S]}$ and $[z_{l,\underline{i}}, z_{l,\overline{i}}]$, which would contradict that $[u,v]_{L[S]} = [u,v]_L$. The symmetric cases (on the right) and the mixed cases (left and right) complete the discussion. So (33) defines a partition β.

To see that β is a congruence, we use Lemma 3.7. To verify $(C_\vee)$, let $a \prec b$, $a \prec c \in L[S]$, $b \neq c$, and $a, b \in L$, $a \equiv b \pmod{\beta}$. We want to prove that $c \equiv b \vee c \pmod{\beta}$. By (33), either $a \equiv b \pmod{\alpha}$ or $a, b \in [z_{l,\underline{i}}, z_{l,\overline{i}}]$ for some $i = 1, \dots, n_l$ (or symmetrically).

First, let $a, b \in L$ and $a \equiv b \pmod{\alpha}$. If $c \in L$, then $c \equiv b \vee c \pmod{\alpha}$ since α is a congruence, so by (33), $c \equiv b \vee c \pmod{\beta}$.

If $c \notin L$, that is, if $c \in F[S]$, then $c \neq m$. Indeed, $c = m$ cannot happen, since m does not cover an element not in $F[S]$. Therefore, $c = z_{l,i}$ for some $i = 1, \dots, n_l$ (or symmetrically). Since $a \prec c = z_{l,i}$ and $a \in L$, we get that $a = y_{l,i}$, $i > 1$ and $b = y_{l,i-1}$. It easily follows that $c \equiv c \vee b \pmod{\alpha}$.

Second, let $a, b \in [z_{l,\underline{i}}, z_{l,\overline{i}}]$ for some $i = 1, \dots, n_l$ (or symmetrically). Then $a = z_{l,j}$ and $b = z_{l,j-1}$ with $z_{l,j} \equiv z_{l,j-1} \pmod{\alpha}$ by (33), and $c = x_{l,j}$. So $c \equiv b \vee c \pmod{\beta}$ by (33), completing the verification of $(C_\vee)$.

To verify $(C_\wedge)$, let $a \succ b$, $a \succ c \in L[S]$, $b \neq c$, and $a, b \in L$, $a \equiv b \pmod{\beta}$. We want to prove that $c \equiv b \wedge c \pmod{\beta}$. By (33), either $a, b \in L$ and $a \equiv b \pmod{\alpha}$ or $a, b \in [z_{l,\underline{i}}, z_{l,\overline{i}}]$ for some $i = 1, \dots, n_l$ (or symmetrically).

First, let $a, b \in L$ and $a \equiv b \pmod{\alpha}$. If $c \in L$, then $c \equiv b \wedge c \pmod{\alpha}$ since α is a congruence, so by (33), $c \equiv b \wedge c \pmod{\beta}$. Let $c \notin L$, that is, if $c \in F[S]$. If $c = m$, then $a \geq t$ and $b \geq a_l$, or symmetrically. But then $a \equiv b \pmod{\alpha}$ would contradict the assumption that $\alpha]S = \mathbf{0}_S$.

Therefore, $c \neq m$, that is, $c = z_{l,i}$ for some $i = 1, \dots, n_l$ (or symmetrically). Since $a \succ c = z_{l,i}$ and $a \in L$, we get that $a = x_{l,i}$, $i > 1$. Now if $b = x_{l,i+1}$, then $c \equiv b \wedge c \pmod{\beta}$ easily follows from (33). So we can assume that $b \neq x_{l,i+1}$. Then by Lemma 4.3(ii), the elements b, $x_{l,i+1}$, and c generate an S_7 sublattice; therefore, $c \equiv b \wedge c \pmod{\beta}$ is easily computed in the S_7 sublattice.

Second, let $a, b \in [z_{l,\underline{i}}, z_{l,\overline{i}}]$ for some $i = 1, \dots, n_l$ (or symmetrically). Then $a = z_{l,j}$, $b = z_{l,j+1}$ and $c = y_{l,j}$. By the definition of $\underline{i}$ and $\overline{i}$, it follows that

$z_{l,j} \equiv z_{l,j+1} \pmod{\alpha}$, which trivially implies that $y_{l,j} \equiv y_{l,j+1} \pmod{\alpha}$, that is, $c \equiv b \wedge c \pmod{\alpha}$, and so $c \equiv b \wedge c \pmod{\beta}$.

Clearly, $\beta = \overline{\alpha}$. $\square$

Note the similarity between the proofs of the conditions $(C_\vee)$ and $(C_\wedge)$. Unfortunately, there is no duality.

Proof of Theorem 24.14.(iii). We need two examples. The first is trivial. Let

$$L = S = C_2^2.$$

Then $L[S] = S_7$ and all congruences of L extend to $L[S]$.

For the second, see Figure 24.13. Let α let be the congruence of L collapsing two opposite sides of the covering square S. This congruence has exactly two nontrivial classes, marked in Figure 24.13 by bold lines. In particular, $a \equiv b \pmod{\alpha}$ fails.

Figure 24.13 also shows the congruence $\overline{\alpha}$ of $L[S]$. Note that $a \equiv b \pmod{\overline{\alpha}}$ in $L[S]$, so α has no extension to $L[S]$. $\square$

Next, we deal with Theorem 24.15.

Proof of Theorem 24.15. Since S is wide, the element t covers an element a in L, with $a \neq a_l, a_r$. Since L is slim, either a is to the left of a_l or to the right of a_r; let us assume the latter, see Figure 24.14. By Lemma 4.3, the set $\{a_l, a_r, a\}$ generates an S_7 sublattice in L.

Then $\mathrm{con}(m, t) \leq \mathrm{con}(a_r, t)$, computed in the S_7 sublattice generated by the set $\{a_l, m, a_r\}$, and $\mathrm{con}(m, t) \geq \mathrm{con}(a_r, t)$, computed in the S_7 sublattice generated by $\{m, a_r, a\}$, so we conclude that $\gamma(S) = \overline{\alpha}_r(S)$.

By (31), $\gamma(S)$ is generated by a congruence of L, namely by $\mathrm{con}(t, a_r)$. $\square$

A simple application of the Swing Lemma describes $\gamma(S)$ for a tight S.

Let L be an SPS lattice with the covering square S. We use the notation of Section 4.2, see Figure 4.4. The crucial new join-irreducible congruence on $L[S]$ is $\gamma(S) = \mathrm{con}_{L[S]}(m, t)$. We now give a complete description of this congruence for tight covering squares.

Theorem 24.17. *Let L be an SPS lattice with a tight covering square S. Let $\mathfrak{q} \neq [m, t]$ be a prime interval in $L[S]$. Then $\mathfrak{q}$ is collapsed by $\gamma(S)$ in $L[S]$ iff there exists a (reduced) sequence of pairwise distinct prime intervals*

(34) $$[m, t] = \mathfrak{r}_0, \mathfrak{r}_1, \ldots, \mathfrak{r}_n = \mathfrak{q}$$

such that $\mathfrak{r}_0$ is down perspective to $\mathfrak{r}_1$, $\mathfrak{r}_0$ externally swings to $\mathfrak{r}_2$, and so on; down perspectivities and externally swings alternate.

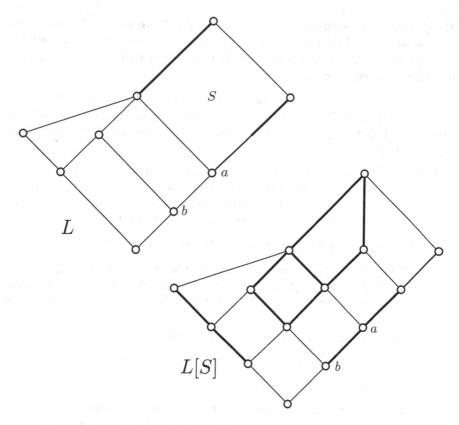

Figure 24.13: A congruence of L that does not extend to $L[S]$.

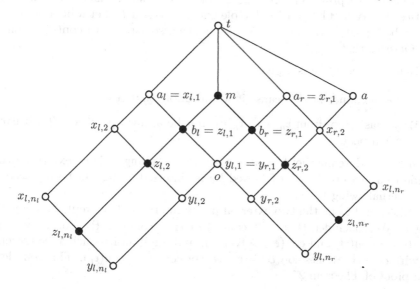

Figure 24.14: Wide square.

Proof. Since, by assumption, $\operatorname{con}(m,t) \geq \operatorname{con}(\mathfrak{q})$, we can apply the Swing Lemma to $L[S]$ and the prime intervals $[m,t]$ and $\mathfrak{q}$, to obtain the sequence (5). Since m is meet-irreducible and $[m,t] \stackrel{\text{up}}{\sim} \mathfrak{r}$, it follows that $[m,t] = \mathfrak{r}$, thereby obtaining the sequence (34). $\square$

Finally, we prove Theorem 24.16. Let L be an SPS lattice. Let $S = \{o, a_l, a_r, t\}$ be a tight square of L. Let $\mathfrak{p}$ be a prime interval of $L[S]$ and assume that $\operatorname{con}_{L[S]}(\mathfrak{p})$ is *not generated* by a congruence of L. Note that $\mathfrak{p} \not\subseteq L$. We distinguish two cases.

(i) $\mathfrak{p} \subseteq F[S]$. By symmetry, we can assume that $\mathfrak{p} \subseteq \{z_{l,n_l}, \ldots, z_{l,n_l} = b_l, m\}$. Then $\mathfrak{p}$ is perspective to a prime interval $\mathfrak{q} \subseteq \{x_{l,n_l}, \ldots, x_{l,n_l} = a_l\} \subseteq L$, hence $\operatorname{con}_{L[S]}(\mathfrak{p}) = \operatorname{con}_{L[S]}(\mathfrak{q})$ with $\mathfrak{q} \subseteq L$, contradicting the assumption that $\operatorname{con}_{L[S]}(\mathfrak{p})$ is *not generated* by a congruence of L.

(ii) Let $\mathfrak{p}$ be neither in L nor in $F[S]$. We need two subcases.

(a) $\mathfrak{p} = [x_{l,i}, z_{l,i}]$ for some $i = 1, \ldots, n_l$. Then $\operatorname{con}_{L[S]}(\mathfrak{p}) = \operatorname{con}_{L[S]}(a_r, t)$, a congruence of $L[S]$ generated by a congruence $\operatorname{con}_L(a_r, t)$ of L, contradicting the assumption that it is *not generated* by a congruence of L.

(b) $\mathfrak{p} = [y_{l,i}, z_{l,i}]$ for some $i = 1, \ldots, n_l$. Then

$$\operatorname{con}_{L[S]}(\mathfrak{p}) = \operatorname{con}_{L[S]}(m,t) = \gamma(S),$$

as claimed in the theorem.

Note that for a wide covering square S, there is a similar description of $\gamma(S)$, except that the sequence (34) may start with an internal swing.

Finally, we prove Theorem 25.2, let L be a slim, planar, semimodular lattice. Let $\alpha > \beta$ be join-irreducible congruences of L. Let $\mathfrak{u}$ be a peak of α and $\mathfrak{v}$ be a peak of β. Let (4) be a sequence associated with $\operatorname{con}(\mathfrak{u}) > \operatorname{con}(\mathfrak{v})$.

Observe that

(i) $\mathfrak{u} = \mathfrak{r}$, because $\mathfrak{u}$ is a peak;

(ii) $\mathfrak{u} \stackrel{\text{dn}}{\sim} \mathfrak{v}$ cannot hold because it would imply that $\alpha = \beta$;

(iii) the last step, from $\mathfrak{r}_{n-1}$ to $\mathfrak{r}_n$ cannot be a down perspectivity, because $\mathfrak{v}$ is a peak.

So $n > 1$, and we may get from $\mathfrak{u}$ to $\mathfrak{v}$ is n steps, ending up in an external swing. (Indeed, an external swing followed by an internal swing can be replaced by an external swing.)

Let $\mathfrak{e}_l$ and $\mathfrak{e}_r$ be the two external prime intervals of the multifork containing $\mathfrak{v}$. We concluded that $\alpha \geq \operatorname{con}(\mathfrak{e}_l)$ or $\alpha \geq \operatorname{con}(\mathfrak{e}_r)$. By Corollary 24.12, $\operatorname{con}(\mathfrak{e}_l) > \operatorname{con}(\beta)$ and $\operatorname{con}(\mathfrak{e}_r) > \operatorname{con}(\beta)$, and any join-irreducible congruence α with $\alpha > \beta$ satisfies $\operatorname{con}(\alpha) \geq \operatorname{con}(\mathfrak{e}_l)$ or $\operatorname{con}(\alpha) \geq \operatorname{con}(\mathfrak{e}_r)$. This completes the proof of Theorem 25.2.

24.6. Discussion

G. Grätzer, H. Lakser, and E. T. Schmidt [105] started the study of planar semimodular lattices and their congruences. This field is surveyed in Chapters 3 and 4 (G. Czédli and G. Grätzer [33] and G. Grätzer [63]) in LTS1.

Theorem 10.1 states that every finite distributive lattice D can be represented as the congruence lattice of a finite planar semimodular lattice L. It is crucial that in this result the lattice L is not assumed to be slim. The M_3 sublattices play a central role in the constructions. So what happens if we cannot have M_3 sublattices?

Problem 24.1. Characterize the congruence lattices of SPS lattices.

Problem 24.2. Characterize the congruence lattices of slim patch lattices.

The following problem is closely related to the two previous ones.

Problem 24.3. Characterize the congruence lattices of patch lattices.

Look at Figure 24.15. It shows the diagrams of $\mathrm{Con}_J L$ and $\mathrm{Con}_J L[S]$ for the lattice L and covering square S. We see that $\mathrm{Con}_J L[S] = \mathrm{Con}_J L \cup \{\gamma(S)\}$.

Problem 24.4. How do $\mathrm{Con}_J L$ and $\mathrm{Con}_J L[S]$ interrelate?

Figure 24.15: Not suborder

Some Applications
of the Swing Lemma

25.1.   The Trajectory Theorem for SPS Lattices

We start with an important definition of G. Czédli [18].

For the trajectories $\mathcal{P} \neq \mathcal{Q}$, let $\mathcal{P} \leq_C \mathcal{Q}$ if $\mathcal{P}$ is a hat trajectory, $1_{\mathrm{top}(\mathcal{P})} \leq 1_{\mathrm{top}(\mathcal{Q})}$, and $0_{\mathrm{top}(\mathcal{P})} \not\leq 0_{\mathrm{top}(\mathcal{Q})}$, see Figure 25.1. Czédli defines $\leq_T$ as the reflexive and transitive closure of $\leq_C$. (The notation in G. Czédli [18] is different.) So for a trajectory $\mathcal{P}$, we can define the closure, $\widehat{\mathcal{P}}$, of $\mathcal{P}$: $\mathcal{Q} \in \widehat{\mathcal{P}}$ iff $\mathcal{P} \leq_C \mathcal{Q}$ and $\mathcal{Q} \leq_C \mathcal{P}$.

Observe that if $\mathcal{P}, \mathcal{P}' \in \widehat{\mathcal{T}}$, then $\mathcal{P} \leq_C \mathcal{Q}$ iff $\mathcal{P}' \leq_C \mathcal{Q}$; similarly, if $\mathcal{Q}, \mathcal{Q}' \in \widehat{\mathcal{T}}$, then $\mathcal{P} \leq_C \mathcal{Q}$ iff $\mathcal{P} \leq_C \mathcal{Q}'$. It follows that, by a slight abuse of terminology, we can use $\leq_T$ as an ordering on

$$\widehat{\mathrm{Traj}}\, L = \{\, \widehat{\mathcal{T}} \mid T \in \mathrm{Traj}\, L\,\}.$$

For a trajectory $\mathcal{T}$, we can define $\mathrm{con}(\widehat{\mathcal{T}}) = \mathrm{con}(\mathcal{T})$. Indeed, let $\mathcal{P}, \mathcal{Q} \in \widehat{\mathcal{T}}$. Then $\mathcal{P} \leq_C \mathcal{Q}$ and $\mathcal{Q} \leq_C \mathcal{P}$, therefore, $1_{\mathrm{top}(\mathcal{P})} \leq 1_{\mathrm{top}(\mathcal{Q})}$ and $1_{\mathrm{top}(\mathcal{Q})} \leq 1_{\mathrm{top}(\mathcal{P})}$, and so $1_{\mathrm{top}(\mathcal{P})} = 1_{\mathrm{top}(\mathcal{Q})}$. Hence, $\mathrm{top}(\mathcal{P})$ and $\mathrm{top}(\mathcal{Q})$ are interior edges of the multifork at $1_{\mathrm{top}(\mathcal{P})} = 1_{\mathrm{top}(\mathcal{Q})}$ and so $\mathrm{con}(\mathrm{top}(\mathcal{P})) = \mathrm{con}(\mathrm{top}(\mathcal{Q}))$, from which $\mathrm{con}(\mathcal{P}) = \mathrm{con}(\mathcal{Q})$ follows.

We have seen that $\widehat{\mathrm{Traj}}\, L$ is an ordered set under the ordering $\leq_T$ and that all the prime intervals $\mathfrak{p}$ in a trajectory $\mathcal{P} \in \widehat{\mathcal{T}}$ generate the same join-irreducible congruence $\mathrm{con}(\mathfrak{p})$ of L. The join-irreducible congruences of L form

© Springer International Publishing Switzerland 2016

G. Grätzer, *The Congruences of a Finite Lattice*,

DOI 10.1007/978-3-319-38798-7_25

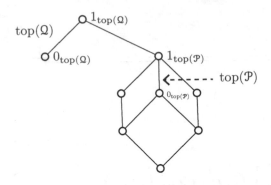

Figure 25.1: $\mathcal{P} \leq_C \mathcal{Q}$.

an ordered set $J(\operatorname{Con} L)$. It is the main result of this section that these two ordered sets are isomorphic, see G. Grätzer [74]. For rectangular lattices, this result was proved in G. Czédli [18]. One can also use Theorem 4.7 (G. Czédli and E. T. Schmidt [36]) to generalize Czédli's result to arbitrary SPS lattices.

Theorem 25.1 (Trajectory Theorem for SPS Lattices). *Let L be an SPS lattice. Then the ordered set $\widehat{\operatorname{Traj}} L$ is isomorphic to the ordered set $J(\operatorname{Con} L)$ under the isomorphism $\widehat{\mathfrak{J}} \mapsto \operatorname{con}(\widehat{\mathfrak{J}})$.*

We are going to prove this result in this section.

First, we prove that

(1) $\mathcal{P} \leq_T \mathcal{Q}$ implies that $\operatorname{con}(\mathcal{P}) \leq \operatorname{con}(\mathcal{Q})$.

Since $\leq_T$ is the reflexive and transitive closure of $\leq_C$, it is sufficient to prove (1) for $\mathcal{P} \leq_C \mathcal{Q}$. So assume the following: $\mathcal{P} \neq \mathcal{Q}$, $\mathcal{P}$ is a hat trajectory, $1_{\operatorname{top}(\mathcal{P})} \leq 1_{\operatorname{top}(\mathcal{Q})}$, and $0_{\operatorname{top}(\mathcal{P})} \not\leq 0_{\operatorname{top}(\mathcal{Q})}$, see Figure 25.1. Then

$$0_{\operatorname{top}(\mathcal{Q})} \equiv 1_{\operatorname{top}(\mathcal{Q})} \pmod{\operatorname{con}(\mathcal{Q})},$$

so

$$0_{\operatorname{top}(\mathcal{Q})} \wedge 1_{\operatorname{top}(\mathcal{Q})} \equiv 1_{\operatorname{top}(\mathcal{Q})} \wedge 1_{\operatorname{top}(\mathcal{Q})} = 1_{\operatorname{top}(\mathcal{Q})} \pmod{\operatorname{con}(\mathcal{Q})}.$$

Let $0_{\operatorname{top}(\mathcal{Q})} \wedge 1_{\operatorname{top}(\mathcal{Q})} \leq a \prec 1_{\operatorname{top}(\mathcal{Q})}$. We conclude that

$$\operatorname{con}(\mathcal{Q}) = \operatorname{con}(\operatorname{top}(\mathcal{Q})) \geq \operatorname{con}(a, 1_{\operatorname{top}(\mathcal{Q})}) \geq \operatorname{con}(\operatorname{top}(\mathcal{Q})) = \operatorname{con}(\mathcal{P}),$$

verifying (1).

Let $a = 0_{\operatorname{top}(\mathcal{Q})} \wedge 1_{\operatorname{top}(\mathcal{P})}$, and remember that $\mathcal{P}$ is a hat trajectory by definition. Since $a < 1_{\operatorname{top}(\mathcal{P})}$, there is a prime interval $\mathfrak{r}$ in the multifork with top $1_{\operatorname{top}(\mathcal{P})}$ such that $a \leq 0_{\mathfrak{r}}$. Hence, $\operatorname{top}(\mathcal{Q})$ is down-congruence perspective to $\mathfrak{r}$, and we have $\operatorname{con}(\mathcal{Q}) \geq \operatorname{con}(\mathfrak{r})$. Since $\operatorname{top}(\mathcal{P})$ is an interior member of our multifork, $\mathrm{con}(\mathfrak{r}) \geq \mathrm{con}(\mathrm{top}(\mathcal{P})) = \mathrm{con}(\mathcal{P})$. Thus, $\mathrm{con}(\mathcal{Q}) \geq \mathrm{con}(\mathcal{P})$, verifying (1).

Second, we prove the converse of (1):

$$(2) \qquad\qquad \mathrm{con}(\mathcal{P}) \leq \mathrm{con}(\mathcal{Q}) \text{ implies that } \mathcal{P} \leq_T \mathcal{Q}.$$

Let $\mathfrak{r} = \mathrm{top}(\mathcal{P})$ and $\mathfrak{q} = \mathrm{top}(\mathcal{Q})$.

By the Swing Lemma and Observation 1, we get the sequence (10) of binary relations.

Note that

(a) trajectories are closed with respect to up and down perspectivities;

(b) the equivalence class $\widehat{P}$ of a trajectory P is closed with respect to interior swings;

(c) whenever r_{i-1} externally swings to r_i, then r_i is the top of a hat trajectory R_i and (denoting the trajectory of r_{i-1} by R_{i-1}), we clearly have that $R_{i-1} \geq_C R_i$.

This completes the proof of (2).

25.2.   The Two-cover Theorem

In this section, we prove an interesting property of the ordered set of join-irreducible congruences of an SPS lattice.

Theorem 25.2 (Two-cover Theorem). *In the ordered set of join-irreducible congruences of an SPS lattice L, an element has at most two covers.*

Contrast this with Theorem 10.1, which states that the ordered set of join-irreducible congruences of a planar semimodular lattice has no special properties.

We start with some easy consequences of the Swing Lemma.

Lemma 25.3. *Let L be an SPS lattice and let $\mathfrak{p}$ and $\mathfrak{q}$ be prime intervals in L with $\mathfrak{p} \overset{ex}{\backsim} \mathfrak{q}$. Then $\mathrm{con}(\mathfrak{p}) > \mathrm{con}(\mathfrak{q})$ in $\mathrm{J}(\mathrm{Con}\,L)$.*

Proof. Since $\mathfrak{p} \overset{ex}{\backsim} \mathfrak{q}$, it follows that $\mathrm{con}(\mathfrak{p}) \geq \mathrm{con}(\mathfrak{q})$. By way of contradiction, assume that $\mathrm{con}(\mathfrak{p}) = \mathrm{con}(\mathfrak{q})$. Then $\mathrm{con}(\mathfrak{q}) \geq \mathrm{con}(\mathfrak{p})$, which is witnessed by a sequence as in (4). However, since $0_{\mathfrak{q}}$ is meet-irreducible and thus no up-perspectivity can start at $\mathfrak{q}$, it follows from (5) that the sequence cannot terminate at $\mathfrak{p}$, a contradiction. $\square$

Finally, we prove Theorem 25.2. Let L be an SPS lattice. Let $S = \{o, a_l, a_r, t\}$ be a tight square of L. Let $\mathfrak{p}$ be a prime interval of $L[S]$ and assume that $\mathrm{con}_{L[S]}(\mathfrak{p})$ is *not generated* by a congruence of L. Note that $\mathfrak{p} \not\subseteq L$. We distinguish two cases.

(i) $\mathfrak{p} \subseteq F[S]$.

By symmetry, we can assume that $\mathfrak{p} \subseteq \{z_{l,n_l}, \ldots, z_{l,n_l} = b_l, m\}$. Then $\mathfrak{p}$ is perspective to a prime interval $\mathfrak{q} \subseteq \{z_{l,n_l}, \ldots, z_{l,n_l} = a_l\} \subseteq L$, hence $\mathrm{con}_{L[S]}(\mathfrak{p}) = \mathrm{con}_{L[S]}(\mathfrak{q})$ with $\mathfrak{q} \subseteq L$, contradicting the assumption that $\mathrm{con}_{L[S]}(\mathfrak{p})$ is *not generated* by a congruence of L.

(ii) Let $\mathfrak{p}$ be neither in L nor in $F[S]$.

We need two subcases.

(a) $\mathfrak{p} = [x_{l,i}, z_{l,i}]$ for some $i = 1, \ldots, n_l$. Then $\mathrm{con}_{L[S]}(\mathfrak{p}) = \mathrm{con}_{L[S]}(a_r, t)$, a congruence of $L[S]$ generated by a congruence $\mathrm{con}_L(a_r, t)$ of L, contradicting the assumption that it is *not generated* by a congruence of L.

(b) $\mathfrak{p} = [z_{l,i}, y_{l,i}]$ for some $i = 1, \ldots, n_l$. Then

$$\mathrm{con}_{L[S]}(\mathfrak{p}) = \mathrm{con}_{L[S]}(m, t) = \gamma(S),$$

as claimed in the theorem.

25.3.   A counterexample

Theorem 10.1 proves that that every finite distributive lattice can be represented as the congruence lattice of a planar semimodular lattice. The proof relies heavily on M_3 sublattices.

I raised in [63] the problem of characterizing congruence lattices of SPS lattices. We proved in Section 25.2 the following necessary condition:

(C1) an element is covered by at most two elements.

It follows from G. Czédli [18, Lemma 2.2] that the ordered set of join-irreducible congruences of an SPS lattice L also satisfies the following condition:

(C2) for any nonmaximal element a, there are (at least) two distinct maximal elements m_1 and m_2 above a.

G. Czédli [20] proved (using his Trajectory Coloring Theorem [18], with a long and deep proof) that the converse does not hold by verifying the following result.

Theorem 25.4. *The eight element distributive lattice D_8 of Figure 25.2 cannot be represented as the congruence lattice of an SPS lattice L.*

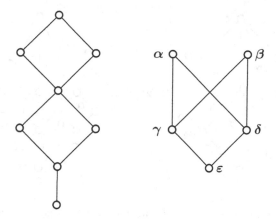

Figure 25.2: The lattice D_8 and the ordered set $P = J(D_8)$.

We prove this result based on my paper [73]. This proof does not utilize the Trajectory Coloring Theorem [18].

Let L be a finite SPS lattice whose join-irreducible congruences form an order P as in Figure 25.2. By G. Grätzer and E. Knapp [87], L has a congruence-preserving extension to a rectangular lattice, so we can assume that L is a rectangular lattice.

Case 1: L is a patch lattice.

By the Structure Theorem for SPS Lattices, we can obtain L from C_2^2 by inserting forks. Let $\mathsf{C}_2^2 = L_0, L_1, \ldots, L_n = L$ be a sequence of fork insertions from C_2^2 to L.

There is only one way to insert a fork into C_2^2; so L_1 is the lattice of Figure 25.3. There is one more join-irreducible congruence in L_1, the congruence γ. We "color" the diagram of L_1 with the join-irreducible congruences generated by the edges.

To get L_2, we pick a covering square S_1 in L_1 and insert a fork into L_1 at S_1. If the top element of S_1 is t, then S_1 is a wide square, so we get no new congruence by Theorem 24.15, see Figure 25.3. The next step, again with no new congruence is shown in the same figure. Note that all covering squares of the lattices L_1, L_2, L_3 satisfy the following condition:

(Col) all covering squares are colored by γ by itself or with α or with β with one exception: the bottom covering square is colored by α and β.

We proceed thus in $k - 1$ steps to get L_k, where $k \geq 1$ is the largest number with the property that the number of join-irreducible congruences do not change from L_{k-1} to L_k. Note that $k < n$ because $L_n = L$ has more join-irreducible congruences. Clearly, L_k also satisfies condition (Col).

We proceed to L_{k+1} by inserting a fork into a covering square S of L_k. There are five cases:

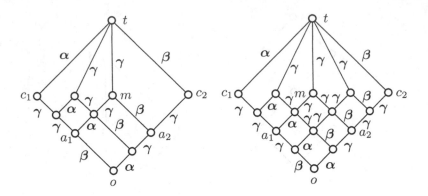

Figure 25.3: The lattices L_2 and L_3 with no new congruence.

(i) the top element of S is t;

(ii) the top element of S is not t and S is "monochromatic", colored by γ;

(iii) the top element of S is not t and S is colored by $\{\gamma, \alpha\}$;

(iv) the top element of S is not t and S is colored by $\{\gamma, \beta\}$;

(v) S is the bottom covering square colored by $\{\alpha, \beta\}$.

Case (i) cannot happen, it would contradict Theorem 24.14 and the definition of k.

If Case (ii) holds, S is tight, so by Theorem 24.15, we add a join-irreducible congruence $\gamma' < \gamma$. By Figure 25.2, we must have $\gamma' = \varepsilon$. This is a contradiction because no fork insertion can add an element δ between two existing elements.

Case (iii) proceeds the same was as Case (ii).

Case (iv) is symmetric to Case (iii).

So we are left with Case (v). In this case, the top element of S is not t, so S is tight. By Theorem 24.16, we get a new join-irreducible congruence δ in L_{k+1} satisfying that $\delta < \alpha$ and $\delta < \beta$, see Figure 25.4.

Let $k + 1 \leq m < n$ be the largest integer so that the number of join-irreducible congruences does not change from L_{k+1} to L_m—we insert forks into wide squares. Then the lattices $L_{k+1}, \ldots, L_m$ share the following property of L_{k+1}:

(P) there are only "monochromatic" squares and $\{\alpha, \beta\}$, $\{\alpha, \gamma\}$, $\{\alpha, \delta\}$, $\{\beta, \gamma\}$, $\{\beta, \delta\}$ squares, but there is no $\{\gamma, \delta\}$ square.

We obtain the last join-irreducible congruence, ε, in L_{m+1}. By property (P), we cannot have $\varepsilon \prec \gamma$ and $\varepsilon \prec \delta$.

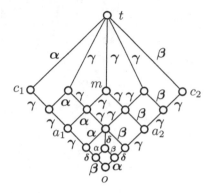

Figure 25.4: The lattice L_{k+1} for case (v).

Case 2: L is not a patch lattice. It follows that the grid D is not a covering square. The lattice D has three join-irreducible congruences forming an antichain of 3 elements. But the ordered set of Figure 25.2 has only one- and two-element antichains.

This completes the proof of Theorem 25.4.

25.4. Discussion

G. Czédli and I started our research into congruences of SPS lattices at about the same time. It was not joint research, but "parallel" research. He read everything I wrote, and was very patient correcting my mistakes.

The first major result was Czédli's Trajectory Theorem for SPS lattices. Its proof was complicated, and almost 30 pages long.

My first research in this field centered on congruences of fork extensions. At the beginning, the paper [71] gave direct (and long and tedious) proofs for the results of Section 24.5. The idea of the Swing Lemma came from this research.

Now the Trajectory Theorem and the results on congruences of fork extensions are presented as applications of the Swing Lemma.

Bibliography

[1] K. Adaricheva and G. Czédli, Note on the description of join-distributive lattices by permutations, *Algebra Universalis,* **72** (2014), 155–162.

[2] J. Anderson and N. Kimura, The tensor product of semilattices, *Semigroup Forum* **16** (1968), 83–88.

[3] A. Arber, *The Mind and the Eye, A Study of the Biologist's Standpoint,* Cambridge University Press, London, 1954.

[4] V. A. Baranskiĭ, On the independence of the automorphism group and the congruence lattice for lattices, Abstracts of lectures of the 15th All-Soviet Algebraic Conference, Krasnojarsk, vol. 1, 11, July 1979.

[5] ———, Independence of lattices of congruences and groups of automorphisms of lattices (Russian), *Izv. Vyssh. Uchebn. Zaved. Mat.* 1984, no. 12, 12–17, 76. English translation: Soviet Math. (Iz. VUZ) **28** (1984), no. 12, 12–19.

[6] G. M. Bergman and G. Grätzer, Isotone maps on lattices. *Algebra Universalis* **68** (2012), 17–37.

[7] J. Berman, On the length of the congruence lattice of a lattice, *Algebra Universalis* **2** (1972), 18–19.

[8] G. Birkhoff, *Universal Algebra,* Proc. First Canadian Math. Congress, Montreal, 1945. University of Toronto Press, Toronto, 1946, 310–326.

© Springer International Publishing Switzerland 2016
G. Grätzer, *The Congruences of a Finite Lattice,*
DOI 10.1007/978-3-319-38798-7

[9] ——, On groups of automorphisms, (Spanish) *Rev. Un. Math. Argentina* **11** (1946), 155–157.

[10] G. Birkhoff, *Lattice Theory.* Third edition. American Mathematical Society Colloquium Publications, vol. XXV. American Mathematical Society, Providence, RI, 1967. vi+418 pp.

[11] K. P. Bogart, R. Freese, and J. P. S. Kung (editors), *The Dilworth Theorems. Selected papers of Robert P. Dilworth*, Birkhäuser Boston, Inc., Boston, MA, 1990. xxvi+465 pp. ISBN: 0-8176-3434-7

[12] B. Chornomaz, A non-capped tensor product of lattices, *Algebra Universalis* **72** (2014), 323–348.

[13] P. Crawley and R. P. Dilworth, *Algebraic Theory of Lattices.* Prentice-Hall, Englewood Cliffs, NJ, 1973. vi+201 pp. ISBN: 0-13-022269-0

[14] G. Czédli, 2-uniform congruences in majority algebras and a closure operator, *Algebra Universalis* **57** (2007), 63–73.

[15] ——, Idempotent Mal'cev conditions and 2-uniform congruences, *Algebra Universalis* **59**, (2008) 303–309.

[16] ——, The matrix of a slim semimodular lattice, *Order* **29** (2012), 85–103.

[17] ——, Representing homomorphisms of distributive lattices as restrictions of congruences of rectangular lattices, *Algebra Universalis* **67** (2012), 313–345.

[18] ——, Patch extensions and trajectory colorings of slim rectangular lattices, *Algebra Universalis* **72** (2014), 125–154.

[19] ——, Coordinatization of join-distributive lattices, *Algebra Universalis* **71** (2014), 385–404.

[20] ——, A note on congruence lattices of slim semimodular lattices, *Algebra Universalis* **72** (2014), 225–230.

[21] ——, The asymptotic number of planar, slim, semimodular lattice diagrams, *Order.*

[22] ——, The ordered set of principal congruences of a countable lattice, *Algebra Universalis* **75** (2016), 351–380.

[23] ——, Diagrams and rectangular extensions of planar semimodular lattices, *Algebra Universalis.*

[24] ——, Quasiplanar diagrams and slim semimodular lattices, *Order.*

[25] _____, Diagrams and rectangular extensions of planar semimodular lattices, *Algebra Universalis*.

[26] _____, Representing a monotone map by principal lattice congruences, *Acta Mathematica Hungarica* **147** (2015), 12–18.

[27] _____, Representing some families of monotone maps by principal lattice congruences, *Algebra Universalis*.

[28] _____, An independence theorem for ordered sets of principal congruences and automorphism groups of bounded lattices, *Acta Sci. Math (Szeged)*.

[29] _____, Large sets of lattices without order embeddings, *Communications in Algebra* **44** (2016), 668–679.

[30] G. Czédli, T. Dékány, L. Ozsvárt, N. Szakács, B. Udvari, On the number of slim, semimodular lattices, *Mathematica Slovaca*.

[31] G. Czédli and G. Grätzer, Lattice tolerances and congruences, *Algebra Universalis* **66** (2011) 5–6.

[32] _____, Notes on planar semimodular lattices. VII. Resections of planar semimodular lattices, *Order* **30** (2013), 847–858.

[33] _____, Planar Semimodular Lattices: Structure and Diagrams. Chapter 3 in LTS1.

[34] G. Czédli, L. Ozsvárt, and B. Udvari, How many ways can two composition series intersect?, *Discrete Math.* **312** (2012), 3523–3536.

[35] G. Czédli and E. T. Schmidt, How to derive finite semimodular lattices from distributive lattices? *Acta Mathematica Hungarica* **121** (2008), 277–282.

[36] _____, Some results on semimodular lattices, Contributions to General Algebra 19. Proceedings of the Olomouc Conference 2010 (AAA 79+ CYA 25), Verlag Johannes Hein, Klagenfurt 2010, 45–56. ISBN 978-3-7084-0407-3.

[37] _____, A cover-preserving embedding of semimodular lattices into geometric lattices, *Advances in Mathematics* **225** (2010), 2455–2463.

[38] _____, The Jordan-Hölder theorem with uniqueness for groups and semimodular lattices, *Algebra Universalis* **66** (2011), 69–79.

[39] _____, Finite distributive lattices are congruence lattices of almost-geometric lattices, *Algebra Universalis* **65** (2011), 91–108.

[40] _____, Slim semimodular lattices. I. A visual approach, *Order* **29** (2012), 481–497.

[41] _____, Slim semimodular lattices. II. A description by patchwork systems, *Order* **30** (2013), 689–721.

[42] _____, Composition series in groups and the structure of slim semimodular lattices, *Acta Sci. Math. (Szeged)* **79** (2013), 369–390.

[43] B. A. Davey and H. A. Priestley, *Introduction to Lattices and Order, Second Edition,* Cambridge University Press, NY, 2002. xii+298 pp. ISBN: 0-521-78451-4

[44] R. P. Dilworth, The structure of relatively complemented lattices, *Ann. of Math.* (2) **51** (1950), 348–359.

[45] H. Dobbertin, Vaught measures and their applications in lattice theory, *J. Pure Appl. Algebra* **43** (1986), 27–51.

[46] G. A. Fraser, The semilattice tensor product of distributive semilattices, *Trans. Amer. Math. Soc.* **217** (1976), 183–194.

[47] R. Freese, Congruence lattices of finitely generated modular lattices, Proceedings of the Ulm Lattice Theory Conference, pp. 62–70, Ulm, 1975.

[48] _____, Computing congruence lattices of finite lattices, *Proc. Amer. Math. Soc.* (1997) **125**, 3457–3463.

[49] R. Freese, G. Grätzer, and E. T. Schmidt, On complete congruence lattices of complete modular lattices, *Internat. J. Algebra Comput.* **1** (1991), 147–160.

[50] R. Freese, J. Ježek, and J. B. Nation, *Free lattices,* Mathematical Surveys and Monographs, vol. 42, American Mathematical Society, Providence, RI, 1995. viii+293 pp.

[51] R. Frucht, Herstellung von Graphen mit vorgegebener abstrakter Gruppe, *Compos. Math.* **6** (1938), 239–250.

[52] _____, Lattices with a given group of automorphisms, *Canad. J. Math.* **2** (1950), 417–419.

[53] N. Funayama and T. Nakayama, On the congruence relations on lattices, *Proc. Imp. Acad. Tokyo* **18** (1942), 530–531.

[54] G. Grätzer, *Universal Algebra,* The University Series in Higher Mathematics, D. van Nostrand Co. Inc., Princeton, N.J., Toronto, Ont., London, 1968. xvi+368 pp.

[55] G. Grätzer, *Lattice Theory. First Concepts and Distributive Lattices*, W. H. Freeman and Co., San Francisco, Calif., 1971. xv+212 pp.

[56] G. Grätzer, *General Lattice Theory*, Pure and Applied Mathematics, vol. 75, Academic Press, Inc. (Harcourt Brace Jovanovich, Publishers), New York-London; Lehrbücher und Monographien aus dem Gebiete der Exakten Wissenschaften, Mathematische Reihe, Band 52. Birkhäuser Verlag, Basel-Stuttgart; Akademie Verlag, Berlin, 1978. xiii+381 pp. ISBN: 0-12-295750-4

(Russian translation: *Obshchaya teoriya reshetok*, translated from the English by A. D. Bol'bot, V. A. Gorbunov, and V. I. Tumanov. Translation edited and with a preface by D. M. Smirnov. "Mir", Moscow, 1982. 454 pp.)

[57] G. Grätzer, *Universal Algebra, second edition*, Springer-Verlag, New York–Heidelberg, 1979. xviii+581 pp. ISBN: 3-7643-5239-6

[58] G. Grätzer, The complete congruence lattice of a complete lattice, Lattices, semigroups, and universal algebra. Proceedings of the International Conference held at the University of Lisbon, Lisbon, June 20–24, 1988. Edited by Jorge Almeida, Gabriela Bordalo and Philip Dwinger, pp. 81–87. Plenum Press, New York, 1990. x+336 pp. ISBN: 0-306-43412-1

[59] G. Grätzer, *General Lattice Theory, second edition*, new appendices by the author with B. A. Davey, R. Freese, B. Ganter, M. Greferath, P. Jipsen, H. A. Priestley, H. Rose, E. T. Schmidt, S. E. E. T. Schmidt, F. Wehrung, and R. Wille. Birkhäuser Verlag, Basel, 1998. xx+663 pp. ISBN: 0-12-295750-4; ISBN: 3-7643-5239-6 *Softcover edition*, Birkhäuser Verlag, Basel–Boston–Berlin, 2003. ISBN: 3-7643-6996-5

[60] G. Grätzer, *The Congruences of a Finite Lattice, A* Proof-by-Picture *Approach*. Birkhäuser Boston, 2006. xxiii+281 pp. ISBN: 0-8176-3224-7.

[61] G. Grätzer, Two problems that shaped a century of lattice theory. *Notices Amer. Math. Soc.* **54** (2007), 696–707.

[62] G. Grätzer, *Lattice Theory: Foundation*. Birkhäuser Verlag, Basel, 2011. xxix+613 pp. ISBN: 978-3-0348-0017-4.

[63] ―――, Planar Semimodular Lattices: Congruences. Chapter 4 in LTS1.

[64] ―――, Notes on planar semimodular lattices. VI. On the structure theorem of planar semimodular lattices. *Algebra Universalis* **69** (2013), 301–304.

328   Bibliography

[65] _____, The order of principal congruences of a lattice. *Algebra Universalis* **70** (2013), 95–105.

[66] _____, Planar Semimodular Lattices: Congruences. Chapter 4 in LTS1.

[67] _____, Sectionally Complemented Lattices. Chapter 5 in LTS1.

[68] _____, Two Topics Related to Congruence Lattices of Lattices. Chapter 10 in LTS1. 54 pp. plus bibliography.

[69] _____, Congruences and prime-perspectivities in finite lattices. *Algebra Universalis* **74** (2015), 351–359.

[70] _____, Congruences in slim, planar, semimodular lattices: The Swing Lemma. *Acta Sci. Math. (Szeged)* .

[71] _____, Congruences of fork extensions of slim, planar, semimodular lattices. *Algebra Universalis*.

[72] _____, A technical lemma for congruences of finite lattices. *Algebra Universalis*.

[73] _____, On a result of Gábor Czédli concerning congruence lattices of planar semimodular lattices. *Acta Sci. Math. (Szeged)* .

[74] _____, Congruences and trajectories in planar semimodular lattices. *Algebra Universalis*.

[75] _____, Homomorphisms and principal congruences of bounded lattices. Acta Sci. Math (Szeged).

[76] _____, Principal congruences and automorphisms of bounded lattices. *Algebra Universalis*.

[77] _____, Homomorphisms and principal congruences of bounded lattices. II. Sketching the proof for sublattices. *Algebra Universalis*.

[78] G. Grätzer and M. Greenberg, Lattice tensor products. I. Coordinatization, *Acta Math. Hungar.* **95** **(4)** (2002), 265–283.

[79] _____, Lattice tensor products. II. Ideal lattices. Acta Math. Hungar. **97** (2002), 193–198.

[80] _____, Lattice tensor products. III. Congruences, *Acta Math. Hungar.* **98** (2003), 167–173.

[81] _____, Lattice tensor products. IV. Infinite lattices, *Acta Math. Hungar.* **103** (2004), 17–30.

[82] G. Grätzer, M. Greenberg, and E. T. Schmidt, Representing congruence lattices of lattices with partial unary operations as congruence lattices of lattices. II. Interval ordering, *J. Algebra* **286** (2005), 307–324.

[83] G. Grätzer, D. S. Gunderson, and R. W. Quackenbush, The spectrum of a finite pseudocomplemented lattice. *Algebra Universalis* **61** (2009), 407–411.

[84] G. Grätzer and David Kelly, A new lattice construction, *Algebra Universalis* **53** (2005), 253–265.

[85] G. Grätzer and E. Knapp, Notes on planar semimodular lattices. I. Construction. *Acta Sci. Math. (Szeged)* **73** (2007), 445–462.

[86] ———, Notes on planar semimodular lattices. II. Congruences. *Acta Sci. Math. (Szeged)* **74** (2008), 37–47.

[87] ———, Notes on planar semimodular lattices. III. Rectangular lattices. *Acta Sci. Math. (Szeged)* **75** (2009), 29–48.

[88] ———, Notes on planar semimodular lattices. IV. The size of a minimal congruence lattice representation with rectangular lattices. *Acta Sci. Math. (Szeged)* **76** (2010), 3–26.

[89] G. Grätzer and H. Lakser, Extension theorems on congruences of partial lattices, *Notices Amer. Math. Soc.* **15** (1968), 732, 785.

[90] ———, Homomorphisms of distributive lattices as restrictions of congruences, *Can. J. Math.* **38** (1986), 1122–1134.

[91] ———, Congruence lattices, automorphism groups of finite lattices and planarity, *C. R. Math. Rep. Acad. Sci. Canada* **11** (1989), 137–142. Addendum, **11** (1989), 261.

[92] ———, On complete congruence lattices of complete lattices, *Trans. Amer. Math. Soc.* **327** (1991), 385–405.

[93] ———, Congruence lattices of planar lattices, *Acta Math. Hungar.* **60** (1992), 251–268.

[94] ———, On congruence lattices of m-complete lattices, *J. Austral. Math. Soc. Ser. A* **52** (1992), 57–87.

[95] ———, Homomorphisms of distributive lattices as restrictions of congruences. II. Planarity and automorphisms, *Canad. J. Math.* **46** (1994), 3–54.

[96] _____, Notes on sectionally complemented lattices. I. Characterizing the 1960 sectional complement. *Acta Math. Hungar.* **108** (2005), 115–125.

[97] _____, Notes on sectionally complemented lattices. II. Generalizing the 1960 sectional complement with an application to congruence restrictions. *Acta Math. Hungar.* **108** (2005), 251–258.

[98] _____, Representing homomorphisms of congruence lattices as restrictions of congruences of isoform lattices. *Acta Sci. Math. (Szeged)* **75** (2009), 393–421.

[99] G. Grätzer, H. Lakser, and R. W. Quackenbush, The structure of tensor products of semilattices with zero, *Trans. Amer. Math. Soc.* **267** (1981), 503–515.

[100] _____, Congruence-preserving extensions of congruence-finite lattices to isoform lattices. *Acta Sci. Math. (Szeged)* **75** (2009), 13–28.

[101] G. Grätzer, H. Lakser, and M. Roddy, Notes on sectionally complemented lattices. III. The general problem, *Acta Math. Hungar.* **108** (2005), 325–334.

[102] G. Grätzer, H. Lakser, and E. T. Schmidt, Congruence lattices of small planar lattices, *Proc. Amer. Math. Soc.* **123** (1995), 2619–2623.

[103] _____, Congruence representations of join-homomorphisms of distributive lattices: A short proof, *Math. Slovaca* **46** (1996), 363–369.

[104] _____, Isotone maps as maps of congruences. I. Abstract maps, *Acta Math. Acad. Sci. Hungar.* **75** (1997), 105–135.

[105] _____, Congruence lattices of finite semimodular lattices, *Canad. Math. Bull.* **41** (1998), 290–297.

[106] _____, Congruence representations of join-homomorphisms of finite lattices: size and breadth, *J. Austral Math. Soc.* **68** (2000), 85–103.

[107] _____, Isotone maps as maps of congruences. II. Concrete maps, *Acta Math. Acad. Sci. Hungar.* **92** (2001), 233–238.

[108] G. Grätzer, H. Lakser, and F. Wehrung, Congruence amalgamation of lattices, *Acta Sci. Math. (Szeged)* **66** (2000), 3–22.

[109] G. Grätzer, H. Lakser, and B. Wolk, On the lattice of complete congruences of a complete lattice: On a result of K. Reuter and R. Wille, *Acta Sci. Math. (Szeged)* **55** (1991), 3–8.

[110] G. Grätzer and J. B. Nation, A new look at the Jordan-Hölder theorem for semimodular lattices. *Algebra Universalis* **64** (2011), 309–311.

[111] G. Grätzer and R. W. Quackenbush, The variety generated by planar modular lattices. *Algebra Universalis* **63** (2010), 187–201.

[112] ———, Positive universal classes in locally finite varieties. *Algebra Universalis* **64** (2010), 1–13.

[113] G. Grätzer, R. W. Quackenbush, and E. T. Schmidt, Congruence-preserving extensions of finite lattices to isoform lattices, *Acta Sci. Math. (Szeged)* **70** (2004), 473–494.

[114] G. Grätzer, I. Rival, and N. Zaguia, Small representations of finite distributive lattices as congruence lattices, *Proc. Amer. Math. Soc.* **123** (1995), 1959–1961. Correction: **126** (1998), 2509–2510.

[115] G. Grätzer and M. Roddy, Notes on sectionally complemented lattices. IV. Manuscript.

[116] G. Grätzer and E. T. Schmidt, Ideals and congruence relations in lattices, *Acta Math. Acad. Sci. Hungar.* **9** (1958), 137–175.

[117] ———, On congruence lattices of lattices, *Acta Math. Acad. Sci. Hungar.* **13** (1962), 179–185.

[118] ———, Characterizations of congruence lattices of abstract algebras, *Acta Sci. Math. (Szeged)* **24** (1963), 34–59.

[119] ———, "Complete-simple" distributive lattices, *Proc. Amer. Math. Soc.* **119** (1993), 63–69.

[120] ———, Another construction of complete-simple distributive lattices, *Acta Sci. Math. (Szeged)* **58** (1993), 115–126.

[121] ———, Congruence lattices of function lattices, *Order* **11** (1994), 211–220.

[122] ———, Algebraic lattices as congruence lattices: The m-complete case, Lattice theory and its applications. In celebration of Garrett Birkhoff's 80th birthday. Papers from the symposium held at the Technische Hochschule Darmstadt, Darmstadt, June 1991. Edited by K. A. Baker and R. Wille. Research and Exposition in Mathematics, 23. Heldermann Verlag, Lemgo, 1995. viii+262 pp. ISBN 3-88538-223-7

[123] ———, A lattice construction and congruence-preserving extensions, *Acta Math. Hungar.* **66** (1995), 275–288.

332   Bibliography

[124] _____, Complete congruence lattices of complete distributive lattices, *J. Algebra* **171** (1995), 204–229.

[125] _____, Do we need complete-simple distributive lattices? *Algebra Universalis* **33** (1995), 140–141.

[126] _____, The Strong Independence Theorem for automorphism groups and congruence lattices of finite lattices, *Beiträge Algebra Geom.* **36** (1995), 97–108.

[127] _____, Complete congruence lattices of join-infinite distributive lattices, *Algebra Universalis* **37** (1997), 141–143.

[128] _____, Representations of join-homomorphisms of distributive lattices with doubly 2-distributive lattices, *Acta Sci. Math. (Szeged)* **64** (1998), 373–387.

[129] _____, Congruence-preserving extensions of finite lattices into sectionally complemented lattices, *Proc. Amer. Math. Soc.* **127** (1999), 1903–1915.

[130] _____, On finite automorphism groups of simple arguesian lattices, *Studia Sci. Math. Hungar.* **35** (1999), 247–258.

[131] _____, Regular congruence-preserving extensions, *Algebra Universalis* **46** (2001), 119–130.

[132] _____, Congruence-preserving extensions of finite lattices to semimodular lattices, *Houston J. Math.* **27** (2001), 1–9.

[133] _____, Complete congruence representations with 2-distributive modular lattices, *Acta Sci. Math. (Szeged)* **67** (2001), 289–300.

[134] _____, On the Independence Theorem of related structures for modular (arguesian) lattices, *Studia Sci. Math. Hungar.* **40** (2003), 1–12.

[135] _____, Representing congruence lattices of lattices with partial unary operations as congruence lattices of lattices. I. Interval equivalence, *J. Algebra.* **269** (2003), 136–159.

[136] _____, Finite lattices with isoform congruences, *Tatra Mt. Math. Publ.* **27** (2003), 111–124.

[137] _____, Congruence class sizes in finite sectionally complemented lattices, *Canad. Math. Bull.* **47** (2004), 191–205.

[138] _____, Finite lattices and congruences. A survey, *Algebra Universalis* **52** (2004), 241–278.

[139] _____, A short proof of the congruence representation theorem for semimodular lattices. *Algebra Universalis* **71** (2014), 65–68.

[140] _____, An extension theorem for planar semimodular lattices. Periodica Mathematica Hungarica (2014) **69** (2014), 32–40.

[141] G. Grätzer, E. T. Schmidt, and K. Thomsen, Congruence lattices of uniform lattices, *Houston J. Math.* **29** (2003), 247–263.

[142] G. Grätzer and J. Sichler, On the endomorphism semigroup (and category) of bounded lattices, Pacif. J. Math. **3** (1970), 639–647.

[143] G. Grätzer and D. Wang, A lower bound for congruence representations, *Order* **14** (1997), 67–74.

[144] G. Grätzer and T. Wares, Notes on planar semimodular lattices. V. Cover-preserving embeddings of finite semimodular lattices into simple semimodular lattices. *Acta Sci. Math. (Szeged)* **76** (2010), 27–33.

[145] G. Grätzer and F. Wehrung, Proper congruence-preserving extensions of lattices, *Acta Math. Hungar.* **85** (1999), 175–185.

[146] _____, A new lattice construction: the box product, *J. Algebra* **221** (1999), 315–344.

[147] _____, Tensor products and transferability of semilattices, *Canad. J. Math.* **51** (1999), 792–815.

[148] _____, Tensor products of lattices with zero, revisited, *J. Pure Appl. Algebra* **147** (2000), 273–301.

[149] _____, The Strong Independence Theorem for automorphism groups and congruence lattices of arbitrary lattices, *Adv. in Appl. Math.* **24** (2000), 181–221.

[150] _____, A survey of tensor products and related constructions in two lectures, *Algebra Universalis* **45** (2001), 117–134.

[151] _____, On the number of join-irreducibles in a congruence representation of a finite distributive lattice, *Algebra Universalis* **49** (2003), 165–178.

[152] G. Grätzer and F. Wehrung eds., *Lattice Theory: Special Topics and Applications. Volume 1.* Birkhäuser Verlag, Basel, 2014.

[153] G. Grätzer and F. Wehrung eds., *Lattice Theory: Special Topics and Applications. Volume 2.* Birkhäuser Verlag, Basel, 2016.

334   Bibliography

[154] Z. Hedrlin and J. Sichler, Any boundable binding category contains a proper class of mutually disjoint copies of itself. *Algebra Universalis* **1** (1971), 97–103.

[155] C. Herrmann, On automorphism groups of arguesian lattices, *Acta Math. Hungar.* **79** (1998), 35–38.

[156] A. P. Huhn, Schwach distributive Verbände. I, *Acta Sci. Math.* (Szeged) **33** (1972), 297–305.

[157] _____, Two notes on n-distributive lattices, Lattice theory (Proc. Colloq., Szeged, 1974), pp. 137–147. *Colloq. Math. Soc. János Bolyai,* vol. 14, North-Holland, Amsterdam, 1976.

[158] _____, On the representation of distributive algebraic lattices, I. *Acta Sci. Math.* (*Szeged*) **45** (1983), 239–246.

[159] _____, On the representation of distributive algebraic lattices. II, *Acta Sci. Math.* **53** (1989), 3–10.

[160] _____, On the representation of distributive algebraic lattices. III, *Acta Sci. Math.* **53** (1989), 11–18.

[161] M. F. Janowitz, Section semicomplemented lattices, *Math. Z.* **108** (1968), 63–76.

[162] J. Jakubík, Congruence relations and weak projectivity in lattices, (Slovak) *Časopis Pěst. Mat.* **80** (1955), 206–216.

[163] K. Kaarli, Finite uniform lattices are congruence permutable, *Acta Sci. Math.* **71** (2005), 457–460.

[164] F. Maeda, *Kontinuierliche Geometrien.* Die Grundlehren der mathematischen Wissenschaften in Einzeldarstellungen mit besonderer Berücksichtigung der Anwendungsgebiete, Bd. 95. Springer-Verlag, Berlin-Göttingen-Heidelberg, 1958. x+244 pp.

[165] R. N. McKenzie, G. F. McNulty, and W. F. Taylor, *Algebras, lattices, varieties, vol. I.* The Wadsworth & Brooks/Cole Mathematics Series. Wadsworth & Brooks/Cole Advanced Books & Software, Monterey, CA, 1987. xvi+361 pp. ISBN: 0-534-07651-3

[166] E. Mendelsohn, Every group is the collineation group of some projective plane. Foundations of geometry (Proc. Conf., Univ. Toronto, Toronto, Ont., 1974), pp. 175–182. Univ. Toronto Press, Toronto, Ont., 1976.

[167] O. Ore, Theory of equivalence relations, *Duke Math. J.* **9** (1942), 573–627.

[168] P. Pudlák, On congruence lattices of lattices, *Algebra Universalis* **20** (1985), 96–114.

[169] P. Pudlák and J. Tůma, Every finite lattice can be embedded into a finite partition lattice, *Algebra Universalis* **10** (1980), 74–95.

[170] A. Pultr and V. Trnková, *Combinatorial, algebraic and topological representations of groups, semigroups and categories,* North-Holland Mathematical Library, vol. 22. North-Holland Publishing Co., Amsterdam-New York, 1980. x+372 pp. ISBN: 0-444-85083-X.

[171] K. Reuter and R. Wille, Complete congruence relations of complete lattices, *Acta. Sci. Math. (Szeged)*, **51** (1987), 319–327.

[172] G. Sabidussi, Graphs with given infinite groups, *Monatsh. Math.* **64** (1960), 64–67.

[173] E. T. Schmidt, Zur Charakterisierung der Kongruenzverbände der Verbände, *Mat. Časopis Sloven. Akad. Vied.* **18** (1968), 3–20.

[174] _____, Every finite distributive lattice is the congruence lattice of some modular lattice, *Algebra Universalis* **4** (1974), 49–57.

[175] _____, On the length of the congruence lattice of a lattice, *Algebra Universalis* **5** (1975), 98–100.

[176] _____, Remark on generalized function lattices, *Acta Math. Hungar.* **34** (1979), 337–339.

[177] _____, The ideal lattice of a distributive lattice with 0 is the congruence lattice of a lattice, *Acta Sci. Math. (Szeged)* **43** (1981), 153–168.

[178] _____, Congruence lattices of complemented modular lattices, *Algebra Universalis* **18** (1984), 386–395.

[179] _____, Homomorphism of distributive lattices as restriction of congruences, *Acta Sci. Math. (Szeged)* **51** (1987), 209–215.

[180] _____, Congruence lattices of modular lattices, *Publ. Math. Debrecen* **42** (1993), 129–134.

[181] _____, On finite automorphism groups of simple arguesian lattices, *Publ. Math. Debrecen* **42** (1998), 383–387.

[182] Z. Shmuley, The structure of Galois connections, *Pacific J. Math.* **54** (1974), 209–225.

[183] S.-K. Teo, Representing finite lattices as complete congruence lattices of complete lattices, *Ann. Univ. Sci. Budapest. Eötvös Sect. Math.* **33** (1990), 177–182.

336   Bibliography

[184] _____, On the length of the congruence lattice of a lattice, *Period. Math. Hungar.* **21** (1990), 179–186.

[185] M. Tischendorf, The representation problem for algebraic distributive lattices, Ph. D. thesis, TH Darmstadt, 1992.

[186] J. Tůma, On the existence of simultaneous representations, *Acta Sci. Math. (Szeged)* **64** (1998), 357–371.

[187] J. Tůma and F. Wehrung, A survey of recent results on congruence lattices of lattices, *Algebra Universalis* **48** (2002), 439–471.

[188] A. Urquhart, A topological representation theory for lattices, *Algebra Universalis* **8** (1978), 45–58.

[189] F. Wehrung, Non-measurability properties of interpolation vector spaces, *Israel J. Math.* **103** (1998), 177–206.

[190] _____, A uniform refinement property of certain congruence lattices, *Proc. Amer. Math. Soc.* **127** (1999), 363–370.

[191] _____, Solutions to five problems on tensor products of lattices and related matters, *Algebra Universalis* **47** (2002), 479–493.

[192] _____, E. T. Schmidt and Pudlák's approaches to CLP. Chapter 7 in LTS1.

[193] _____, Congruences of Lattices and Ideals of Rings. Chapter 8 in LTS1.

[194] _____, Liftable and Unliftable Diagrams. Chapter 9 in LTS1.

[195] Y. Zhang, A note on "Small representations of finite distributive lattices as congruence lattices", *Order* **13** (1996), 365–367.

Index

Symbols
0-separating homomorphism, 17
$\{0,1\}$-embedding, 13
$\{0,1\}$-homomorphism, 13
$\{0,1\}$-sublattice, 14
1/3-boolean triple, 221
2-distributive lattice, 138, 152,
    236, 251, 332
2/3-boolean triple, 209, 210
4-cell, 47
4-cell lattice, 47

A
$\{a, b\}$-fork, 181
Absorption identity, 9
Adaricheva, K., 49, 51, 323
Algebra, 9
    partial, 57
Anderson, J., 271, 323
Antichain, 4, 321
Antisymmetric, 3
Antitone map, 5
Arber, A., xxx, 323
Arguesian identity, 138
Arguesian lattice, 138
Associative identity, 9
Atom, 19
    dual, 19
Atom Lemma, 64, 162

Atomistic lattice, 19
Automorphism, 12

B
Balanced triple, 68
Baranskiĭ, V. A., xxxii, 137,
    189, 251, 323
Baranskiĭ-Urquhart Theorem, xxxii,
    251
Base congruence, 22
Bergman, G. M., 323
Berman, J., 110, 323
Bijective, 12
Binary
    operation, 9
    partial operation, 57
    relation, 3, 317
        product, 16
    term, 23
Birkhoff's Subdirect Representation
    Theorem, 22
Birkhoff, G., xxxiv, 22, 108, 189, 199,
    323, 324
Block, 7, 15
    trivial, 7
Bogart, K. P., 324
Bol'bot, A. D., 327
Boolean lattice, 25
Boolean sublattice, 47

© Springer International Publishing Switzerland 2016
G. Grätzer, *The Congruences of a Finite Lattice*,
DOI 10.1007/978-3-319-38798-7

Printed in the United States
By Bookmasters